Numerical Simulation of Wind Turbines

Numerical Simulation of Wind Turbines

Editors

Alessandro Bianchini
Giovanni Ferrara

MDPI • Basel • Beijing • Wuhan • Barcelona • Belgrade • Manchester • Tokyo • Cluj • Tianjin

Editors

Alessandro Bianchini	Giovanni Ferrara
Department of Industrial Engineering	Department of Industrial Engineering
Università degli Studi di Firenze	Università degli Studi di Firenze
Firenze	Firenze
Italy	Italy

Editorial Office
MDPI
St. Alban-Anlage 66
4052 Basel, Switzerland

This is a reprint of articles from the Special Issue published online in the open access journal *Energies* (ISSN 1996-1073) (available at: www.mdpi.com/journal/energies/special_issues/Numerical_Simulation_Wind_Turbine).

For citation purposes, cite each article independently as indicated on the article page online and as indicated below:

LastName, A.A.; LastName, B.B.; LastName, C.C. Article Title. *Journal Name* **Year**, *Volume Number*, Page Range.

ISBN 978-3-0365-1165-8 (Hbk)
ISBN 978-3-0365-1164-1 (PDF)

Contents

About the Editors

Giovanni Ferrara

Giovanni Ferrara is full professor at the Deparment of Industrial Engineering at Università degli Studi di Firenze, and he is the coordinator of the REASE research group. He is an active member of the EU wind energy community with many years of experience, and he has authored more than 80 international publications in wind energy related to his many scientific contributions in the area of machinery and energy engineering.

Alessandro Bianchini

Alessandro Bianchini is an assistant professor at the Department of Industrial Engineering at Università degli Studi di Firenze (UNIFI). He has authored more than 90 publications in wind energy, and he is presently serving as the representative of UNIFI within the board of the European Academy of Wind Energy (EAWE), as the Chair of the "Wind Energy Committee"of the ASME Turbo Expo Conference, and as an associate editor of the open-access journals *Wind Energy Science* and *Frontiers in Energy*. He is an expert in aerodynamic simulation of wind turbines, from engineering codes to computational fluid dynamics.

Preface to "Numerical Simulation of Wind Turbines"

It is undisputed that wind energy has become a consolidated industry, probably representing the leading form of energy conversion from renewables. Wind turbines are, however, not only the largest rotating machines on Earth, but also very complex systems, whose design and optimization require the contribution of several disciplines, from aerodynamics to structural analyses, from control to electric engineering. At the moment, however, a truly holistic approach to wind turbine design is still missing, with different groups of experts often working not sufficiently close to each other. To achieve it, it is apparent that simulation will play a key role since it represents the common framework, within which the different disciplines and sciences can communicate. Developments are needed in each of them, from more accurate, but computationally affordable, simulation techniques, to innovative control logics or innovative design solutions. The contributions from the research works included in the Special Issue *Simulation of Wind Turbines* offer new data, information, and findings to continue the R&D effort in wind turbine simulation, with the aim of stimulating the research community to further contribute to the development of the field.

Alessandro Bianchini, Giovanni Ferrara
Editors

energies

MDPI

Editorial

Special Issue "Numerical Simulation of Wind Turbines"

Giovanni Ferrara * and **Alessandro Bianchini**

Department of Industrial Engineering, Università degli Studi di Firenze, via di Santa Marta 3, 50139 Firenze, Italy; alessandro.bianchini@unifi.it
* Correspondence: giovanni.ferrara@unifi.it

Citation: Ferrara, G.; Bianchini, A. Special Issue "Numerical Simulation of Wind Turbines". *Energies* **2021**, *14*, 1616. https://doi.org/10.3390/en14061616

Received: 15 February 2021
Accepted: 11 March 2021
Published: 15 March 2021

To fulfill global needs for a more sustainable energy, a further development of wind energy is fostered. Wind turbines represent the largest rotating machines on Earth and their upscaling trend is expected to continue in upcoming years. On the other hand, even small wind turbines [1,2] can play a role in a future energy scenario of distributed energy production, especially in combination with other renewable energy sources in proximity of populated areas [3] served by a smart-grid logic: in these applications, novel concepts and designs are also under development [4–7].

Overall, it is undisputed that wind energy has become a consolidated industry, with the connected benefits and drawbacks. Among the latter, it is apparent that the complexity of wind turbines and their installation in challenging, open environments imply the need of accounting for a variety of "side-issues" that go beyond the pure engineering design of the turbine, such as, for example, the interaction with the atmosphere and the wake of other machines [2], the grid integration, or the effects of environmental conditions [8,9]. As a matter of fact, as the development of wind energy requires a multi-disciplinary approach, it is the combination of several areas that make things happen and that often defines new scientific challenges [10].

At the moment, however, a truly holistic approach to wind turbine design is still missing. To achieve it, it is apparent that simulation will play a key role since it represents the common framework, within which the different disciplines and sciences can communicate. Developments are needed in each of them, from more accurate but computationally affordable simulation techniques [11,12], to innovative control logics [13] or effective methods for noise prediction [14]. Focusing more specifically on aerodynamics, it is apparent that the use of high-fidelity computational techniques is key to properly model the new generation of long and flexible blades, where aero-elastically-tailored designs are needed, and different flow control devices are often used to alleviate the loads [15,16]. As discussed, all these new methods and tools, however, should be developed synergically by the different groups of specialists, in view of that integration that represents the only way of properly addressing the reliable modeling of such a complex system like a wind turbine.

The contributions from the research works included in this Special Issue offer new data, information, and findings to continue the R&D effort in wind turbine simulation, with the aim of stimulating the research community to further contribute to the development of the field. Sincere thanks are, therefore, due to all the authors that contributed with their works to this Special Issue.

Funding: This research received no external funding.

Conflicts of Interest: The authors declare no conflict of interest.

References

1. Papi, F.; Nocentini, A.; Ferrara, G.; Bianchini, A. On the Use of Modern Engineering Codes for Designing a Small Wind Turbine: An Annotated Case Study. *Energies* **2021**, *14*, 1013. [CrossRef]
2. Yang, X.; Foti, D.; Kelley, C.; Maniaci, D.; Sotiropoulos, F. Wake Statistics of Different-Scale Wind Turbines under Turbulent Boundary Layer Inflow. *Energies* **2020**, *13*, 3004. [CrossRef]
3. Lipian, M.; Czapski, P.; Obidowski, D. Fluid-Structure Interaction Numerical Analysis of a Small, Urban Wind Turbine Blade. *Energies* **2020**, *13*, 1832. [CrossRef]
4. Lanzafame, R.; Mauro, S.; Messina, M.; Brusca, S. Development and Validation of CFD 2D Models for the Simulation of Micro H-Darrieus Turbines Subjected to High Boundary Layer Instabilities. *Energies* **2020**, *13*, 5564. [CrossRef]
5. Balduzzi, F.; Zini, M.; Molina, A.C.; Bartoli, G.; De Troyer, T.; Runacres, M.C.; Ferrara, G.; Bianchini, A. Understanding the Aerodynamic Behavior and Energy Conversion Capability of Small Darrieus Vertical Axis Wind Turbines in Turbulent Flows. *Energies* **2020**, *13*, 2936. [CrossRef]
6. Mansour, H.; Afify, R. Design and 3D CFD Static Performance Study of a Two-Blade IceWind Turbine. *Energies* **2020**, *13*, 5356. [CrossRef]
7. Sobczak, K.; Obidowski, D.; Reorowicz, P.; Marchewka, E. Numerical Investigations of the Savonius Turbine with Deformable Blades. *Energies* **2020**, *13*, 3717. [CrossRef]
8. Papi, F.; Cappugi, L.; Salvadori, S.; Carnevale, M.; Bianchini, A. Uncertainty Quantification of the Effects of Blade Damage on the Actual Energy Production of Modern Wind Turbines. *Energies* **2020**, *13*, 3785. [CrossRef]
9. Santo, G.; Peeters, M.; Van Paepegem, W.; Degroote, J. Fluid–Structure Interaction Simulations of a Wind Gust Impacting on the Blades of a Large Horizontal Axis Wind Turbine. *Energies* **2020**, *13*, 509. [CrossRef]
10. Van Kuik, G.A.; Peinke, J.; Nijssen, R.; Lekou, D.; Mann, J.; Sørensen, J.N.; Ferreira, C.; van Wingerden, J.W.; Schlipf, D.; Gebraad, P.; et al. Long-term research challenges in wind energy—A research agenda by the European Academy of Wind Energy. *Wind Energy Sci.* **2016**, *1*, 1–39. [CrossRef]
11. Ma, Z.; Lei, L.; Dowell, E.; Zeng, P. An Experimental Study on the Actuator Line Method with Anisotropic Regularization Kernel. *Energies* **2020**, *13*, 977. [CrossRef]
12. Li, Z.; Yang, X. Evaluation of Actuator Disk Model Relative to Actuator Surface Model for Predicting Utility-Scale Wind Turbine Wakes. *Energies* **2020**, *13*, 3574. [CrossRef]
13. Perez-Becker, S.; Marten, D.; Nayeri, C.N.; Paschereit, C.O. Implementation and Validation of an Advanced Wind Energy Controller in Aero-Servo-Elastic Simulations Using the Lifting Line Free Vortex Wake Model. *Energies* **2021**, *14*, 783. [CrossRef]
14. Tang, H.; Lei, Y.; Li, X. An Acoustic Source Model for Applications in Low Mach Number Turbulent Flows, Such as a Large-Scale Wind Turbine Blade. *Energies* **2019**, *12*, 4596. [CrossRef]
15. Sy, M.S.; Abuan, B.E.; Danao, L.A.M. Aerodynamic Investigation of a Horizontal Axis Wind Turbine with Split Winglet Using Computational Fluid Dynamics. *Energies* **2020**, *13*, 4983. [CrossRef]
16. Wiśniewski, P.; Balduzzi, F.; Buliński, Z.; Bianchini, A. Numerical Analysis on the Effectiveness of Gurney Flaps as Power Augmentation Devices for Airfoils Subject to a Continuous Variation of the Angle of Attack by Use of Full and Surrogate Models. *Energies* **2020**, *13*, 1877. [CrossRef]

energies

MDPI

Article

On the Use of Modern Engineering Codes for Designing a Small Wind Turbine: An Annotated Case Study

Francesco Papi , Alberto Nocentini, Giovanni Ferrara and Alessandro Bianchini *

Department of Industrial Engineering, Università Degli Studi di Firenze, via di Santa Marta 3, 50139 Firenze, Italy; fr.papi@unifi.it (F.P.); alberto.nocentini@stud.unifi.it (A.N.); giovanni.ferrara@unifi.it (G.F.)
* Correspondence: alessandro.bianchini@unifi.it

Abstract: While most wind energy comes from large utility-scale machines, small wind turbines (SWTs) can still play a role in off-grid installations or in the context of distributed production and smart energy systems. Over the years, these small machines have not received the same level of aerodynamic refinement of their larger counterparts, resulting in a notably lower efficiency and, therefore, a higher cost per installed kilowatt. In an effort to reduce this gap during the design of a new SWT, the scope of the study was twofold. First, it aimed to show how to combine and best exploit the modern engineering methods and codes available in order to provide the scientific and industrial community with an annotated procedure for a full preliminary design process. Secondly, special focus was put on the regulation methods, which are often some of the critical points of a real design. A dedicated sensitivity analysis for a proper setting is provided, both for the pitch-to-feather and the stall regulation methods. In particular, it is shown that stall regulation (which is usually preferred in SWTs) may be a cost-effective and simple solution, but it can require significant aerodynamic compromises and results in a lower annual energy output in respect to a turbine making use of modern stall-regulation strategies. Results of the selected case study showed how an increase in annual energy production (AEP) of over 12% can be achieved by a proper aerodynamic optimization coupled with pitch-to-feather regulation with respect to a conventional approach.

Keywords: wind turbine; pitch; stall; aerodynamics; engineering codes

Citation: Papi, F.; Nocentini, A.; Ferrara, G.; Bianchini, A. On the Use of Modern Engineering Codes for Designing a Small Wind Turbine: An Annotated Case Study. *Energies* **2021**, *14*, 1013. https://doi.org/10.3390/en14041013

Academic Editor: David Wood
Received: 13 January 2021
Accepted: 10 February 2021
Published: 15 February 2021

1. Introduction

To fulfill global energy needs, manufacturers and most of the wind turbine industry have concentrated their efforts on large utility-scale machines [1]. The standard design for horizontal-axis turbines consists of a three-blade, upwind rotor featuring an active yaw and pitch regulation. Such machines benefit from large levels of aerodynamic optimization, often using purposely developed airfoils featuring twisted and tapered blades and large resources for development and testing. On the other hand, small wind turbines (SWTs) often do not feature the same level of optimization, with low power coefficients often resulting from unoptimized designs [2]. Such sub-optimal aerodynamic designs have been identified amongst the issues that hamper the diffusion and economic feasibility of SWTs [3,4], with larger SWTs suffering the most from the often used simplistic approaches [4]. This type of turbine, which marked the dawn of wind energy, is still used in a variety of applications, from rural areas to off-grid applications [5]; notwithstanding this, their high levelized cost of energy [6] has thus far hampered an effective diffusion. On the other hand, interest has been rising lately again, as testified by the creation of a dedicated technical committee for SWTs by the European Academy of Wind Energy (EAWE) [7]; this is mainly due to the role that distributed production, even with small rated power, could have in the transition towards smart energy systems [8]. In doing so, the "old generation" of turbines seems unsuitable in terms of efficiency and flexibility, and so better designs are about to be explored.

The present article aimed to analyze the main issues causing low power output in an SWT while also detailing how an effective preliminary design can be achieved by using and properly integrating current industry best-practices and open-source tools. In particular, even though the latter are indeed familiar to the wind energy community, their conscious use is not trivial, and organic design guidelines are often not available. Effectively and economically designing an SWT is not a trivial task, and many hurdles must be overcome in terms of aerodynamics, materials, structural resistance, and economics.

While a good overview of these issues can be found in [4], the aim of the present study was twofold. On the one hand, it aspired to provide the reader with an organic overview of the steps that need to be followed for a first turbine design, suggesting how to integrate existing engineering open-source tools and how to tune them, especially in cases of the realistic turbulent inflow conditions that are required by standards (summarized in the chart in Appendix A). This, while not completely novel from a scientific point of view, is thought to be of industrial relevance and significance for newcomers. Guidelines and general indications on blade design can be in fact found in the available literature [9,10]. For instance, various aspects of aerodynamic design and optimization were discussed in [11]. Such studies, however, do not account for control or dynamic inflow conditions [12]. The study instead specifically focused on the implication of control for small wind turbines. In particular, it is shown that using modern control strategies, which have rarely been applied to SWTs, can lead to much more efficient design and more convenient loading. The importance of making an early decision regarding control in the design phase was assessed, as this aspect significantly influences aerodynamic design, and controller tuning and optimization should go hand in hand with aerodynamic optimization. Most small wind turbines indeed use a stall as their main power-limiting strategy. This involves controlling the rotor speed so that, as the wind speed increases, the turbine gradually enters the stall, the lift decreases, and the drag increases, thus effectively regulating the power output. Fixed-speed stall-controlled turbines were the de-facto standard in the nineties [2], and successful applications of this design can be found [13]; however, most stall-regulated turbines, including commercially available products, now feature variable speed generators [14]. By adopting variable-speed control, a turbine is able to operate at or near the design tip–speed ratio (TSR) at a low wind speed, greatly improving energy capture. Even when adopting variable speed control, however, significant compromises must be made in order to ensure good stall regulation, from setting the blades to a manual fixed pitch angle to varying the twist and chord distributions of the blade. Such compromises can be avoided if pitch regulation is employed. Two kinds of pitch regulation strategies are possible: pitch-to-stall and pitch-to-feather. As noted in [15], the pitch-to-stall strategy is able to provide effective regulation, though it increases most design loads. Moreover, given that the pitch-to-feather strategy is the most widely adopted control method in modern utility-scale turbines, this kind of pitch regulation is discussed in this paper. While the benefits of this control strategy are apparent, it does not come without drawbacks, mainly connected to its added complexity and, especially, cost. However, examples of manufacturers proposing this kind of solution can be found, as is the case with the line of products by Tozzi Nord [16]. For all of these reasons combined, two SWT designs are compared in this paper—a variable-speed stall-regulated turbine and a variable-speed, pitch-regulated turbine.

The selected testcase for the entire analysis was a 50 kW machine with a 200 m^2 swept rotor area, which is in line with the definition of an SWT according to International Electrotechnical Commission (IEC) 61400-2 [12]. The authors indeed had a direct industrial experience with this size of machine, and this experience mainly drove the present study. Due to the industrial non-disclosure agreement with the partner, however, all analyses were repeated on a purely theoretical case study. Notwithstanding this, the results are fully representative of those found in reality.

2. Methods

In this section, the numerical tools used throughout the study are briefly presented. The methods to determine a preliminary chord and twist distribution are explained. Then, the airfoils considered in the design process are discussed, and the modifications done to the ideal design in order to meet the desired targets in terms of power output are analyzed in detail.

2.1. Numerical Tools

The aerodynamic design process described in the previous sections was conducted using open-source, industry-standard computational tools. It is worth pointing out again that these methods are indeed not novel and well-known to the scientific community. Additionally, they have been extensively validated on a many study cases, both of small and utility-scale wind turbines, assessing their suitability for the scope. However, while approaching the industrial design of a new SWT, the authors found that detailed guidelines on how to consciously, organically, and in an integrated fashion use these methods were missing. On these bases, this study would like to represent a support to the industrial and scientific community in an overall attempt of improving the future design of this class of machines. A brief overview of the tools used in the study is provided in this paragraph.

Lift and drag airfoil characteristics were obtained with XFoil [17]. The tool is based on an inviscid panel method, and it has been used in the design of a vast amount of airfoils for all sorts of engineering uses, including the families used herein [18]. Even though a recent study pointed out that this method may have some issues in case of low Reynolds numbers and high angles of attack [19], its use in horizontal-axis rotors from the present power output and above can be considered as a solid choice, especially for the first design of blades, when multiple design variations need to be compared quickly.

For the present study, the characteristics were calculated using 200 panels per airfoil and setting a trailing edge gap of 2%. A Reynolds number of 1×10^6 was used. The boundary layer transition was calculated with the N_{crit}-based shear layer transition method [20], and a value of $N_{crit} = 9$ was used. The Reynolds number matched the final operating Re number fairly well, ranging between 0.8×10^6 and 1.3×10^6 depending on operating conditions, and was therefore considered acceptable; however, if this is not the case, a few design iterations might be required to ensure that lift and drag polars were suitable for the test case. The full-blade aerodynamic design was conducted in OpenFAST [21]. This open-source modular tool was developed by the National Renewable Energy Laboratory (NREL) and can model the full response of wind turbines, accounting for a wide variety of effects such as aerodynamics, elastodynamics, control-dynamics, and, for offshore installations, hydrodynamics. The code has been widely adopted, validated, and used in the design of multiple, industry-standard, reference wind turbines [22,23]. In the present study, only the aerodynamic and control-dynamics perspectives were explored in detail. The aerodynamic module AeroDyn [24] allows for the simulation of dynamic inflow conditions in the presence of atmospheric turbulence. Blade element momentum (BEM)-based aerodynamics also include corrections for wind shear, yaw misalignment, tip and hub losses, and tower-shadow effects. Dynamic stall is treated with the Beddoes–Leishman dynamic stall model included in AeroDyn. This correction is especially relevant for a stall-operated turbine operating in turbulent conditions. The turbine controller was integrated through the ServoDyn module. For the pitch-controlled turbine, an external routine was used, as detailed in the next section.

2.2. Ideal Blade Shape

A tentative blade design could be determined using BEM theory. A detailed explanation of the equations and their derivation can be found in [9], as only the aspects relative to blade design are briefly presented herein. By manipulating BEM equations to express the power coefficient for each radial section without considering drag [9], one gets:

$$C_P(r) = \frac{8}{\lambda^2}\lambda_r^3 a'(1-a) \tag{1}$$

The equation can be rearranged and written in terms of the flow angle φ:

$$C_P(r,\lambda,\varphi) = \frac{8}{\lambda^2}\sin^2\varphi(\cos\varphi - \lambda_r\sin\varphi)(\sin\varphi + \lambda_r\cos\varphi)\lambda_r^2 \tag{2}$$

The flow angle distribution along the span that maximizes the power coefficient (Cp) can be found by setting the partial derivative of Equation (2) equal to zero:

$$\frac{\partial}{\partial\varphi}(C_P(r,\lambda,\varphi)) = 0 \tag{3}$$

Solving Equation (3), one then gets:

$$\varphi = \frac{2}{3}\tan^{-1}\left(\frac{r}{R}\lambda\right) \tag{4}$$

The local blade twist can be calculated based on the flow angle as:

$$\gamma = \varphi - \theta - \alpha_{des} \tag{5}$$

where θ is the blade pitch angle and α_{des} is the local design angle of attack. The local blade chord can also be expressed as [9]:

$$c = \frac{8\pi r}{BC_l}(1-\cos\varphi) \tag{6}$$

The twist and chord distributions obtained from Equations (5) and (6) do not account for drag and tip losses, and so the design angle of attack α_{des} should be selected as the angle of attack that maximizes the glide ratio of the airfoil employed at the selected local radius. As shown in [10], when the airfoil glide ratio exceeds 40, the assumption of neglecting drag can be reasonably assumed. Moreover, the proposed design method determines the ideal blade shape in design conditions, with set tip–speed ratio (TSR) and pitch angle. It is then apparent that the design conditions should be chosen carefully. Rather than choosing the rotor and wind speed at rated conditions, a sounder choice would be to choose operating conditions based on the design wind speed distribution. In the present test case, a design wind speed of 8.5 m/s was chosen as the mean wind speed of a class IIA wind speed distribution. Another good choice could be the mode of the wind speed distribution. When designing a fixed-speed wind turbine, the mode of the wind speed distribution (i.e., the most frequent wind speed) should be chosen as a design point in order to ensure the turbine is operating at its design TSR most of the time. A variable speed wind turbine, on the other hand, can vary rotor speed to maintain a nominal TSR as wind speed varies, and the mean wind speed is therefore also a good choice because it ensures that the rotor speed is closer to the nominal value at the design point. The design TSR must also be chosen carefully, as this will contribute to determining rotor speed. Modern rotors generally operate between TSRs of 4 and 10 [25]. Higher tip–speed ratios decrease blade solidity and increase aerodynamic noise [26,27]. Therefore, based on these considerations and similar existing turbine designs [14,28–30], a medium-low TSR of 5.7 was selected here.

2.3. Airfoil Families

In order to obtain smooth chord and twist distributions, airfoils from the same family must be used along the entire blade. Several airfoil families have been designed specifically for wind turbines over the years by laboratories, scientists, and technical institutions such as NREL (USA), Risø (Denmark), and Delft (The Netherlands). [18,31–33]. The selection of the required airfoils plays a crucial role in the aerodynamic design process. The shape of the selected airfoils is a compromise between performance, regulation characteristics (especially important in stall-regulated wind turbines), and structural stiffness. The mid and outer sections of a wind turbine blade are typically optimized for high aerodynamic performance, while the inner sections are designed to provide the required structural integrity and stiffness for the blade. Suggesting a family of airfoils is definitely not an easy task, since any of them have specific benefits and drawbacks that may be relevant to each different application; also, companies sometimes are willing to design proprietary airfoils tailored for their machine. However, the scope of the present work was to show how, even in case where one selects a very well-known family of "standard" airfoils, effective turbine designs can be achieved. In detail, research into airfoil families that would be suitable for a 50 kW wind turbine with a rotor diameter of 16 m led to the selection of two different airfoil families belonging to the S800 group developed and tested by NREL [18] for medium-size turbines rated at 20–150 kW with blades from 5 to 10 m in length, which was the size category of our interest.

The first family considered (Figure 1a) [18,34], with thin tip airfoils, was designed in 1987 and includes the S805A, S806A, S807, and S808 airfoils. This airfoil family was designed to have a low tip maximum lift coefficient (Cl_{max}) (1.0) for a Reynolds number just over $1*10^6$, and it is suitable for stall-regulated blades. The "A" designation stands for an improved version of an airfoil, based on wind-tunnel test results for a similar airfoil.

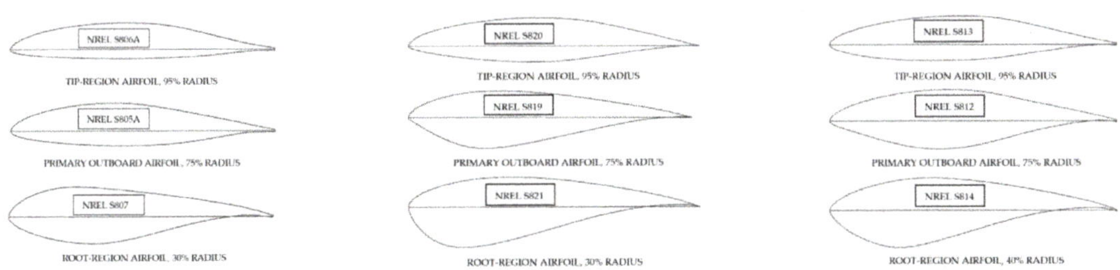

(a) (b) (c)

Figure 1. (**a**) Thin-airfoil family for medium blades (low tip Cl_{max}), (**b**) thick-airfoil family for medium blades (low tip Cl_{max}), (**c**) thick-airfoil family for large blades (low tip Cl_{max}).

The second family (Figure 1b) [18,35], having thick tip airfoils, was designed in 1993 and consists of the S819, S820, and S821 airfoils. This family was designed to have performance characteristics similar to the previous family. The greater tip-region thickness helps accommodate overspeed-control mechanisms for stall-regulated rotors at the expense of a slightly higher drag [36,37]. Though these mechanisms are not used in modern turbines and were thus not included in this case study, the increased thickness is structurally beneficial. The S821 blade-root airfoil was designed to have restrained maximum lift coefficients, and have low profile-drag coefficients, and to be as insensitive as possible to roughness.

The low design lift coefficient of these airfoil families is indeed a design trait [33,36,38]. Specifically, on an SWT, the operating Reynolds number must be as high as possible to achieve the best aerodynamic performance. Based on Equation (6), decreasing the design lift coefficient implies an increase of the chord size required to reach a certain performance level. In turn, this increases the operating Reynolds number, which helps to lower drag and increase the glide ratio of the blade [10].

Finally, the S812, S813, and S814 (Figure 1c) [18,39–41] airfoil family was designed for large rotors rated at 100–400 kW with blades 10–15 m in length. Though this family of airfoils did not seem to fit the specification of the test case, it has been used successfully on the Atlantic Orient AOC 15/50 three-bladed wind turbine. The designation 15/50 refers to the 15 m diameter rotor and its rated output of 50 kW [18,28,42]. This rated output is achieved at 12 m/s by the 50 Hz version and at 11.3 m/s by the 60 Hz version. This airfoil family was therefore also taken into consideration.

2.4. Preliminary Performance Curves

Following the steps to determine the blade design presented in Section 2.1, the first steady analyses on OpenFAST for the three airfoil families were carried out. The focus in this phase was put on the stall-regulated turbine, as in this case, the aerodynamic design also influenced the regulation characteristics. As discussed previously, the design point was chosen to be 8.5 m/s, which is the mean wind speed of a class IIA (see Table 1).

Table 1. Turbine specifications. IEC: International Electrotechnical Commission.

Parameter	Value
IEC wind class Rotation axis	IIA Horizontal
Number of blades	3
Rotor diameter Hub radius	16 m 0.5 m
Rated Power	50 kW
Cut-in/cut-out wind speed	2–20 m/s
Rated wind speed	12 m/s
Hub height	20.5 m
Airfoil family	NREL S821-19-20

Figure 2 shows the aerodynamic power produced as a function of the wind speed. These power curves were obtained without stall delay correction, and—only afterward once the most promising design was chosen—were the polars 3D-corrected and further refinements done (see Section 2.5). The regulation method for this preliminary design was variable speed stall regulation. In particular, the powers produced at 12 m/s were as follows: 50.4 kW for the S819–21 family, 52.6 kW for the S812–14 family, and 52.1 for the S805–8 family.

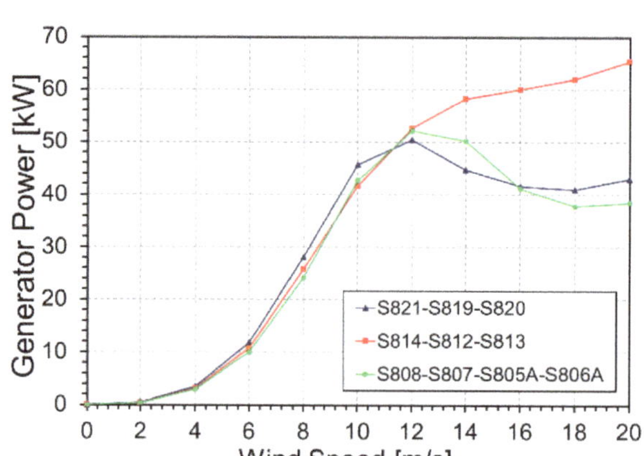

Figure 2. Turbine curves of generator power from the steady simulations in OpenFAST. Maximum rotor speed was set to 60 rpm.

Upon examination of the performance comparison in terms of power output, one could notice that the S821, S819, and S820 family was preferable to the others. In fact, it allowed us to reach the set power target of 50 kW at 12 m/s with good stall regulation, and it generated more power for below-rated wind speeds than the S805, S806, S807, and S808 family. The main preliminary characteristics of the turbine after the first preliminary design phase are shown in Table 1.

2.5. Modifications to Ideal Design

The ideal blade design seen so far did not account for various practical aspects that have to be considered in a real design [43]. The ideal blade first needs to be tapered at the tip. In this area, tip losses, which are not considered during preliminary design, would greatly decrease the energy extracted at blade tip [19,38]. In the present study, the tip region was tapered from 95% blade span outwards empirically. Since BEM methods were used in the present preliminary design study, although Prandtl's tip and hub-loss corrections were included in this work, accurate tip-loss evaluation was not possible and the influence of different tapering strategies was hard to assess. In fact, despite the fact that BEM methods are able to capture the primary effects of blade tapering by resolving blade–chord variations, the chord variations at the blade tip also influence tip-vortex strength and, as a consequence, blade loading. While some tip-loss correction models are somewhat sensible for tip chord distribution [44], Prandtl's model is not [24]. Moreover, while these corrections may be more sophisticated, they remain unlinked to the underlying physics; therefore, to properly study the effects of this phenomena, more sophisticated aerodynamic models are required. Decreasing the chord at the tip region also decreases aerodynamic loading, which is beneficial from a load standpoint and has little aerodynamic penalty due to the presence of the aforementioned tip-losses.

The blade also needs to be tapered at the blade root, where it is connected to the rotor hub. In this area, the local tip–speed ratio is very low, the local radius is short, and the produced torque is thus very low. On this basis, it is common practice to taper a blade empirically.

The lift and drag coefficients used throughout the blade must also be corrected to account for 3D flow effects. 3D effects were first noted by Himmelskamp [45] and tend to greatly increase the high-angle of attack lift of the inboard sections of a rotating blade [46]. These effects are present in rotating blades and although the underlying physics are not fully understood to this day, they seem to be caused by complex flow interactions in the boundary layer. In practical terms, radial pressure nonuniformities along the rotor blade crease radial flow patterns, which have the main effect of delaying the stall. A brief explanation of the phenomena, as well as additional references, can be found in Chapter 3 of [10]. In the present study, the model proposed by Bak [47] was adopted and suggested. This model corrects both the lift and drag coefficients, and it can be relatively easily applied as an empirical correction step before the aerodynamic simulations are performed. The inclusion of 3D-effects was found to have a notable impact on turbine performance, especially for the stall-regulated turbine, as shown in Figure 3. Power was found to vary quite noticeably. At 12 m/s, the blades with a fixed -6.5° pitch angle produced 49.4 kW (3D) and 48.4 kW (2D), while the blades with a -5° pitch angle produced 50.4 kW (2D) and 59.8 kW (3D).

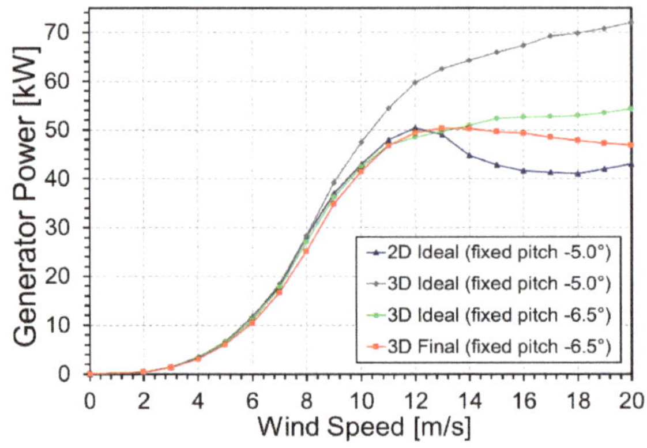

Figure 3. Generator power as a function of wind speed for the stall-regulated ideal blade design with 2D lift and drag coefficients, the ideal design with 3D effects, and the final design with 3D effects.

For this reason, the fixed blade pitch of the stall-regulated turbine was further tuned, and the twist of the inboard sections of the blade, which are most affected by stall delay, were modified to ensure the desired regulation characteristics, as shown in Figure 4. Reducing the twist angle increases the angle of attack, therefore pushing this part of the blade towards the stall. In fact, the angle of attack can be found from the flow angle φ, twist angle δ, and pitch angle θ as:

$$\alpha = \varphi - \delta - \theta \tag{7}$$

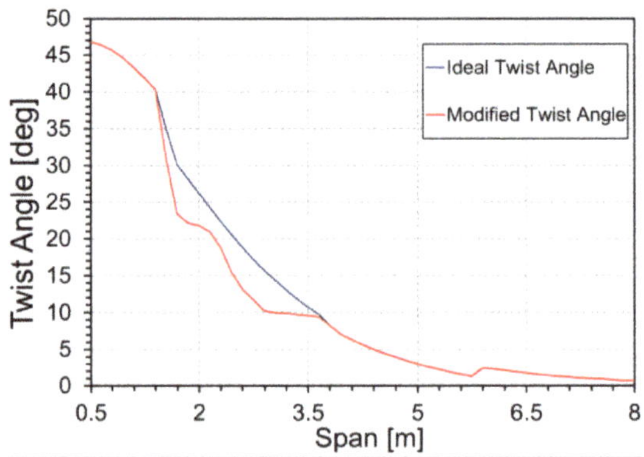

Figure 4. Changes to the optimal twist distribution to ensure good stall regulation.

Therefore, reducing the twist angle increases angle of attack, although it should be noted that changing blade twins influences axial and tangential induction, therefore changing the induced velocities and affecting the flow angle in Equation (7). The overall trend, however, remains valid, although some trial and error might be necessary. In other words, changing the twist angle can be seen as partial compensation for the stall delay effect, which, in contrast and as the name suggests, tends to delay the point of the stall, thus negatively affecting the blade's regulation capacity.

3. Turbine Control

In this section, the two adopted regulation methods, as well as the benefits and possible drawbacks of each solution, are explained in detail. This is probably the key element of the study since modern control strategies were often not applied to SWTs in the past. Throughout this paper, however, it will be proven that their use can be largely beneficial also in these rotors, leading to more effective designs.

3.1. Pitch Control

Blade pitch control as a means of power curtailment is the modern control method, adopted on all utility-scale wind turbines. While two methods of pitch control are available, i.e., pitch-to-stall and pitch-to-feather, only the latter is used because it allows for much lower out-of-plane loads at high wind speeds.

The open-source NREL ROSCO controller [48] was used for this test case and is suggested as a valuable tool for a first analysis. The variable-speed pitch controller was developed based on the work of Mulders et al. [49], and it is able to regulate generator torque and blade pitch. It also allows for yaw control and individual pitch control (IPC).

Below the rated wind speed, a blade pitch is kept constant at fine pitch; in this case, it was set to 0°. The generator torque is calculated as in Equation (8):

$$\tau_g = K\omega_g^2 \tag{8}$$

where ω_g is the generator speed. As also shown in [50], this simple formula is the result of the fact that in order to ensure maximum performance, the turbine must operate at peak Cp for all below-rated wind speeds. In the absence of pitch control, not active in this region, Cp is a function of the tip–speed-ratio alone that must therefore be kept constant. Therefore, as the theoretical available power in proportional to the cube of the wind speed, the generated power must be proportional to the cube of the rotor speed. The power maximizing the generator's torque constant (Equation (9)) can be calculated as [48]:

$$K = \frac{\rho\pi R^5 C_P}{2\lambda^3 N_g} \tag{9}$$

where N_g is the generator drive ratio. The relation can be easily derived from the expression of the rotor power coefficient by imposing generator torque, as in Equation (8).

Above rated power, the generator torque is fixed to the design torque $\tau_g = P_r/\omega_r$ and the blade pitch is controlled with a Proportional-Integral (PI) controller (Equation (10)):

$$\Delta\theta(t) = K_p\Delta\omega_g(t) + K_i\int \Delta\omega_g\,dt \tag{10}$$

where θ is the blade pitch. The proportional and integral gains Kp and Ki, respectively, depend on the blade pitch angle; in particular, as the blade pitch increases, rotor speed variations are more sensitive to small pitch variations. PI control in Equation (10) is derived from a more general Proportional-Integral-Derivative (PID) control strategy with the derivative term (D-term) set to zero. This is common practice in wind turbine pitch controllers. In most cases, the controller is able to adequately control rotor speed without the D-term, making controller tuning easier because there is one less parameter to tune. Furthermore, the D-term is very sensitive to high frequency fluctuations of the rotor speed, and an ill-chosen D-term could therefore introduce instability in a controller. Traditional tuning techniques involve the linearization of the system around an operating point to find controller gains. The linearization procedure must be repeated several times in the operating range. Alternatively, various authors have proposed methods to empirically calculate the gains [51].

In the present testcase, the open-source ROSCO toolbox [52] was used to tune the controller. The gains were analytically calculated and depended on the design natural frequency ω_{des} and damping ratio ξ_{des}. In general, increased values of ω_{des} decrease rotor speed response time, while increased values of ξ_{des} decrease the amount of rotor speed overshoot. For the present testcase, the values of 0.82 and 1.4 were empirically selected for ω_{des} and ξ_{des}, respectively. These values were substantially higher than those found in much larger reference wind turbines, where values of ω_{des} of 0.2–0.3 and ξ_{des} of 0.7–0.9 are common [16,42] and are needed to effectively regulate a small wind turbine with low rotor inertia. As noted in [48], there is a limit to how fast rotor speed can be controlled (how high ω_{des} can be) without incurring in erratic blade pitch behavior, and the value of 0.82 adopted for this case was found to be at the upper limit of this range. The controller response was tested with wind-step simulations below at and above rated speeds, as well as in turbulent wind. While response to turbulent wind is discussed in the following section, response to wind increments of 2 m/s are shown in Figure 5. The erratic blade pitch behavior can be clearly seen at 300 and 350 s.

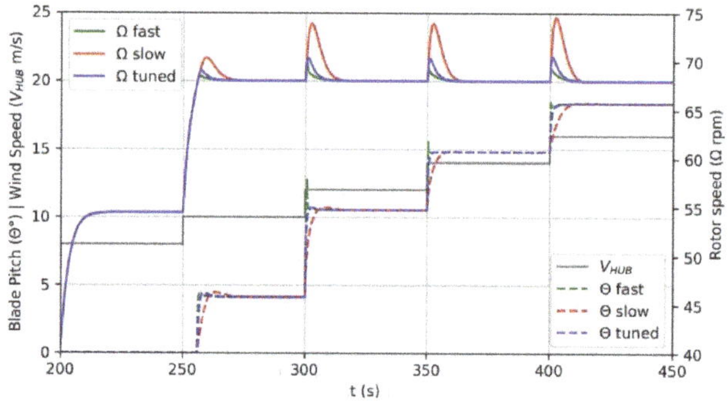

Figure 5. Turbine response to step-wind profiles for different values of controller natural frequency and damping ratio. Design natural frequency $(\omega_{des}) = 0.82$ and damping ratio $(\xi_{des}) = 1.4$ for the tuned case, $\omega_{des} = 1.2$ and $\xi_{des} = 2$ for the "fast" case, and $\omega_{des} = 0.4$ and $\xi_{des} = 0.8$ for the "slow" case. Values for the "slow" case were still greater than those typically employed on utility-scale machines.

3.2. Stall Control

In this section, a variable-speed, stall-regulated strategy that eliminates the need for ancillary aerodynamic control systems is evaluated.

The variable-speed operation of wind turbines presents certain advantages over constant speed operation [50,53]. The primary advantage claimed for variable-speed turbines is the increased energy capture during partial load operation. Variable-speed operation allows the turbine to operate at near optimum Cp and to maximize power over a range of wind speeds. Moreover, variable-speed wind turbines use the inertia of the rotating mechanical parts of the system as a flywheel; this helps to smooth power fluctuations and reduces the drive train mechanical stress. Secondary benefits are acoustic signature and power quality [51]. The control logic is described in detail in [54], but the main details are explained herein as regulation strategy that significantly influences turbine regulation and, consequently, aerodynamic choices.

Typical variable-speed wind turbines have different regions of operation, as shown in Figure 5, where the generator torque as a function of the generator speed is shown. The turbine startup occurs in region 1, where the generator torque is zero. Once the generator speed has reached cut-in speed and power is produced normally, the turbine is operating in region 2. In this region, the generator torque control is used to vary the speed of the turbine to maintain the constant TSR corresponding to optimum Cp, thus maximizing the energy capture. In region 2, the torque curve is calculated as in Equation (7) and intersects the rated torque at a rotor speed that is significantly higher than the rated speed. It would of course be beneficial to operate the turbine on region 2 at an optimum Cp curve up to where it intersects the rated torque, but the operation of the turbine at these high rotor speeds would result in a high blade tip speed and unacceptable noise emissions [31]. Therefore, a transition region is included between regions 2 and 3 (region $2\frac{1}{2}$). Region $2\frac{1}{2}$ depends linearly on rotor speed, starting at a rotor speed lower than the rated speed ω_1 and reaching the rated torque at, or slightly below, the rated speed ω_2. The generator torque for this region can be expressed as Equation (11):

$$\tau_g(\omega) = \tau_1 + \frac{\tau_{rated} - \tau_1}{\omega_2 - \omega_1}(\omega - \omega_1) \tag{11}$$

where ω is rotor speed, τ_1 is the generator torque at the rotor speed in which this region starts (ω_1), τ_{rated} is rated torque, and ω_2 is the rotor speed at which we reach rated torque. Above the rated speed, the generator torque is set equal to the rated torque τ_{rated}.

In region 3, generator torque is simply held constant at rated torque (see Figure 6).

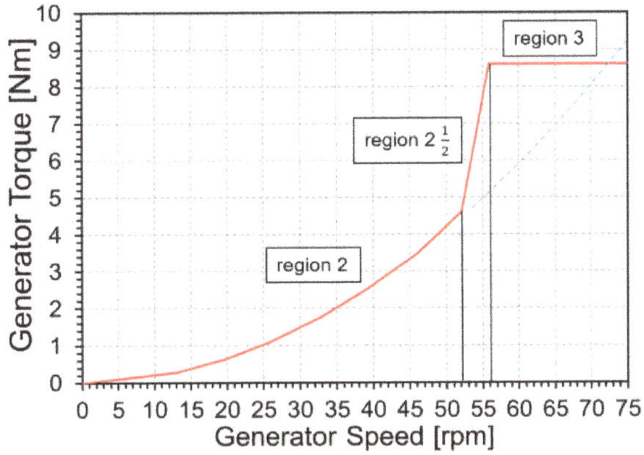

Figure 6. Variable-speed turbine operating regions.

Adequately tuning the slope and position of region $2\frac{1}{2}$ ensures effective turbine regulation. Through region $2\frac{1}{2}$, the turbine is controlled to limit its rotational speed and, consequently, output power. In fact, limiting rotor speed decreases the TSR and forces the rotor into an aerodynamically stalled condition. This is usually called the "soft-stall" approach because it allows for the introduction of rather benign stall characteristics for the purposes of controlling maximum power.

3.3. Control Input Parameters

In Table 2, the main parameters used to set the torque-control strategy of the two turbines are shown. These are a result of a (in most cases) necessary sensitivity analysis. This paragraph hopefully helps the interested reader understand the influence of some of the main control parameters and how they can be tuned to reach the desired turbine performance. A baseline for these control parameters can be determined using the methods detailed in Sections 3.1 and 3.2 Several common techniques to ensure that power is correctly regulated using both control schemes were adopted for this study. The rated rotational speed of the stall-regulated turbine (i.e., the beginning of region 3 in Figure 6) was limited to 60 rpm because the turbine was designed to operate at a nominal TSR at 8.5 m/s wind speed and to enter off-design conditions as wind speed increases to force the blade to stall and the power to be regulated. Therefore, to effectively regulate power with a stall control scheme, the turbine needs to be forced to enter off-design conditions before the rated wind speed. A nominal rotor speed for the pitch-regulated turbine was chosen so that the design TSR could be maintained up to 10 m/s wind speed. The high value of rated generator torque for the stall-regulated turbine was set to avoid rotor overspeed in high wind speed turbulent scenarios. In practice, this means that, even at a rated power, the turbine would operate in region $2\frac{1}{2}$. Operating in this region would ensure that the rotor does not speed-up as a response to steep wind speed increases. On the contrary, if the rotor is allowed to speed-up, the TSR and, consequently, the power increase, causing the rotor to quickly become uncontrolled and reach its terminal velocity. This also highlights the importance of considering dynamic inflow conditions early on in the design stage. Finally, the value of K (Equation (8)), was different in the two cases because the peak C_P design TSR were different for the two turbines.

Table 2. OpenFAST variable-speed generator model inputs.

Parameter	Stall-Regulation	Pitch-Regulation
Rated generator speed	60 rpm	68 rpm
Rated generator torque	12,878.58 Nm	7727.02 Nm
K (see Equation (7))	1.645 Nm/rpm^2	1.694 Nm/rpm^2
Slip % in region $2\frac{1}{2}$	26%	n/a

4. Simulation Set-Up

Once preliminary steady-state performance curves are obtained, it is important to account for more realistic environmental cases early in the design process. The reasons are twofold: first, it is important to assess turbine behavior in dynamic conditions, and secondly, the turbine will have to be certified in the later stages of the design process. For instance, as mentioned previously, it is crucial to verify that adequate turbine control is achieved in dynamic conditions. Moreover, the design loads calculated by simulating the turbine in dynamic environmental conditions can be used as a base for preliminary structural design. Here, the turbine was simulated in a normal power production situation, corresponding to the IEC Design Load Cases (DLCs) 1.2 [12]. The chosen turbine class was class IIA. This represents a class of turbines designed for medium wind speed (W.S.) and high turbulence sites.

One hundred fourteen 10-min simulations were performed for each turbine, reproducing operating conditions specified by the IEC 61400-2 power production DLC-group, including wind shear, yaw misalignment, and turbulence, as detailed in Table 3.

Table 3. IEC 61400-2 DLC 1.2 main set-up parameters.

Parameter	Value
Type of Evaluation	Fatigue/Ultimate
Simulation Length	600 s
Number of Simulations per W.S. and Yaw Angle	3
Wind Speeds	2–20 m/s increments of 1 m/s
Yaw Angles	0/+8/−8 deg
Vertical Inflow Angle	8 deg
Total Number of Simulations	228
Total Simulated Time	38 h

These simulations had wind speeds between two and twenty meters per second in intervals of one meter per second following the standard and industry-accepted guidelines [12,55]. These design cases were representative of power production under normal wind conditions and would therefore be the most common within the turbine lifespan. Though DLCs are designed with structural certification in mind, they were used in this study to verify the productivity of the turbine, as they allowed us to simulate a normal power production scenario. Each simulation used a different turbulent speed (i.e., different turbulent wind field) in order to more realistically reproduce the conditions the turbine will encounter during operation and to avoid biases that might be introduced by a specific wind pattern.

As an additional verification, the convergence of power and annual energy production (AEP) was evaluated, as shown in Figure 7 and Table 4. This was important to evaluate to make sure that the predicted power curves could be considered independent from specific turbulence characteristics. The convergence of power was evaluated in terms of mean power per wind speed calculated with respect to the case using six turbulence speeds per wind speed (adding up to a total of one-hundred-fourteen simulations), as shown in Figure 7a. The analogous convergence of power standard deviation is shown in Figure 7b. Mean power was sufficiently well-predicted by using four turbulent speeds, with variations in mean power below 3% for all wind speeds. The standard deviations required more simulations to properly converge. As shown in Figure 7b, using five turbulent speeds ensured variations in Standard Deviation (STD) below 5% for all wind speeds.

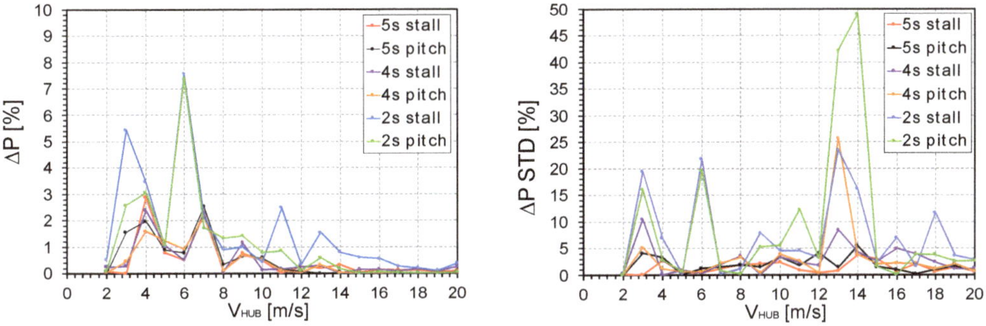

(**a**) (**b**)

Figure 7. Relative error of power (**a**) and power standard deviation (**b**) per wind speed bin with respect to six speeds per wind speed value, using 5 (5 s), 4 (4 s), and 2 (2 s) turbulent speeds per wind speed.

Table 4. Statistical convergence of annual energy production (AEP).

Speeds per WS	**6**	**5**	**4**	**2**
AEP stall	46.06	46.11	46.22	46.32
ΔAEP (%) stall	-	0.117	0.345	0.575
AEP pitch	52.55	52.69	52.65	52.53
ΔAEP (%) pitch	-	0.251	0.176	0.040

AEP already showed strong convergence at two speeds per wind speed and is largely insensitive to increasing the number of speeds. This was in-line with the finding of Bortolotti et al. [56], who noted convergence on predicted fatigue loads and AEP using a small number of turbulent speeds. In conclusion, the minimum requirements of IEC 61400-2 in terms of turbulent speeds were able to guarantee the convergence of power and AEP in the present testcase.

5. Results

In this section, the results of the different design choices discussed so far are critically compared in order to let the reader evaluate their impact on the final performance.

5.1. Steady-State Performance

In order to evaluate general rotor performance, a steady-state performance comparison was carried out. Generator power as a function of windspeed is shown in Figure 8a. Both the stall and the pitch-regulated turbines were able to reach the desired output power of 50 kW. However, the pitch-regulated turbine reached rated power at 10 m/s wind speed, while rated power was not reached until 12 m/s in the stall-regulated turbine. For low wind speeds of up to 8 m/s, the increased power output of the pitch-regulated turbine depended on the increased aerodynamic efficiency of this blade, caused by the fact that blade twist and pitch angle did not need to be compromised for effective stall regulation. From 9 m/s and above, the control strategy also had a direct effect, as the rotor speed was limited for the stall-regulated turbine in order to drive the blades to stall, as shown in Figure 8b and discussed in the previous section.

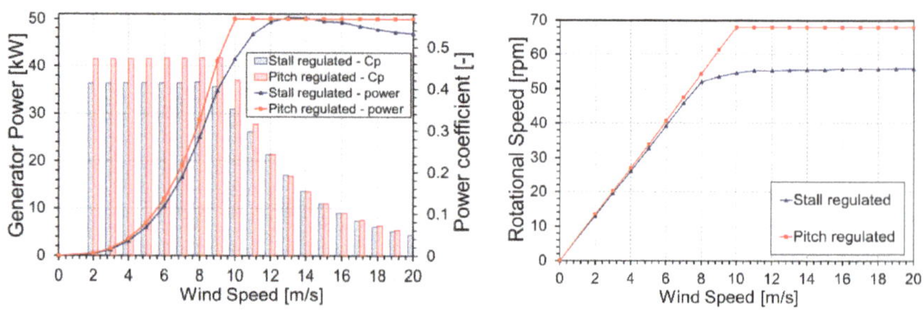

(**a**) (**b**)

Figure 8. (**a**) Generator power and aerodynamic power coefficient; (**b**) rotational speed for pitch- and stall-regulated rotors in steady-state condition.

The observations made from a perusal of Figure 8a are confirmed in Figure 9, where the power coefficient is shown as a function of the tip–speed ratio: the compromises adopted for the stall-regulated turbine resulted in a generally lower power coefficient. Furthermore, the shape of the curve was very different, with the stall-regulated turbine presenting a pronounced peak in Cp, unlike the pitch-regulated turbine that could operate near peak-Cp for a broad range of TSRs. It is also interesting to note that both turbines effectively operated in the area of the Cp–TSR curve that is on the left of the Cp peak. This was crucial, especially for a stall-regulated turbine, where tuning the shape of a Cp–TSR curve and forcing the rotor to operate in off-design conditions are the sole ways turbine control can be properly ensured.

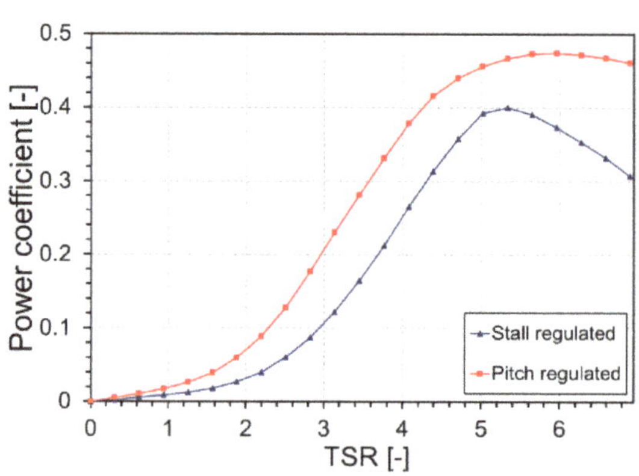

Figure 9. Aerodynamic power coefficient as a function of tip–speed ratio (TSR) in steady-state conditions.

5.2. Aerodynamic Performance

In this section, the aerodynamic performance in dynamic conditions is discussed. Figure 10 shows the generator power as a function of the wind speed for the two turbines in the conditions specified by IEC 61400-2 DLC 1.2 [12] and discussed in Section 4. The error bars show the maximum and the minimum calculated values, while the filled areas represent the standard deviation. The analysis of the mean values shows that the power produced by the pitch-regulated turbine was higher than that generated by the stall-regulated turbine. This was largely because the pitch-regulated turbine was more efficient below the rated wind speed (Figure 8a). Furthermore, the standard deviation was generally lower for the pitch-controlled turbine at all wind speeds, and power output seemed to be better controlled, especially at high wind speeds. With their respective differences, these results show how the control systems of both turbines were able to adequately regulate the turbine in turbulent inflow conditions. The oscillation of the minimum power values for both turbines was due to the strong wind speed oscillations during turbulent simulations. This is in fact a key aspect of SWTs, whose installation contexts are often characterized by very turbulent winds.

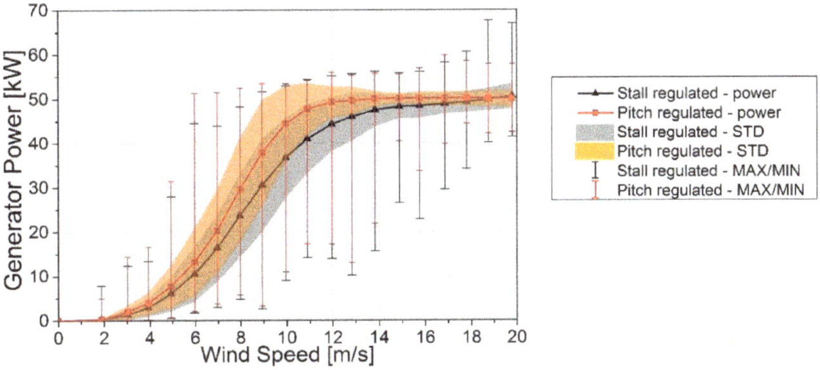

Figure 10. Generator power mean, standard deviation (shaded areas), and maxima and minima.

When comparing the power predicted in steady and dynamic conditions, some interesting considerations can be drawn. Figure 11a,b compares the power curves in steady and dynamic conditions for the pitch- and stall-regulated turbines. The effects of vertical up-flow, yaw-misalignment, and turbulence intensity can be globally evaluated.

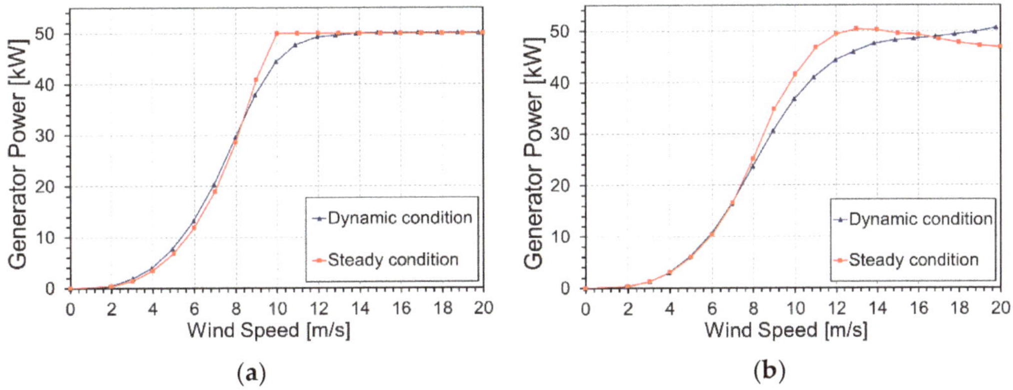

Figure 11. Power curves of the pitch- (**a**) and stall-controlled (**b**) variable speed concept.

For the pitch-regulated turbine, especially, there was a tendency to increase power output below rated power, while both the pitch- and stall-regulated turbines drastically decreased power around the rated wind speed. This was mainly an effect of turbulence intensity, as many authors have shown [57–60], and underlines the importance of taking realistic operating conditions into account in the design process of a wind turbine (while in the past, this was discarded in many small wind turbines). For instance, in Figure 11b, one can notice how in turbulent flow conditions, the rated power was not reached until 20 m/s average wind speed; this could lead the designer to modify the turbine design, e.g., by compensating for this effect by reducing the fixed pitch angle.

By comparing the performance obtained in steady conditions with that in dynamic conditions, it can be noticed how the gap between the pitch-regulated and stall-regulated power curves widened in turbulent wind. This is very visible in Figure 12. Referring to this figure, the area between the curves in steady conditions was 27.06 kW*m/s (Figure 12a) and 55.26 kW*m/s in dynamic conditions (Figure 12b). This was a consequence of the flatter TSR–Cp curve of the stall-regulated case, as shown above in Figure 9. For this reason, the pitch-regulated turbine was less sensitive to variations in TSR and could operate near peak Cp for longer time. This is again a consideration that was often unclear in the old generation of stall-regulated SWTs, and it seems to suggest that the real benefits of pitch regulation are higher than expectations and thus possibly able to compensate for the increased cost.

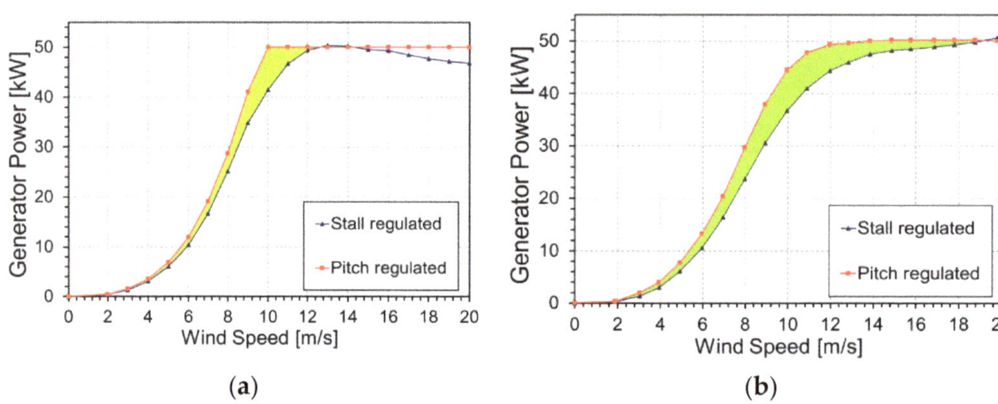

Figure 12. Power curves for the pitch- and stall-regulated turbines in steady conditions (**a**) and dynamic inflow conditions (**b**).

In above-rated flow speed operation, the stall-regulated turbine was able to self-regulate power output, as previously shown in Figure 7a. Unfortunately, as shown in Figure 13, the result was an increase in axial load for the stall-regulated rotor—more force was transferred into axial loading rather than into rotating the blade. In fact, as wind speed increased in the stall-regulated rotor, the force vector rotated downwind to decrease the torque component and increase the thrust component. Therefore, a potential advantage of pitch regulation over its stall counterpart is decreased peak axial loads, which decrease rotor structural requirements and may lower the risk of failure during high-wind events.

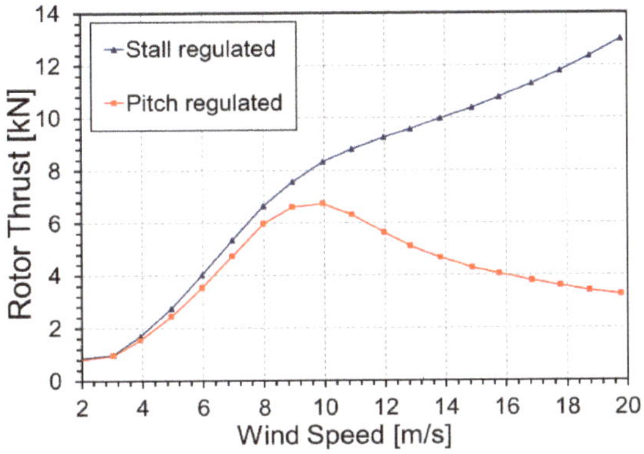

Figure 13. Average rotor thrust curves in dynamic conditions.

5.3. Annual Energy Production (AEP)

Differences in power delivery and efficiency discussed in the previous sections result in different annual energy production values. The AEP was calculated according to IEC 61400-2 standard turbine classes from the results of the dynamic simulations. A Weibull wind speed distribution with shape factor of 2 and an average value of 8.5 m/s was used to model sites of IEC wind class IIA, with medium wind speed and high turbulence intensities.

The results of AEP estimations are displayed in Figure 14. It can be noted that the energy capture was very low at low wind and high wind speeds, though for different reasons. At low wind speeds, a wind turbine cannot deliver enough power, while high wind speeds occur only for short times during a year.

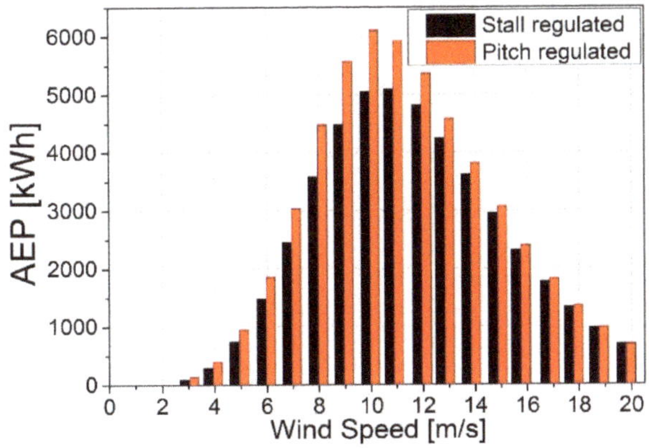

Figure 14. Annual energy production per wind average wind speed in dynamic conditions.

In the analyzed case study, the pitch-regulated turbine produced 12.36% more energy (kWh) annually than the same stall-regulated turbine. The annual energy production calculated for the stall-regulated turbine was 46.058 MWh/year, while the pitch-regulated turbine produced 52.554 MWh/year.

5.4. Results in the Time Domain

In order to do a comparative analysis between pitch and stall control strategies, it was also useful to look at time characteristics. The first reason for this is to show that the controllers and simulation models worked properly. The second is to show the impact of the two control methods on power output, which also has an effect on global energy capture. Finally, the third is to get an impression of the power quality of the different controls.

In the following, below, above, and at around rated wind speed simulations are discussed, and the results for a 600 s time interval are shown. For the partial load time characteristics, an average wind speed of 7 m/s was selected. In this scenario, wind speed rarely reached its rated value, and power limiting did not occur. In this area, the main goal was to maximize energy harvesting.

In Figure 15a, the generated power is shown for the two turbines. In the 600 s time interval, the power of the pitch-regulated turbine was always slightly higher than that produced by the stall-regulated turbine, as was expected given the higher Cp. The power output was globally similar for the two turbines, as power regulation did not kick in until higher wind speeds were reached.

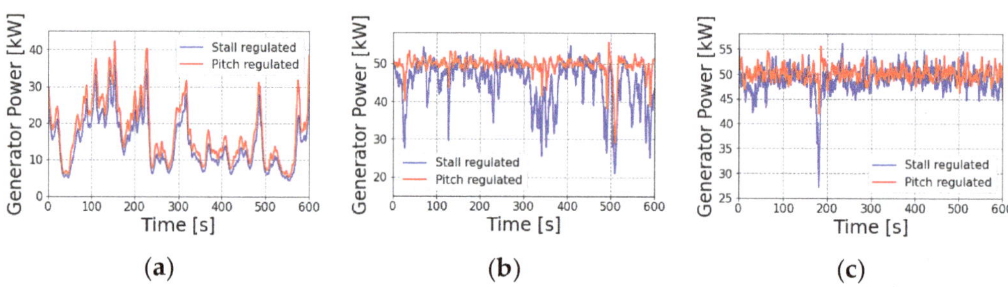

Figure 15. Generator power for a (**a**) 7 m/s average wind speed simulation, (**b**) a 12 m/s average wind speed simulation, and (**c**) a 16 m/s average wind speed simulation.

For the near rated wind speed time characteristics, an average wind speed of 12 m/s was selected. The main interest lies in the transitions from partial load to rated power and vice versa. In this area, a smooth transition between the power maximization and the power limiting was of interest. In Figure 15b, the power output for the two turbines at 12 m/s average wind speed is shown.

The stall-controlled variable-speed concept showed very steep power changes when entering and leaving region 3. While power overshoots were similar in magnitude for the two regulation concepts, power output dropped significantly as wind speed dropped below rated for the stall-regulated variant. When the turbine was operating at rated power, the blades were in partial or total stall; therefore, due to dynamic effects, power dropped significantly as the blade gradually exits stalled.

In Figure 15c, the behavior of the two different concepts at wind speeds above rated wind speed is shown; in particular, an average wind speed of 16 m/s was selected. At these wind speeds, there is always much more power in the wind than the wind turbine can handle. Therefore, the power output must be curtailed. Overall, the pitch-controlled turbine appeared to be able to regulate power more efficiently, although both control systems provided satisfactory results. As noted also when analyzing operation around rated wind speed, the generator power dropped significantly more on the stall-regulated turbine at the 180 s mark, an effect that could be again related to the stall state the blade is in.

Some interesting trends can also be inferred from the rotor speed of the same simulations at 7, 12, and 16 m/s shown in Figure 15; these trends are shown in Figure 16.

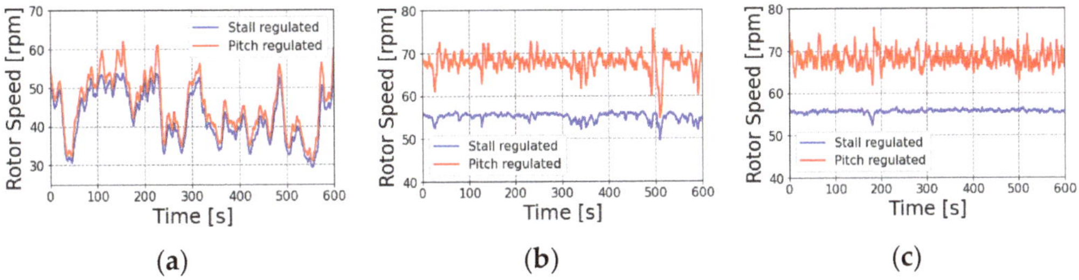

Figure 16. Generator speed for a (**a**) 7 m/s average wind speed simulation, (**b**) a 12 m/s average wind speed simulation, and (**c**) a 16 m/s average wind speed simulation. Results for the same simulations shown in Figure 15.

At 7 m/s, the two turbines behaved similarly, with the stall-regulated turbine producing more power and operating at a higher rotor speed. The stall-regulated turbine operated at a lower TSR, as intended and shown in Figure 9. At 12 and 16 m/s, the average rotor speed was higher for the pitch-regulated turbine and both turbines were operating at their nominal rotor speed, thus indicating that the controllers were performing as intended. It can be noted how the stall-regulated turbine was able to maintain a nearly constant rotor speed. The differences between control systems can be explained as follows: the pitch controller employed in this study maintained a constant torque above rated and regulated rotor speed and power trough blade-pitch feathering. Thus, fluctuations in power were caused by variations in rotor speed and vice-versa. The stall controller, on the other hand, was set to operate in region 2.5 at a wind speed above rated (further details are discussed previously in Section 3.3), and controlled rotor speed at the expense of fluctuations in torque and power. Keeping rotor speed in check is very important for a stall controller because if rotor speed could increase, the turbine would increase its TSR and quickly accelerate out of control.

6. Conclusions

In this study, the design process of a 50 kW turbine from blade selection to performance assessment was used to show how modern engineering codes and recent tools for turbine control can be effectively used to design an efficient small wind turbine.

Focusing first on aerodynamic design, it must be noted that he intended final control strategy (i.e., pitch or stall control) needs to be defined early on in the design process because the resulting final blade shape may be significantly influenced by the choice. In the case of SWTs, it is preferable to use a family of airfoils that targets a high glide ratio with moderate lift coefficients, as this helps to increase blade chord and, hence, operating Reynolds number. Furthermore, although these effects are not fully understood and their inclusion in the design process is somewhat uncertain, it is very important to consider 3D-effects. Such phenomena play a key role in the inner parts of the blade and have been shown, as expected, to significantly influence the stall-regulation capabilities of an SWT. The presented guidelines and aerodynamic design procedure are general and can be applied to all turbine sizes, not only to SWTs. It must be noted, however, that when designing very large wind turbines (10–20 MW), pitch control is the undisputed choice and focus is placed mainly on structural loads. In fact, these rotors operate at extremely high Reynolds numbers, therefore achieving high peak aerodynamic performance almost effortlessly. On the other hand, structural optimization is extremely important to keep blade cost and weight down, as well as to guarantee robust blade design. For this class of rotors, a more integrated design procedure, focusing on loads in addition to aerodynamics and control in the initial stages of rotor design, should be considered.

Focusing on control, the study showed the basic approaches and methods to implement both pitch and stall control in small wind turbines. In this sense, even though existing books and reports very often only focus on the stationary power curve, it has been shown here that in dynamic conditions, i.e., in a power-production DLC case from international design standards, the power curve of the turbine significantly changes, thus indicating the importance of accounting for such conditions in the design process and, especially, in the selection of the best control strategy.

For the stall-regulated turbine, an overall good level of performance was achieved. The peak aerodynamic power coefficient for the selected case study was around 0.4, which is in line with turbines of this class. When adopting a pitch regulation strategy, however, fewer compromises to the blade design have to be made in order to ensure good power regulation; in this case, no fixed blade pitch angle needed to be set, and the ideal blade twist distribution could be used. As a consequence, the aerodynamic power coefficient improved significantly, reaching a value of nearly 0.5, which is in line with most modern utility-scale turbines. Furthermore, rotor thrust continued to increase above rated wind speed for the stall-regulated turbine, as opposed to the trend observed when using pitch regulation. This points to the possibility that pitch regulation also has the added benefit of lowering axial blade loads. In this sense, this work has shown how the use of a modern, pitch-to-feather control strategy has the potential to significantly improve SWT performance through more effective power regulation and due to the fact that many compromises to the aerodynamic design can be avoided.

Author Contributions: Conceptualization, A.B. and F.P.; methodology, F.P.; software, A.N.; validation, A.N., F.P.; investigation, A.N.; data curation, A.N.; writing—original draft preparation, A.N.; writing—review and editing, F.P. and A.B.; supervision, A.B., G.F.; project administration, G.F.; All authors have read and agreed to the published version of the manuscript.

Funding: This research received no external funding.

Informed Consent Statement: Not applicable.

Data Availability Statement: Data available on request due to restrictions.

Conflicts of Interest: The authors declare no conflict of interest.

Appendix A

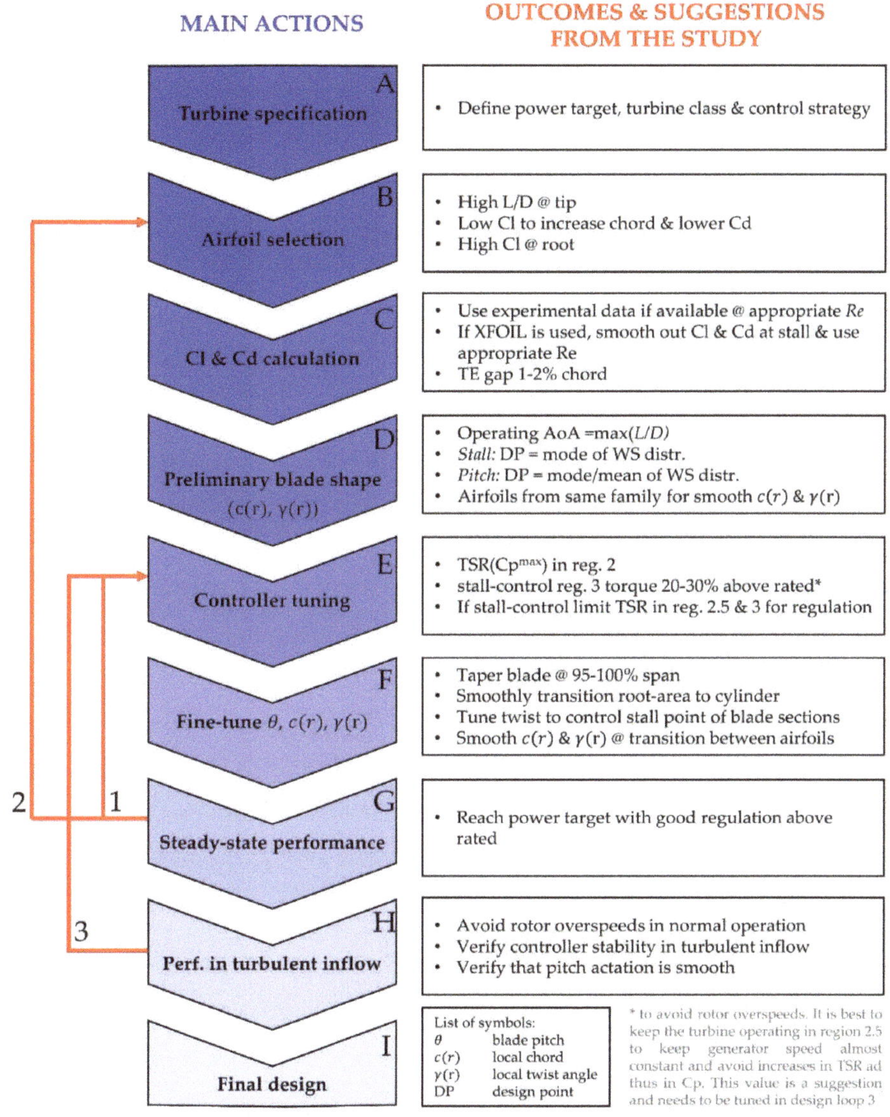

- Design loops 1&2 are necessary to understand if the chosen airfoil families and the controller parameters setting provide satisfactory performance. In many cases satisfactory performance can be reached trough loop 1 and fine-tuning blade shape in step F. If performance is still not satisfactory or stall regulation not as desired different airfoils must be chosen.
- The Final Design Loop allows for modifications to the ideal blade design in order to optimize the final aerodynamic performance. Trough steps E and F controller and blade can be fine-tuned in order to ensure good regulation and stability in turbulent inflow.

Figure A1. Conceptual scheme of the design phases for a SWT design procedure. Outcomes and suggestions from the study are reported for each phase.

References

1. Veers, P.; Dykes, K.; Lantz, E.; Barth, S.; Bottasso, C.L.; Carlson, O.; Clifton, A.; Green, J.; Green, P.; Holttinen, H.; et al. Grand Challenges in the Science of Wind Energy. *Science* **2019**, *366*, eaau2027. [CrossRef] [PubMed]
2. Gardner, P.; Garrad, A.; Hansen, L.F.; Jamieson, P.; Morgan, C.; Murray, F.; Tindal, A. *Wind Energy—The Facts—Part 1: Technology*; EWEA: Brussels, Belgium, 2010.
3. Bukala, J.; Damaziak, K.; Kroszczynski, K.; Malachowski, J.; Szafranski, T.; Tomaszewski, M.; Karimi, H.R.; Jozwik, K.; Karczewski, M.; Sobczak, K. Small Wind Turbines: Specification, Design, and Economic Evaluation. In *Wind Turbines—Design, Control and Applications*; Aissaoui, A.G., Tahour, A., Eds.; InTech: London, UK, 2016; ISBN 978-953-51-2495-5.
4. Clausen, P.D.; Reynal, F.; Wood, D.H. Design, manufacture and testing of small wind turbine blades. In *Advances in Wind Turbine Blade Design and Materials*; Elsevier: Amsterdam, The Netherlands, 2013; pp. 413–431. ISBN 978-0-85709-426-1.
5. Tzen, E. Small Wind Turbines for on Grid and off Grid Applications. *IOP Conf. Ser. Earth Environ. Sci.* **2020**, *410*, 012047. [CrossRef]
6. Simic, Z.; Havelka, J.G.; Bozicevic Vrhovcak, M. Small Wind Turbines—A Unique Segment of the Wind Power Market. *Renew. Energy* **2013**, *50*, 1027–1036. [CrossRef]
7. EAWE Committees. Available online: https://www.eawe.eu/organisation/committees/ (accessed on 13 January 2021).
8. Artal-Sevil, J.S.; Dufo, R.; Dominguez, J.A.; Bernal-Agustin, J.L. Small Wind Turbines in Smart Grids. Transformation of Electrical Machines in Permanent Magnet Synchronous Generators. In Proceedings of the 2018 Thirteenth International Conference on Ecological Vehicles and Renewable Energies (EVER), Monte-Carlo, Monaco, 10–12 April 2018; pp. 1–8.
9. James, F.M.; Jon, G.M.; Anthony, L.R. *Wind Energy Explained: Theory, Design and Application*, 2nd ed.; John Wiley & Sons: Hoboken, NJ, USA, 2010; ISBN 978-0-470-01500-1.
10. Burton, T. (Ed.) *Wind Energy Handbook*; John Wiley & Sons: Hoboken, NJ, USA, 2001; ISBN 978-0-471-48997-9.
11. Chaudhary, M.K.; Roy, A. Design & Optimization of a Small Wind Turbine Blade for Operation at Low Wind Speed. *World J. Eng.* **2015**, *12*, 83–94. [CrossRef]
12. IEC. IEC 61400-2. In *Wind Turbines—Part 2: Design Requirements for Small Wind Turbines*; IEC: Geneva, Switzerland, 2006.
13. Jacobson, R.; Meadors, M.; Jacobson, E.; Link, H. *Power Performance Test Report for the AOC 15/50 Wind Turbine*; NREL: Golden, CO, USA, 2003.
14. Eunice Energy Group (EEG). *EW16 THETIS—Wind Turbine 50kw*; Eunice Energy Group: Athens, Greece, 2020; p. 10.
15. Samani, A.E.; de Kooning, J.D.M.; Kayedpour, N.; Singh, N.; Vandevelde, L. The Impact of Pitch-To-Stall and Pitch-To-Feather Control on the Structural Loads and the Pitch Mechanism of a Wind Turbine. *Energies* **2020**, *13*, 4503. [CrossRef]
16. Small Wind Turbine Generators—Tozzi Green. Available online: www.tozzinord.com (accessed on 12 February 2021).
17. Drela, M. XFOIL: An Analysis and Design System for Low Reynolds Number Airfoils. In *Low Reynolds Number Aerodynamics*; Mueller, T.J., Ed.; Springer: Berlin/Heidelberg, Germany, 1989; Volume 54, pp. 1–12. ISBN 978-3-540-51884-6.
18. Tangler, J.L.; Somers, D.M. *NREL Airfoil Families for HAWTs*; NREL: Golden, CO, USA, 1995.
19. Battisti, L.; Zanne, L.; Castelli, M.R.; Bianchini, A.; Brighenti, A. A Generalized Method to Extend Airfoil Polars over the Full Range of Angles of Attack. *Renew. Energy* **2020**, *155*, 862–875. [CrossRef]
20. Drela, M.; Gilest, M.B. Viscous-Inviscid Analysis of Transonic and Low Reynolds Number Airfoils. *AIAA J.* **1987**, *25*, 1347–1355. [CrossRef]
21. OpenFAST. Available online: https://github.com/OpenFAST (accessed on 31 October 2019).
22. Jonkman, J.; Butterfield, S.; Musial, W.; Scott, G. *Definition of a 5-MW Reference Wind Turbine for Offshore System Development*; NREL/TP-500-38060; NREL: Golden, CO, USA, 2009; p. 947422.
23. Gaertner, E.; Rinker, J.; Sethuraman, L.; Zahle, F.; Anderson, B.; Barter, G.E.; Abbas, N.J.; Meng, F.; Bortolotti, P.; Skrzypinski, W.; et al. *IEA Wind TCP Task 37: Definition of the IEA 15-Megawatt Offshore Reference Wind Turbine*; NREL/TP-5000-75698; NREL: Golden, CO, USA, 2020; p. 1603478.
24. Jonkman, J.M.; Hayman, G.J.; Jonkman, B.J.; Damiani, R.R. *AeroDyn User's Guide and Theory Manual*; NREL: Golden, CO, USA, 2015.
25. Manwell, J.F.; McGowan, J.G.; Rogers, A.L. Aerodynamics of Wind Turbines. In *Wind Energy Explained: Theory, Design and Application*; John Wiley & Sons, Ltd: Chichester, UK, 2010; pp. 91–155. ISBN 978-1-119-99436-7.
26. Peter, J.S.; Richard, J.C. Wind Turbine Blade Design. *Energies* **2012**, *5*, 3425–3449.
27. Gasch, R.; Twele, J. *Wind Power Plants Fundamentals, Design, Construction and Operation*, 2nd ed.; Springer: Berlin/Heidelberg, Germany, 2002; ISBN 978-3-642-22938-1.
28. Atlantic Orient AOC 15/50. Available online: https://it.wind-turbine-models.com/turbines/2164-atlantic-orient-aoc-15-50 (accessed on 12 February 2021).
29. Smith, J.; Huskey, A.; Jager, D.; Hur, J. *Duration Test Report for the Entegrity EW50 Wind Turbine*; NREL: Golden, CO, USA, 2012.
30. E-3120 50kW Wind Turbine. Available online: http://www.burnley.gov.uk/attachments/12_0055_Appendix_Endurance_E-3120_turbine_specifications.pdf (accessed on 12 February 2021).
31. Mamadaminov, U.M. Review of Airfoil Structure for Wind Turbine Blades. *Or. Inst. Technol.* **2013**, *Energy Engineering I*, 1–8.
32. Fuglsang, P.; Bak, C. Development of the Risø wind turbine airfoils. *Wind Energy Int. J. Prog. Appl. Wind Power Convers. Technol.* **2004**, *7*, 145–162. [CrossRef]

33. Timmer, W.A.; Van Rooij, R.P.J.O.M. Summary of the Delft University Wind Turbine Dedicated Airfoils. *J. Sol. Energy Eng.* **2003**, *125*, 488–496. [CrossRef]
34. Somers, D.M. *Design and Experimental Results for the S805 Airfoil*; NREL: Golden, CO, USA, 1997.
35. Somers, D.M. *The S819, S820 and S821 Airfoils: October 1992–November 1993*; NREL: Golden, CO, USA, 2005.
36. Tangler, J.L.; NREL; Somers, D.M.; Airfoils, Inc. *Airfoils for Wind Turbine*; NREL: Golden, CO, USA, 1996.
37. Griffin, D.A.; Lynette, R. *Investigation of Aerodynamic Braking Devices for Wind Turbine Applications*; NREL: Golden, CO, USA, 1997.
38. Islam, M.R.; Bashar, L.B.; Saha, D.K.; Rafi, N. Comparison and Selection of Airfoils for Small Wind Turbine between NACA and NREL's S series Airfoil Families. *Int. J. Res. Electr. Electron. Commun. Eng.* **2019**, *4*, 1–12.
39. Ramsay, R.R.; Hoffmann, M.J.; Gregorek, G.M. *Effects of Grit Roughness and Pitch Oscillations on the S812 Airfoil: Airfoil Performance Report, Revised (12/99)*; Ohio State University: Columbus, OH, USA, 1999.
40. Ramsay, R.R.; Gregorek, G.M. *Effects of Grit Roughness and Pitch Oscillations on the S813 Airfoil: Airfoil Performance Report, Revised (12/99)*; Ohio State University: Columbus, OH, USA, 1996.
41. Somers, D.M.; Tangler, J.L. *Wind-Tunnel Test of the S814 Thick-Root Airfoil*; NREL: Golden, CO, USA, 1995.
42. Dal Monte, A.; de Betta, S.; Castelli, M.R.; Benini, E. Proposal for a coupled aerodynamic—Structural wind turbine bladeoptimization. *Compos. Struct.* **2017**, *159*, 144–156. [CrossRef]
43. Hansen, M.O.L. *Aerodynamics of Wind Turbines*, 2nd ed.; Earthscan: London, UK; Sterling, VA, USA, 2008; ISBN 978-1-84407-438-9.
44. Shen, W.Z.; Zhu, W.J.; Sørensen, J.N. Study of Tip Loss Corrections Using CFD Rotor Computations. *J. Phys. Conf. Ser.* **2014**, *555*, 012094. [CrossRef]
45. Himmelskamp, H. *Profiluntersuchungen an Einem Umlaufenden Propeller*; Report No 1945; Diss. Goettingen; Max-Planck-Inst. fuer Stroemungsforschung: Goettingen, Germany, 1950.
46. Hand, M.M.; Simms, D.A.; Fingersh, L.J.; Jager, D.W.; Cotrell, J.R.; Schreck, S.; Larwood, S.M. *Unsteady Aerodynamics Experiment Phase VI: Wind Tunnel Test Configurations and Available Data Campaigns*; NREL/TP-500-29955; NREL: Golden, CO, USA, 2001; p. 15000240.
47. Bak, C. Three-Dimensional Corrections of Airfoil Characteristics Based on Pressure Distributions. In Proceedings of the European Wind Energy Conference, Athens, Greece, 27 February–2 March 2006.
48. Abbas, N.J.; Wright, A.; Pao, L. An Update to the NREL Baseline Wind Turbine Controller. *Renew. Energy* **2020**, *17*, Preprint.
49. Mulders, S.P.; van Wingerden, J.W. Delft Research Controller: An open-source and community-driven wind turbine baseline controller. *J. Phys. Conf. Ser.* **2018**, *1037*, 032009. [CrossRef]
50. Muljadi, E.; Pierce, K.; Migliore, P. A conservative control strategy for variable-speed stall-regulated wind turbines. In Proceedings of the 2000 ASME Wind Energy Symposium, Reno, NV, USA, 10–13 January 2000.
51. Wright, A.D.; Fingersh, L.J. *Advanced Control Design for Wind Turbines Part I: Control Design, Implementation, and Initial Tests*; NREL: Golden, CO, USA, 2008.
52. *ROSCO Tollbox*, Version 1.0.0. GitHub Repository. GitHub: San Francisco, CA, USA, 2020.
53. Li, D.; Chen, C. Comparison of Energy Efficiency between Fixed-speed and Variable-speed Wind Turbines. *Energy Eng.* **2009**, *101*, 71–80. [CrossRef]
54. Jonkman, J.M.; Buhl, M.L., Jr. *FAST User's Guide*; NREL: Golden, CO, USA, 2005.
55. Hansen, M.H.; Thomsen, K.; Natarajan, A.; Barlas, A. Design Load Basis for Onshore Turbines. In *DTU Wind Energy Technical Report*; Department of Wind Energy: Roskilde, Denmark, 2015.
56. Bortolotti, P.; Canet, H.; Bottasso, C.L.; Loganathan, J. Performance of Non-Intrusive Uncertainty Quantification in the Aeroservoelastic Simulation of Wind Turbines. *Wind Energy Sci.* **2019**, *4*, 397–406. [CrossRef]
57. Bardala, L.M.; Sætrana, L.R. Influence of turbulence intensity on wind turbine power curves. *Energy Procedia* **2017**, *137*, 553–558. [CrossRef]
58. Kaiser, K.; Langreder, W.; Hohlen, H.; Højstrup, J. Turbulence Correction for Power Curves. In *Wind Energy*; Springer: Berlin/Heidelberg, Germany, 2007.
59. Albers, A.; Jakobi, T.; Rolf, R.; Stoltenjohannes, J. Influence of meteorological variables on measured wind turbine power curves. In Proceedings of the European Wind Energy Conference & Exhibition, Milan, Italy, 7–10 May 2007.
60. Wagner, R.; Courtney, M.; Larsen, T.J.; Paulsen, U.S. *Simulation of Shear and Turbulence Impact on Wind Turbine Performance*; Danmarks Tekniske Universitet, Risø Nationallaboratoriet for Bæredygtig Energi: Roskilde, Denmark, 2010.

energies

MDPI

Article

Wake Statistics of Different-Scale Wind Turbines under Turbulent Boundary Layer Inflow

Xiaolei Yang [1,2], Daniel Foti [3], Christopher Kelley [4], David Maniaci [4] and Fotis Sotiropoulos [5,*]

1 The State Key Laboratory of Nonlinear Mechanics, Institute of Mechanics, Chinese Academy of Sciences, Beijing 100190, China; xyang@imech.ac.cn

2 School of Engineering Sciences, University of Chinese Academy of Sciences, Beijing 100049, China

3 Department of Mechanical Engineering, University of Memphis, Memphis, TN 38152, USA; dvfoti@memphis.edu

4 Sandia National Laboratories (Sandia National Laboratories is a multimission laboratory managed and operated by National Technology & Engineering Solutions of Sandia, LLC, a wholly owned subsidiary of Honeywell International Inc., for the U.S. Department of Energy's National Nuclear Security Administration under contract DE-NA0003525.), Albuquerque, NM 87185, USA; clkell@sandia.gov (C.K.); dcmania@sandia.gov (D.M.)

5 Department of Civil Engineering, College of Engineering and Applied Sciences, Stony Brook University, Stony Brook, NY 11790, USA

* Correspondence: fotis.sotiropoulos@stonybrook.edu; Tel.: +1-631-632-8380

Received: 11 May 2020; Accepted: 8 June 2020; Published: 11 June 2020

check for updates

Abstract: Subscale wind turbines can be installed in the field for the development of wind technologies, for which the blade aerodynamics can be designed in a way similar to that of a full-scale wind turbine. However, it is not clear whether the wake of a subscale turbine, which is located closer to the ground and faces different incoming turbulence, is also similar to that of a full-scale wind turbine. In this work we investigate the wakes from a full-scale wind turbine of rotor diameter 80 m and a subscale wind turbine of rotor diameter of 27 m using large-eddy simulation with the turbine blades and nacelle modeled using actuator surface models. The blade aerodynamics of the two turbines are the same. In the simulations, the two turbines also face the same turbulent boundary inflows. The computed results show differences between the two turbines for both velocity deficits and turbine-added turbulence kinetic energy. Such differences are further analyzed by examining the mean kinetic energy equation.

Keywords: turbine wake; turbine size; large-eddy simulation; actuator surface model

1. Introduction

As wind energy grows as a main energy resource for the whole world, further research is still needed to reduce the cost of wind energy to keep it economically competitive [1]. Turbine wakes affect both the power production and operation and maintenance costs of wind energy. Wind tunnel experiments at meter-scale or even smaller wind turbines [2–8] and field measurements of subscale wind turbines (e.g., the SWiFT facility [9–12]) play a vital rule in understanding the dynamics of turbine wakes and provide valuable datasets for validating computational models. However, the sizes of these meter-scale turbines and subscale turbines are often much smaller than a full-scale wind turbine, which calls into question how well these small-scale wind turbines can represent full-scale wind turbines [13].

When designing a meter-scale or a subscale wind turbine, geometric, kinematic and dynamic similarities should be maintained to ensure their equivalence to a full-scale wind turbine. For a meter-scale wind turbine, which can be about 1000 times smaller than a full-scale wind turbine, it is even challenging to ensure only geometric similarity (e.g., the rotor diameter and chord length) and the kinematic similarity (e.g., the tip speed ratio $\lambda = \Omega R/U$, where Ω is the rotor rotating speed, R is the rotor radius and U is the incoming wind speed). As such, researchers often build a meter-scale wind turbine in a way that the power coefficient and thrust coefficient are similar to that of a full-scale wind turbine [6]. For the scale effects on turbine wakes, Howard and Guala [14] compared the velocity deficits of a meter-scale turbine with that of the University of Minnesota 2.5 MW EOLOS turbine and observed significant differences at $x/D = 1.5$ and $x/D = 2.5$ turbine downwind locations but relatively small differences at $x/D = 3$ turbine downwind location, where D is the rotor diameter. Heisel, Hong and Guala [15] further compared the wake meandering from the meter-scale turbine and the EOLOS turbine and observed similar meandering frequencies related to the bluff body shear layer instability.

For a subscale wind turbine, which may be 3 or 4 times smaller than a full-scale wind turbine, the geometric similarity and the kinematic similarity can be ensured relatively easier compared to meter-scale wind turbines. However, because of the difference in Reynolds number, the dynamic similarity, namely the distributions of lift and drag coefficients along the blade for wind turbines, still cannot be guaranteed easily while keeping the geometric and kinematic similarities. Kelley et al. [16] proposed to loosely maintain the dynamic similarity by keeping the same dimensionless bound circulation along the blade by relaxing the constraints on the geometric similarity. Although the aerodynamics of a subscale wind turbine can be made closer to a full-scale wind turbine, an important difference is that the subscale wind turbine is located closer to the ground where the mean shear stress and the turbulence intensity change significantly with distance from the ground. However, how this difference affects the representation of a subscale wind turbine wake to that of a full-scale wind turbine is not clear yet. To address this issue, in this work we simulate a subscale wind turbine and a full-scale wind turbine, which are geometrically and kinematically equivalent, and are dynamically equivalent by applying the same lift and drag coefficients in turbine parameterizations, and under the same turbulent boundary layer inflow.

It is noted that this work is different from the studies in the literature on investigating the effects of inflow turbulence on turbine wakes. In the literature, the turbine wakes under different turbulent inflows, which are due to different ground roughness lengths, were investigated in [2,17] using wind tunnel experiments and large-eddy simulation, respectively. The effects of different inflow turbulence on turbine wakes, which is caused by different thermal stratifications, were studied using wind tunnel experiments and large-eddy simulation in [18,19], respectively. In [20], turbine wakes under different inflow turbulence caused by an upwind hill of different heights were simulated using large-eddy simulation. In [21], the coherent tip vortices of a utility-scale wind turbine were investigated for inflows of different turbulence intensities. In [22], the wake meandering of a utility-scale wind turbine was investigated for inflows of different turbulence intensities. All these studies were focused on the wake of one-size wind turbine. To the best of our knowledge, no studies have been carried out on comparing wakes from turbines of different sizes, which is critical as we use the knowledge from a subscale wind turbine to a full-scale wind turbine.

The rest of the paper is organized as follows: in Section 2 we briefly describe large-eddy simulation with actuator surface models for turbine blades and nacelle; we then compare the results from the two turbines of different sizes in Section 4; at last we draw conclusions from this work in Section 5.

2. Numerical Method

The Virtual Flow Simulator (VFS-Wind) code [23–25] is employed in this work to simulate wakes from the two turbines, which solves the incompressible Navier–Stokes equations in curvilinear coordinates shown as follows:

$$J\frac{\partial U^j}{\partial \xi^j} = 0, \tag{1}$$

$$\frac{1}{J}\frac{\partial U^i}{\partial t} = \frac{\xi_l^i}{J}\left(-\frac{\partial (U^j u_l)}{\partial \xi^j} + \frac{\mu}{\rho}\frac{\partial}{\partial \xi^j}\left(\frac{g^{jk}}{J}\frac{\partial u_l}{\partial \xi^k}\right) - \frac{1}{\rho}\frac{\partial}{\partial \xi^j}\left(\frac{\xi_l^j p}{J}\right) - \frac{1}{\rho}\frac{\partial \tau_{lj}}{\partial \xi^j} + \frac{1}{\rho}f_l\right), \tag{2}$$

where x_i ($i = 1,2,3$) are the Cartesian coordinates, ξ^j ($j = 1,2,3$) are the curvilinear coordinates, U^i denotes the contravariant volume flux, u_i are the velocity vector in Cartesian coordinates, J is the Jacobian of the geometric transformation, ξ^i_l represent the transformation metrics, g^{jk} represents the contravariant metric tensor, ρ is the density, μ is the dynamic viscosity, p is the pressure, $f_l (l = 1,2,3)$ are the body forces from actuator type turbine models, and τ_{ij} is the subgrid scale stress modeled using the dynamic subgrid scale model [26]. The governing equations are discretized in space using a second-order central differencing scheme and in time using a second-order accurate fractional step method. The pressure Poisson equation is solved using an algebraic multigrid acceleration along with GMRES solver. The momentum equation is solved using the matrix-free Newton–Krylov method. The VFS-Wind code has been validated extensively using laboratory and field measurements [21,24] and applied to utility-scale wind turbines [22,27,28].

The actuator surface models for turbine blades and nacelle developed in [25] are employed in this work. In the actuator surface model for turbine blades, the blade is represented with the actuator surface defined by chords at different radial locations. Forces computed using the blade element method are distributed on the actuator surface to represent the effect of blades on the incoming flow. Compared with the actuator line parameterization, the actuator surface can represent better the geometrical effect of the blade in the chordwise direction. A model for the nacelle is also necessary for accurately predicting turbine wakes, which affects the hub vortex and meandering in the far wake [29]. In this work the nacelle is represented using an actuator surface formed by the actual surface of the nacelle. The effects of nacelle on the incoming flow are modeled using distributed forces with the wall-normal component calculated by satisfying the no-flux boundary condition and the wall-tangent component calculated by specifying a friction coefficient.

3. Test Cases and Computational Details

3.1. Specifics of the Employed Turbine Design

A full-scale wind turbine and a subscale wind turbine are studied in this work as shown in Figure 1. The diameters of the two turbines are 80 m and 27 m, respectively. The full-scale turbine T80 is located 93.3 m above the ground, while the subscale turbine T27 is located 31.5 m above the ground. The T27 turbine is designed by Kelley et al. in [30] (Design A). The T80 turbine is designed by simply scaling the T27 turbine, such that the radial distributions of the twist angle and the normalized chord length (as shown in Figure 2) are the same for the two turbines. It is noticed that the T27 turbine has the same rotor diameter and hub height as the SWiFT turbine but with different designs as shown in Figure 2 and Table 1. The scaling ratio is chosen in a way the size of the full-scale turbine is similar to that of a land-based wind turbine. The radial distribution of the airfoil shapes is shown in Table 1. For the present cases, the tip speed ratio is 9. The power coefficient, $C_P = \frac{P}{0.5\rho\pi R^2 U^3}$ (where P is the power, R is the rotor radius, D is the rotor diameter, and U is the incoming wind speed), is 0.47 and 0.49 for the turbine T80 and T27, respectively. The thrust coefficient, $C_T = \frac{T}{0.5\rho\pi R^2 U^2}$ (where T is the thrust), is 0.65 and 0.68 for the turbine T80 and T27, respectively. The incoming wind speed U employed for calculating C_P and C_T is obtained by averaging the time-averaged velocity over a disk of radius R parallel with the rotor rotating plane and located $1D$ upwind of the turbine. The same lift and drag coefficients are employed for both turbines to ensure dynamic similarity and to simplify the problem such that the only two differences are the turbine size and the inflow.

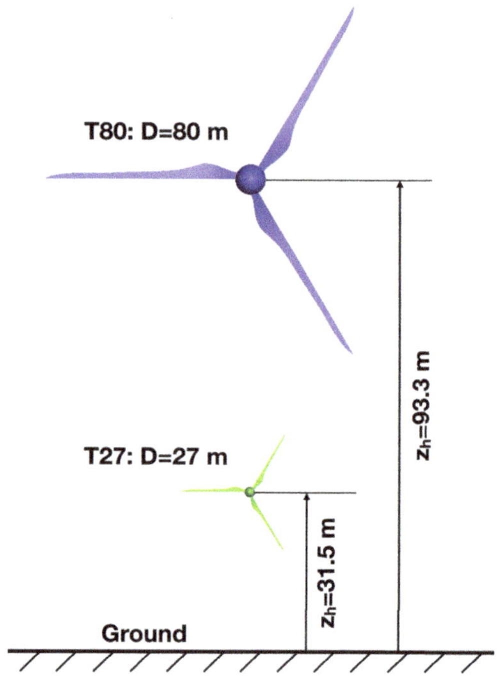

Figure 1. Schematic of the two turbines of different sizes employed in this work. Large blue turbine: T80, diameter $D = 80$ m, hub height $z_h = 93.3$ m; small green turbine: T27, diameter $D = 27$ m, hub height $z_h = 31.5$ m.

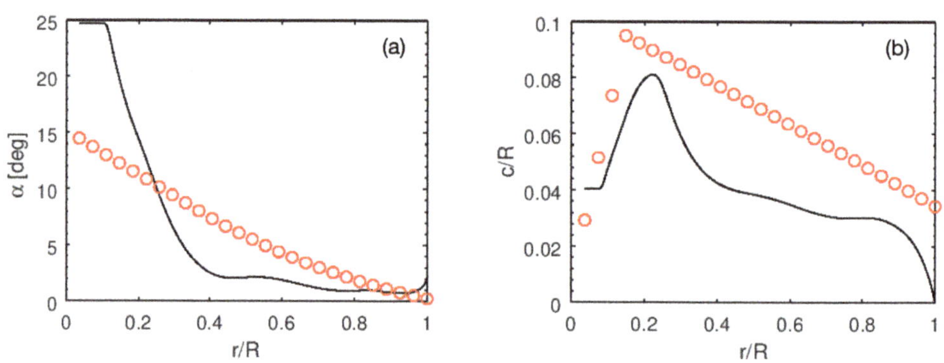

Figure 2. Radial distributions of (**a**) twist angle and (**b**) chord length. The twist angle and the normalized chord length are the same for turbines T80 and T27. Black lines: the employed turbine design; Red circle: SWiFT turbine design.

Table 1. Distribution of airfoil shapes along the radial direction [30]. The information in parentheses shows the airfoil shape distribution of the SWiFT turbine [31].

Section	r/R	Shape
1	$0 \leq r/R < 0.125$ $(0 \leq r/R < 0.038)$	Circle (Circle)
2	$0.125 \leq r/R < 0.225$ $(0.038 \leq r/R < 0.231)$	DU 40 (NACA 63-235)
3	$0.225 \leq /R < 0.375$ $(0.231 \leq r/R < 0.615)$	DU 35 (NACA 63-224)
4	$0.375 \leq /R < 0.475$ $(0.615 \leq r/R < 0.923)$	DU 30 (NACA 63-218)
5	$0.475 \leq /R < 0.675$ $(0.923 \leq r/R < 1)$	DU 25 (NACA 63-214)
6	$0.675 \leq /R < 0.825$	NACA $63_6 - 621$
7	$0.825 \leq /R \leq 1$	NACA $63_6 - 618$

3.2. Case Setup

The sizes of the computational domain are $L_x \times L_y \times L_z = 1.20\delta \times 2.09\delta \times \delta$ and $L_x \times L_y \times L_z = 0.405\delta \times 2.09\delta \times \delta$ with the corresponding numbers of grid nodes $N_x \times N_y \times N_z = 1501 \times 361 \times 282$ and $N_x \times N_y \times N_z = 1501 \times 351 \times 256$ for T80 and T27 turbine simulations, respectively, where x, y and z represent the downwind, lateral and vertical directions, respectively, and $\delta = 1000$ m is the boundary layer thickness. The width (L_y) and the height (L_z) of the computational domain are set being the same as those of the precursory simulation. In the downwind direction, the length of the computational domain (L_x) is 15 rotor diameters for both cases. The turbine is located $2D$ from the inlet plane. The length of the computational domain, although it is much shorter than its width, is typical for simulations of stand-alone wind turbines [4,22], and will not affect incoming large-scale coherent structures as they are generated from a precursory simulation. In the downwind direction, the grid nodes are uniformly distributed with grid spacing $D/100$. In the other two directions, the grids are uniformly distributed in the near turbine region ($|y - y_t| < 0.75D$ and $z < 2.2D$, where y_t is the turbine coordinate in the lateral direction) with the grid spacing $\Delta h = D/100$ and gradually stretched to boundaries of the computational domain. Employing the actuator surface model requires higher spatial resolutions. A grid of spacing $D/160$ was employed in [21], where the same actuator surface model is applied to predict the coherent tip vortices of a utility-scale wind turbine. In [25], on the other hand, a grid of spacing $D/40$ was shown being enough for the same actuator surface model to accurately predict the wake statistics of a hydrokinetic turbine, e.g., the velocity deficit and turbulence kinetic energy, which are also of interest in this work. Based on this study and the other work for a utility-scale wind turbine [27], we believe that the resolution of the employed grid is enough for the quantities of interest in this work. The sizes of time step are $\Delta t U_h/D = 7.9 \times 10^{-4}$ and 5.9×10^{-4} for T80 and T27 cases, respectively, where U_h and D are their own incoming downwind velocity at hub height and diameter, respectively.

The roughness length of the ground is $k_o = 0.01$ m, which represents the field characteristics at the SWiFT site. The thermal stratification is neutral. The free-slip boundary condition and the logarithmic law for rough walls are applied at the top and bottom of the computational domain, respectively. The periodic boundary condition is applied in the lateral direction. The same inflow is applied in the two simulations, which is generated from a precursory simulation. In the precursory simulation, the computational domain is $L_x \times L_y \times L_z = 6.28\delta \times 2.09\delta \times \delta$. The grid nodes are uniformly distributed in the horizontal directions, while are stretched in the vertical direction with the height of the first off wall grid cell $\Delta z_1 = \delta/1000$. Periodic boundary conditions are applied in the horizontal directions. At the ground and top boundary of the domain, boundary conditions the same as those in the wind turbine simulations are employed. The size of time step is $\Delta t U_h/\delta = 8 \times 10^{-4}$. The statistics of the inflow generated from the precursory simulation is shown in Figure 3. In the precursory simulation, the velocity fields on a $y - z$ plane at every time step are saved. Since the spatial and temporal resolutions of the precursory simulation are different from those of the turbine simulations, linear interpolations are employed in both space and time in order to apply the obtained inflow on the inlet plane of the wind turbine simulation.

In both cases, simulations are first carried out to a full developed state, and then further advanced in time for 100 and 70 rotor revolutions to compute time-averaged quantities for T80 and T27 cases, respectively. It is noted in Section 4 that the profiles of turbulence kinetic energy at $1D$ rotor upwind are not smooth enough. Although a similar number of rotor revolutions for temporal averaging has been employed in a hydrokinetic turbine case [32], further averaging in time is probably needed especially to take into account the low frequency variations caused by incoming large eddies. However, it will require about 1000 rotor revolutions [27] or even more, which is time-consuming and not realistic for this study. Taking the T80 case as an example, as one rotor revolution needs about 3.5 h wall clock time by using 640 compute cores, simulating 1000 rotor revolutions will take about 5 months wall clock time. As shown in Section 4, the profiles of turbulence kinetic energy and other quantities in the turbine wake are fairly smooth, so we believe that the main conclusions drawn from this work are not affected by the length of temporal averaging.

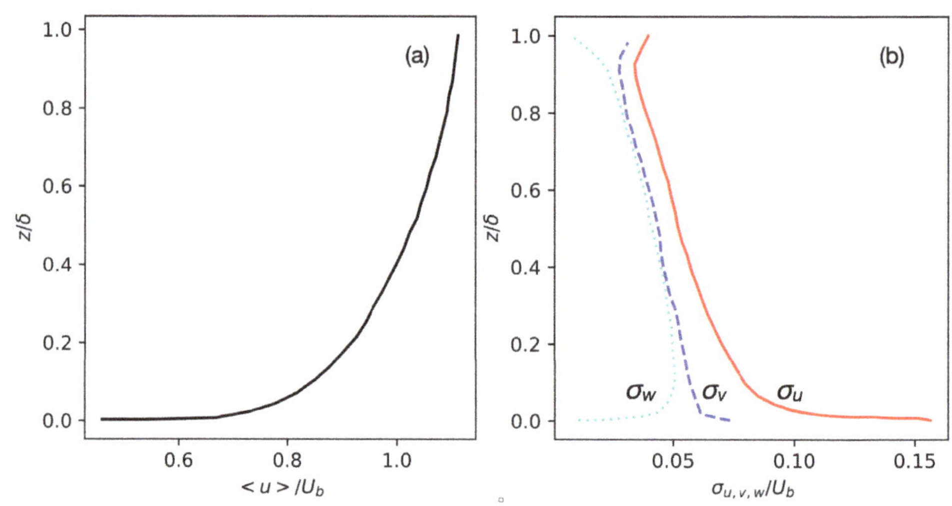

Figure 3. Statistics of inflow employed in the simulation for (**a**) mean downwind velocity and (**b**) standard deviation of velocity fluctuations, where U_b is the bulk velocity.

4. Results

As no measurements are available for the employed turbine designs, we attempt to validate the employed computational setup by comparing the simulated velocity deficit ΔU with the measurements at the SWiFT site considering the T27 turbine and the SWiFT turbine are of the same size and have comparable blade designs, in which the velocity deficit ΔU is defined as

$$\Delta U = U_{in} - U \tag{3}$$

where U is the time-averaged downwind velocity at different downwind locations, U_{in} is the time-averaged incoming downwind velocity (which is taken at $1D$ upwind the turbine). As in [33], a slight offset of $0.1D$ is imposed in the negative y direction to compensate for the wake deflection observed in the measured data. It is seen in Figure 4 that the simulation velocity deficit profiles show an overall good agreement with the measurements considering the complex wind and turbine operating conditions in the field and different turbine designs. It is also noticed the lateral velocity deficit profiles from the T80 case and the T27 case are very similar at the considered turbine downwind locations.

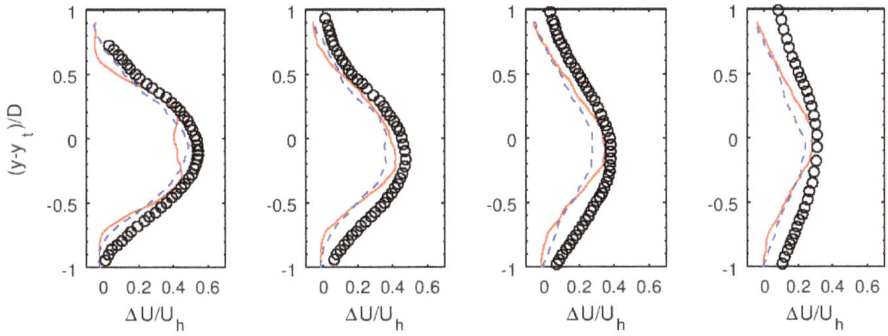

Figure 4. Comparison of the simulated velocity deficit profiles with measurements at the SWiFT site. The measured data are digitized from [33]. Details on the measurements can be found in [34]. Red lines: T80; Blue lines: T27; Circles: measurements.

We then examine the contours of the instantaneous downwind velocity from the two cases in Figure 5. It is seen that the wake remains an annular shape until about $2D$ downwind of the T80 turbine, while becomes unstable immediately downwind of the T27 turbine. At further downwind locations, the flow structures of the T80 turbine's wake remain quite coherent with its center biased towards the ground, which, on the other hand, looks chaotic and meanders significantly in the vertical direction with its center shifting above the centerline of the rotor for the T27 turbine.

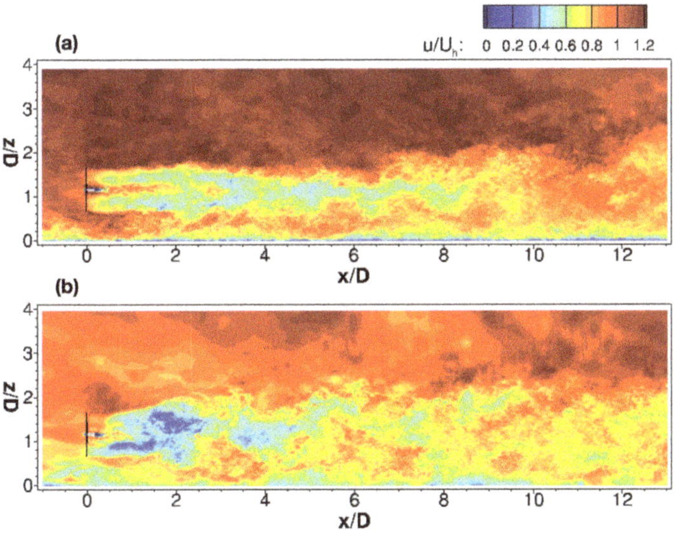

Figure 5. Instantaneous downwind velocity on the x-z plane passing through the rotor center for (**a**) T80 and (**b**) T27, respectively.

Next, we examine the time-averaged quantities from the two cases. In Figure 6 we compare the time-averaged downwind velocity from the two cases. It is seen that the velocity deficits from the T27 case are higher for the upper parts of the wake when $x/D < 2$, which are similar for the T80 case. At far wake locations, the wake center from the T80 case is below the hub height, while above the hub height for the T27 case.

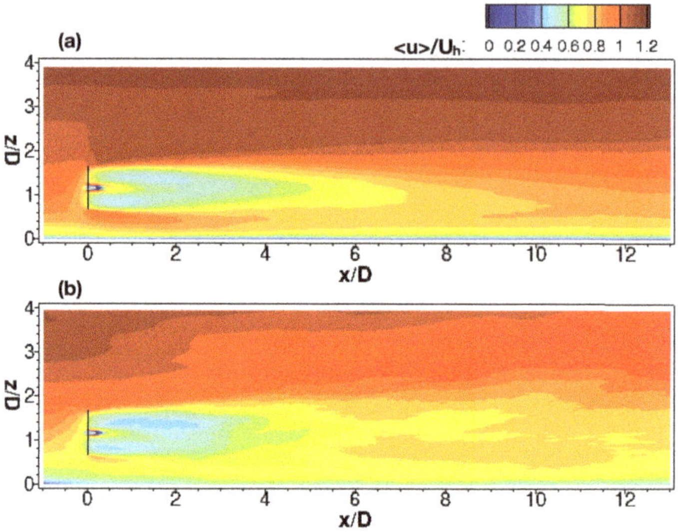

Figure 6. Time-averaged downwind velocity on the x-z plane passing through the rotor center for (**a**) T80 and (**b**) T27, respectively.

In Figure 7 we examine the turbulence kinetic energy (TKE) from the two cases. One similar observation from the two cases is that the high TKE region are in the top shear layer for both cases. One significant difference between the two cases is that the high TKE region persists to about 7*D* downwind the turbine for the T80 case, which only persists to about 4*D* turbine downwind for the T27 case.

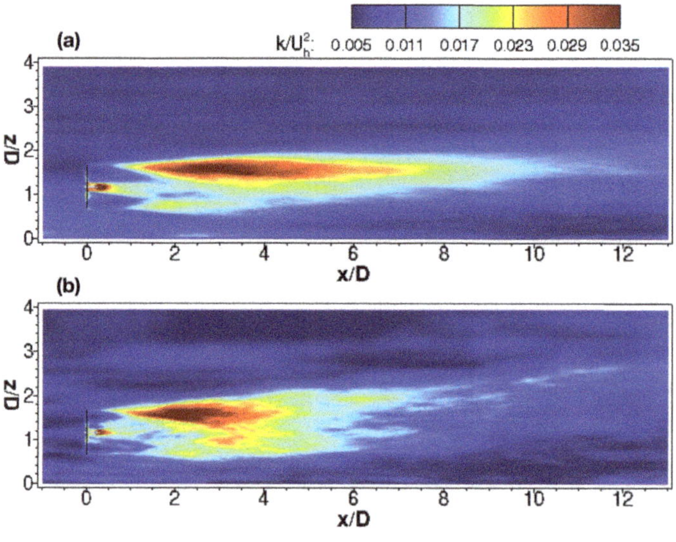

Figure 7. Turbulence kinetic energy (TKE) on the x-z plane passing through the rotor center for (**a**) T80 and (**b**) T27, respectively.

To qualitatively show the differences between the two cases, we examine the time-averaged velocity deficit (Equation (3)) and the turbine-added TKE profiles. The turbine-added TKE (Δk) is defined as

$$\Delta k = k - k_{in} \tag{4}$$

where k is the TKE at different downwind locations, k_in is the TKE of the inflow (which is taken at $1D$ upwind of the turbine for the present cases). It is seen in Figure 8a that incoming time-averaged downwind velocity profiles normalized using the corresponding length and velocity scales are quite similar, with one difference that the incoming velocity is slightly lower for the lower part of the turbine for the T27 case. At $x/D = 2$, the velocity deficit profiles are similar between the two cases with slightly higher velocity deficits from the T27 case. Moving to further downwind locations, the center of the wake gradually shifts to the upper part of the wake, which is above the upper tip of the turbine at $x/D = 10$.

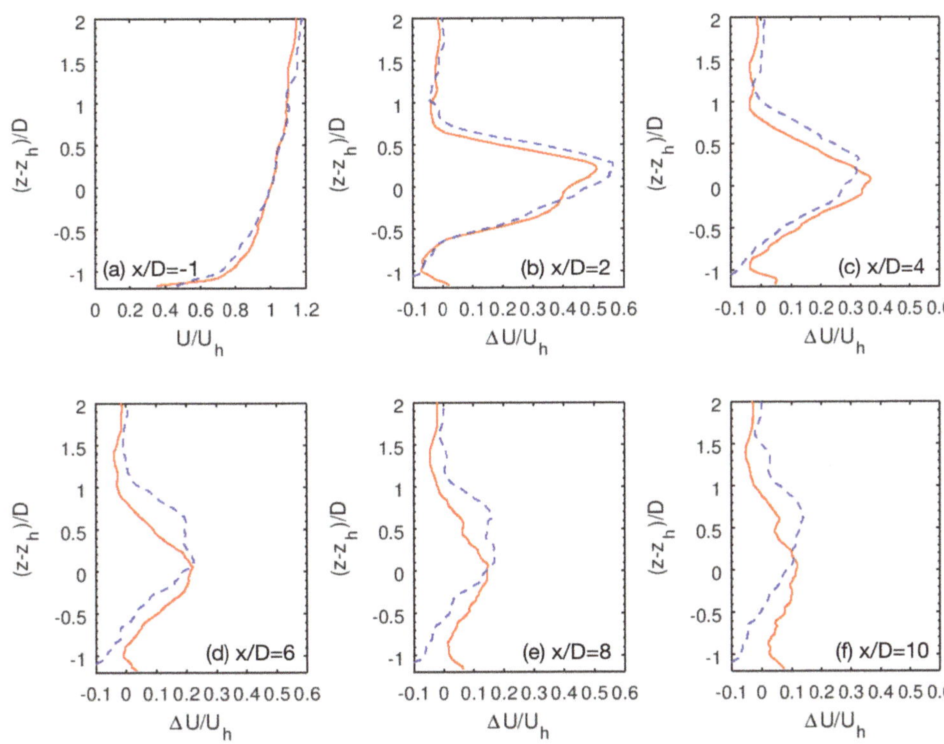

Figure 8. (**a**) Vertical profiles of time-averaged downwind velocity at $1D$ upwind of the turbine; (**b**–**f**) Vertical profiles of time-averaged velocity deficit at different locations downwind of the turbine. Red lines: T80; Blue lines: T27.

In Figure 9 we compare the turbine-added TKE from the two cases. First, we examine the normalized TKE at the $1D$ upwind of the turbine in Figure 9a. Because of different turbine sizes and hub heights, it is seen that the TKE of the inflow in the turbine region are quite different between the two cases, that the inflow TKE for the T27 case is higher than the T80 case for locations above the hub height while slightly lower for locations below the hub height. At $x/D = 2$, the vertical distributions from the two cases are similar, in that two peaks exist within the turbine top and bottom shear layers, respectively, with the one on the top about 1.5 times higher. At this location, it is also noticed that the normalized Δk from the T27 case is higher at almost all z locations. Starting from

$x/D = 4$, one observation is that the bottom peak of Δk disappears for both cases. One major difference is that the top peak of Δk still exists for the T80 case, which becomes flat and with a much wider high Δk region for the T27 case. It is also noticed that the maximum Δk from the T80 case is much higher than that from the T27 case. These observations are consistent with the lower incoming TKE for the upper region of the T80 case allowing the wake to maintain coherent helical wake vorticity structures for longer distance downstream than for the same region of the T27 case, where the higher incoming turbulence likely drives these structures to mix in to more evenly distributed vorticity sooner.

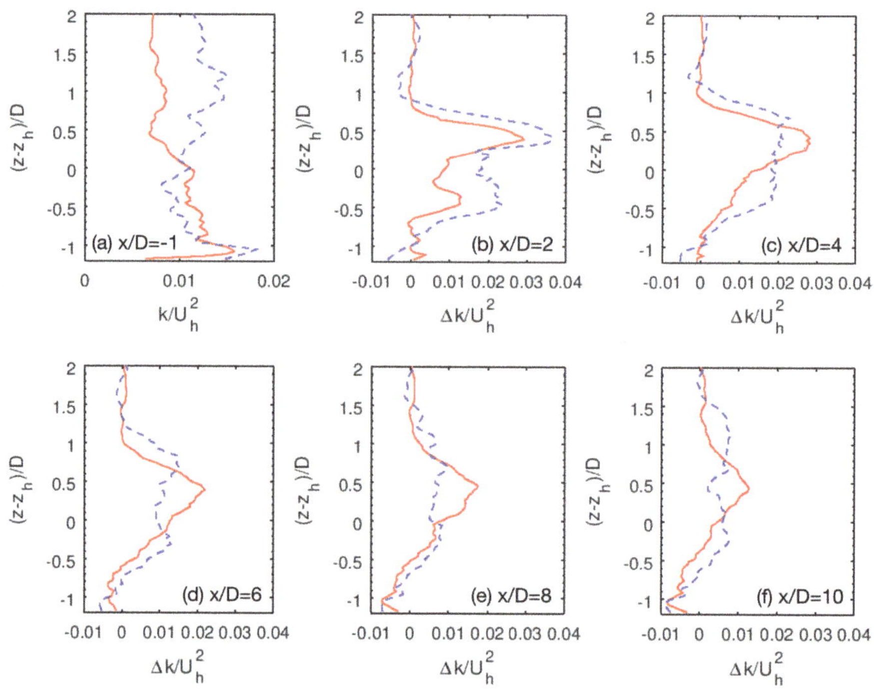

Figure 9. (**a**) Vertical profiles of turbulence kinetic energy at 1D upwind of the turbine; (**b–f**) Vertical profiles of turbine-added turbulence kinetic energy at different locations downwind of the turbine. Red lines: T80; Blue lines: T27.

To explore the reason for different distributions of velocity deficits and turbine-added TKE, we examine the mean kinetic energy (MKE) equation integrated over $y - z$ plane, which is shown as follows:

$$0 = MC + PT + TC + DF + TP + DP, \tag{5}$$

where MC, PT, TC, DF, TP, DP are the convection of the MKE by the mean flow, transport terms due to mean pressure, turbulence fluctuations, the diffusion term, the negative of turbulence production term (which transfers energy from the mean flow to turbulence), and the dissipation term, respectively. The expressions for the various terms in Equation (5) are given as follows:

$$MC = -\int_{y_t-R}^{y_t+R}\int_{z_h-R}^{z_h+R} \langle u_j \rangle \frac{\partial\left(\langle u_i \rangle \langle u_i \rangle /2\right)}{\partial x_j} dzdy \tag{6}$$

$$PT = -\int_{y_t-R}^{y_t+R}\int_{z_h-R}^{z_h+R} \frac{\partial\left(\langle p \rangle \langle u_j \rangle /\rho\right)}{\partial x_j} dzdy \tag{7}$$

$$TC = -\int_{y_t-R}^{y_t+R}\int_{z_h-R}^{z_h+R}\frac{\partial\left(\langle u_i'u_j'\rangle\langle u_i\rangle\right)}{\partial x_j}dzdy \tag{8}$$

$$DF = 2\int_{y_t-R}^{y_t+R}\int_{z_h-R}^{z_h+R}\frac{\partial\left((\nu+\nu_t)\,S_{ij}\langle u_i\rangle\right)}{\partial x_j}dzdy \tag{9}$$

$$TP = 2\int_{y_t-R}^{y_t+R}\int_{z_h-R}^{z_h+R}\langle u_i'u_j'\rangle\frac{\partial\langle u_i\rangle}{\partial x_j}dzdy \tag{10}$$

$$DP = -2\int_{y_t-R}^{y_t+R}\int_{z_h-R}^{z_h+R}(\nu+\nu_t)\,S_{ij}\frac{\partial\langle u_i\rangle}{\partial x_j}dzdy \tag{11}$$

where y_t is the coordinate of the rotor center in the lateral direction, R is the rotor radius, $S_{ij} = \left(\partial\langle u_i\rangle/\partial x_j + \partial\langle u_j\rangle/\partial x_i\right)/2$ is the strain rate tensor. The turbulence convection term TC is further decomposed into three components for the contributions from three directions as follows:

$$TC = TC_x + TC_y + TC_z, \tag{12}$$

where

$$TC_x = -\int_{y_t-R}^{y_t+R}\int_{z_h-R}^{z_h+R}\frac{\partial\left(\langle u_i'u'\rangle\langle u_i\rangle\right)}{\partial x}dzdy, \tag{13}$$

$$TC_y = \left(\int_{z_h-R}^{z_h+R}\left(-\langle u_i'v'\rangle\langle u_i\rangle\right)dz\right)\Big|_{y_t-R}^{y_t+R}, \tag{14}$$

$$TC_z = \left(\int_{y_t-R}^{y_t+R}\left(-\langle u_i'w'\rangle\langle u_i\rangle\right)dy\right)\Big|_{z_h-R}^{z_h+R}. \tag{15}$$

First, we show in Figure 10 different terms in Equation (5) for both cases. It is seen that the mean convection (MC) term is balanced with the pressure transport term in the near wake region where the pressure recovers to the ambient pressure by extracting mean kinetic energy from the wake. In the far wake region, the MC term is mainly balanced with the turbulence convection (TC) term. The above observations are similar to the wake of a model wind turbine located downwind of a three-dimensional hill [20]. The major differences between the T80 and the T27 cases are observed in the TC and TP terms, of which the magnitudes are higher in the near wake for the T27 case.

In Figure 11 we compare the three components of the TC term between the T80 and T27 cases. It is seen that the mean kinetic energy losses are due to the downwind component of the turbulence convection term TC_x term when $x/D < 4$ and $x/D < 2$ for the T80 and T27 cases, respectively. At further downwind locations the effect of the TC_x term on the MKE budget is negligible. On the other hand, the other two components of the turbulence convection terms TC_y and TC_z contributes positively to the MKE budget at almost all downwind locations except in the region immediately downwind the turbine. The TC_y term from the T27 case is higher than that from the T80 case for all downwind locations. On the other hand, the TC_z term from the T27 case is almost the same as that from the T80 case for $x/D < 2$, while lower than that from the T80 case for further downwind locations. That MKE entrainment from the top is lower than that from two sides explains the upward shift of the wake center observed in Figure 8 for the T27 case.

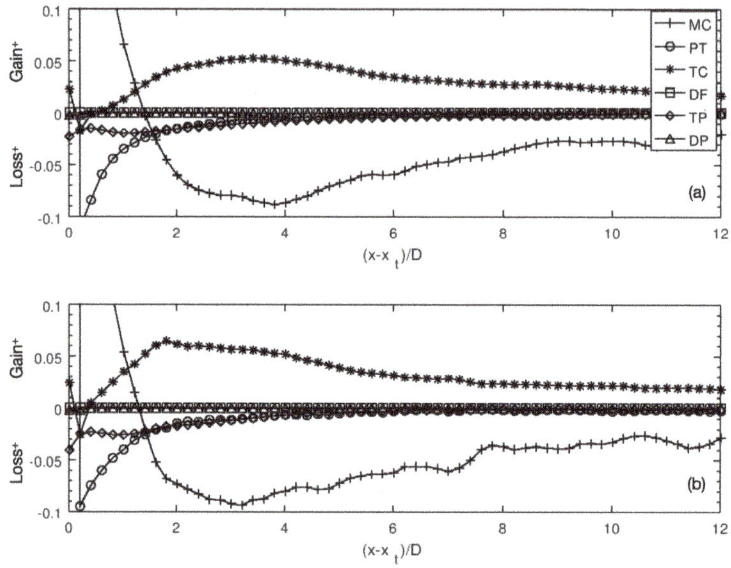

Figure 10. Budget of mean kinetic energy for (**a**) T80 and (**b**) T27, respectively. Line with crosses: MC term (Equation (6)); Line with circles: PT term (Equation (7)); Line with stars: TC term (Equation (8)); Line with squares: DF term (Equation (9)); Line with diamonds: TP term (Equation (10)); Line with triangles: DP term (Equation (11)). Different terms are normalized using $\frac{1}{2}DU_h^3$.

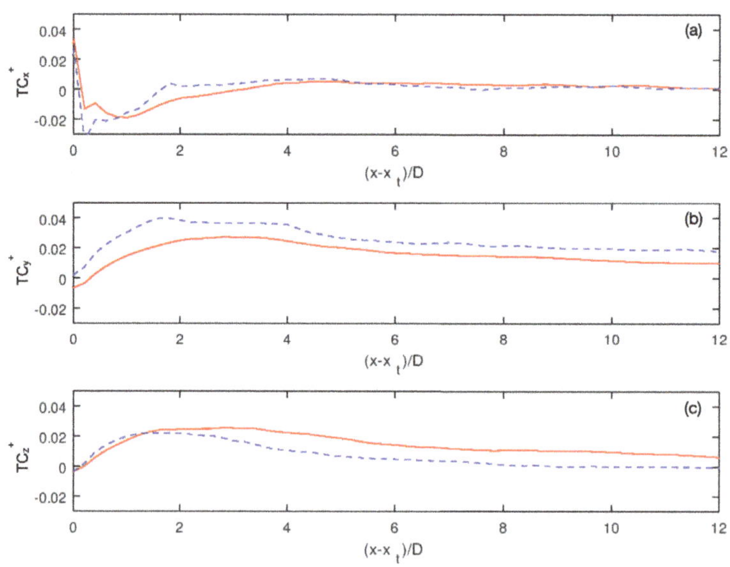

Figure 11. The three components of the turbulence convection (TC) term for (**a**) TC_x (Equation (13)), (**b**) TC_y (Equation (14)) and (**c**) TC_z (Equation (15)), respectively. Red line: T80; Blue line: T27. Different terms are normalized using $\frac{1}{2}DU_h^3$.

To better understand the differences in TC_y and TC_z terms between the two cases, we examine the averaged Reynolds stress and downwind velocity on the control surface from the two cases in Figures 12 and 13, in which different terms are defined as follows:

$$U_{side} = \left(\left(\int_{z_h-R}^{z_h+R} \langle u \rangle dz \right) \Big|_{y_t-R} + \left(\int_{z_h-R}^{z_h+R} \langle u \rangle dz \right) \Big|_{y_t+R} \right) /2, \tag{16}$$

$$U_{bot} = \left(\int_{y_t-R}^{y_t+R} \langle u \rangle dy \right) \Big|_{z_h-R}, \tag{17}$$

$$U_{top} = \left(\int_{y_t-R}^{y_t+R} \langle u \rangle dy \right) \Big|_{z_h+R}, \tag{18}$$

$$\langle u'v' \rangle_{side} = \left(\left(\int_{z_h-R}^{z_h+R} \langle u'v' \rangle dz \right) \Big|_{y_t-R} + \left(\int_{z_h-R}^{z_h+R} \langle u'v' \rangle dz \right) \Big|_{y_t+R} \right) /2, \tag{19}$$

$$\langle u'w' \rangle_{bot} = \left(\int_{y_t-R}^{y_t+R} \langle u'w' \rangle dy \right) \Big|_{z_h-R}, \tag{20}$$

$$\langle u'w' \rangle_{top} = \left(\int_{y_t-R}^{y_t+R} \langle u'w' \rangle dy \right) \Big|_{z_h+R}. \tag{21}$$

As seen in Figure 12a, the averaged downwind velocity on the two sides of the surface are nearly the same for the T80 and T27 cases. The averaged Reynolds stress term $\langle u'v' \rangle_{side}$ from the T27 case, on the other hand, is larger than that from the T80 case as shown in Figure 13a, which is the key reason the TC_y term the T27 case is larger than that from the T80 case (shown in Figure 11b). On the bottom of the control surface, the averaged downwind velocity from the T27 case is larger than that from the T80 case (as shown in Figure 12b), while the $\langle u'w' \rangle_{bot}$ from the T27 case is smaller than that from the T80 case from $2D$ to $5D$ rotor downwind and close to zero at further downwind locations, which is the same as the T80 case (shown in Figure 13b). This makes that the differences in entrainment from the bottom surface are insignificant between the two cases especially at far wake locations. On the other hand, the averaged downwind velocity on the top surface is significantly lower for the T27 case at all downwind locations as shown in Figure 12c, which is caused by the upward shift of the wake center for the T27 case as shown in Figure 8. Meanwhile, the $\langle u'w' \rangle_{top}$ from the T27 case is also significantly lower than that from the T80 case at downwind locations $x/D > 4$. Therefore, this explains why the TC_z term from the T27 case is lower than that from the T80 case at downwind locations $x/D > 2$ (shown in Figure 11c).

To explore the reason for the different downwind variations of the TKE shown in Figure 9, we compare the turbulence production term from the two cases in Figure 14. It is observed that the magnitude of the TP term from the T27 case is significantly larger than that from the T80 case when $x/D < 2$, while is similar to that from the T80 case at further downwind locations. Higher TP term indicates more energy is transferred to TKE from MKE. This is consistent to what we observed in Figure 9 that the TKE from the T27 case is significantly higher than that from the T80 case in the near wake region.

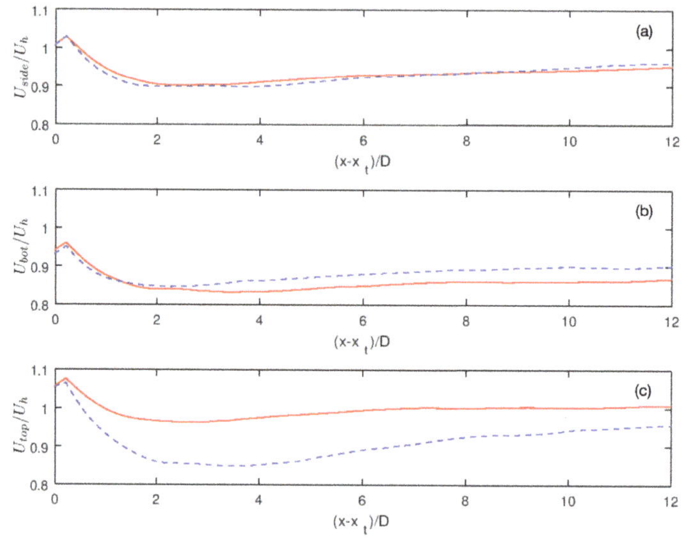

Figure 12. The averaged downwind velocity on the surface of the control volume for (**a**) on the two side surface of the control volume (Equation (16)), (**b**) on the bottom surface of the control volume Equation (17) and (**c**) on the top surface of the control volume Equation (18).

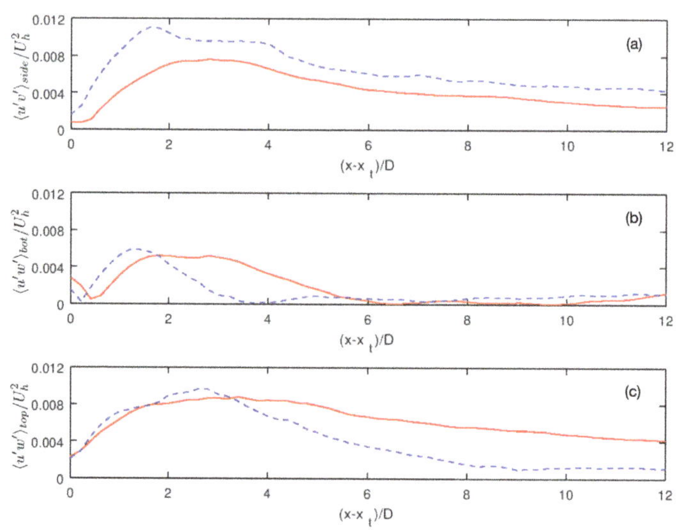

Figure 13. The averaged Reynolds stress on the surface of the control volume for (**a**) on the two side surface of the control volume (Equation (19)), (**b**) on the bottom surface of the control volume Equation (20) and (**c**) on the top surface of the control volume Equation (21).

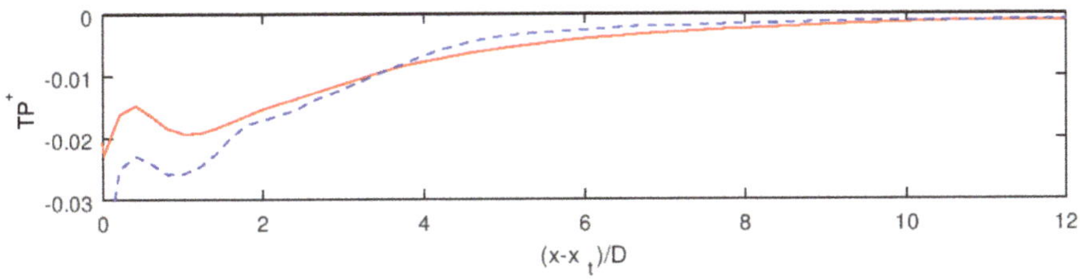

Figure 14. The turbulence production (TP) term (Equation (10)). Red line: T80; Blue line: T27. The TP term is normalized using $\frac{1}{2}DU_h^3$.

5. Conclusions

In this work we investigated the wake from a full-scale turbine (T80) and a subscale turbine (T27) using the VFS-Wind code to carry out large-eddy simulation with actuator surface models for turbine blades and nacelle. The T80 turbine is 2.96 times larger than the T27 turbine in terms of both rotor diameter and hub height. In both cases, the same turbulent inflow is employed. The same lift and drag coefficients are also employed for the two turbines. The key differences between the inflows for the two cases are the different distributions of wind shear and turbulence intensity across the rotor caused by different rotor diameters and different hub heights. The computed results show differences between the two cases for both velocity deficits and turbine-added turbulence kinetic energy. It is observed that the wake center of the T27 turbine shifts upwards in the far wake of the turbine (e.g., more than $0.5D$ above the hub height at $10D$ downwind the rotor), which remains at hub height for the T80 turbine. The maximum turbulence kinetic energy in the wake of the T27 turbine is higher than that from the T80 case in the near wake region (e.g., more than 20% higher at $2D$ downwind the rotor), but decreases rapidly and becomes lower than that from the T80 turbine at further downwind locations (e.g., about 50% lower at $8D$ downwind the rotor). We explore the reason for these differences by examining the budgets for the mean kinetic energy. Compared with the T80 case, we found that the MKE entrainment for the T27 case is higher for the lateral component of the turbulence convection term, while lower for the vertical component of the turbulence convection term. This explains the upward shift of wake center for the T27 case. We also examined the turbulence production term. It is observed that the magnitude of the turbulence production term from the T27 case is significantly higher than that from the T80 case in the near wake region, which explains the higher turbulence kinetic energy in the near wake for the T27 case. Both the inflow turbulence and the size of the turbine relative to the incoming eddies may cause these observed differences in the wake characteristics. Further study (e.g., space-time correlation study [35]) is needed to probe in more depth the underlying cause for the differences between the two turbines of different scales. A systematic study on the wakes from turbines of different scales for different turbulent inflows is also needed to further investigate the phenomenon observed in this work.

Author Contributions: Conceptualization, D.M., X.Y. and F.S.; methodology, X.Y. and D.F.; software, X.Y. and D.F.; validation, X.Y. and D.F.; resources, D.M.; writing—original draft preparation, X.Y.; writing—review and editing, X.Y., D.F., C.K., D.M. and F.S.; visualization, X.Y.; supervision, F.S.; project administration, F.S. and D.M.; funding acquisition, F.S. and D.M. All authors have read and agreed to the published version of the manuscript.

Funding: This work was supported by Department of Energy DOE (DE-EE0002980, DE-EE0005482 and DE-AC04-94AL85000). Computational resources were provided by Sandia National Laboratories and the University of Minnesota Supercomputing Institute.

Conflicts of Interest: The authors declare no conflict of interest.

References

1. Veers, P.; Dykes, K.; Lantz, E.; Barth, S.; Bottasso, C.L.; Carlson, O.; Clifton, A.; Green, J.; Green, P.; Holttinen, H.; et al. Grand challenges in the science of wind energy. *Science* **2019**, *366*, eaau2027. [CrossRef] [PubMed]
2. Chamorro, L.P.; Porté-Agel, F. A wind-tunnel investigation of wind-turbine wakes: Boundary-layer turbulence effects. *Bound. Layer Meteorol.* **2009**, *132*, 129–149. [CrossRef]
3. Chamorro, L.P.; Porte-Agel, F. Turbulent flow inside and above a wind farm: A wind-tunnel study. *Energies* **2011**, *4*, 1916–1936. [CrossRef]
4. Shen, W.Z.; Zhu, W.J.; Sørensen, J.N. Actuator line/Navier—Stokes computations for the MEXICO rotor: Comparison with detailed measurements. *Wind Energy* **2012**, *15*, 811–825. [CrossRef]
5. Krogstad, P.Å.; Eriksen, P.E. "Blind test" calculations of the performance and wake development for a model wind turbine. *Renew. Energy* **2013**, *50*, 325–333. [CrossRef]
6. Howard, K.B.; Singh, A.; Sotiropoulos, F.; Guala, M. On the statistics of wind turbine wake meandering: An experimental investigation. *Phys. Fluids* **2015**, *27*, 075103. [CrossRef]
7. Lignarolo, L.; Ragni, D.; Scarano, F.; Ferreira, C.S.; Van Bussel, G. Tip-vortex instability and turbulent mixing in wind-turbine wakes. *J. Fluid Mech.* **2015**, *781*, 467–493. [CrossRef]
8. Jin, Y.; Liu, H.; Aggarwal, R.; Singh, A.; Chamorro, L.P. Effects of freestream turbulence in a model wind turbine wake. *Energies* **2016**, *9*, 830. [CrossRef]
9. Barone, M.; White, J. *DOE/SNL-TTU Scaled Wind Farm Technology Facility: Research Opportunities for Study of Turbine-Turbine Interaction*; SANDIA Report SAND2011-6522; Sandia National Laboratories: Albuquerque, NM, USA, 2011.
10. Berg, J.; Bryant, J.; LeBlanc, B.; Maniaci, D.C.; Naughton, B.; Paquette, J.A.; Resor, B.R.; White, J.; Kroeker, D. Scaled wind farm technology facility overview. In Proceedings of the 32nd ASME Wind Energy Symposium, National Harbor, MD, USA, 13–17 January 2014; p. 1088.
11. Yang, X.; Boomsma, A.; Barone, M.; Sotiropoulos, F. Wind turbine wake interactions at field scale: An LES study of the SWiFT facility. In *Journal of Physics: Conference Series*; IOP Publishing: Bristol, UK, 2014; Volume 524, p. 012139.
12. Yang, X.; Boomsma, A.; Sotiropoulos, F.; Resor, B.R.; Maniaci, D.C.; Kelley, C.L. Effects of spanwise blade load distribution on wind turbine wake evolution. In Proceedings of the 33rd Wind Energy Symposium, Kissimmee, FL, USA, 5–9 January 2015; p. 0492.
13. Hong, J.; Toloui, M.; Chamorro, L.P.; Guala, M.; Howard, K.; Riley, S.; Tucker, J.; Sotiropoulos, F. Natural snowfall reveals large-scale flow structures in the wake of a 2.5-MW wind turbine. *Nat. Commun.* **2014**, *5*, 4216. [CrossRef]
14. Howard, K.B.; Guala, M. Upwind preview to a horizontal axis wind turbine: A wind tunnel and field-scale study. *Wind Energy* **2016**, *19*, 1371–1389. [CrossRef]
15. Heisel, M.; Hong, J.; Guala, M. The spectral signature of wind turbine wake meandering: A wind tunnel and field-scale study. *Wind Energy* **2018**, *21*, 715–731. [CrossRef]
16. Kelley, C.L.; Maniaci, D.C.; Resor, B.R. Scaled aerodynamic wind turbine design for wake similarity. In Proceedings of the 34th Wind Energy Symposium, San Diego, CA, USA, 4–8 January 2016; p. 1521.
17. Wu, Y.T.; Porté-Agel, F. Atmospheric turbulence effects on wind-turbine wakes: An LES study. *Energies* **2012**, *5*, 5340–5362. [CrossRef]
18. Chamorro, L.P.; Porté-Agel, F. Effects of thermal stability and incoming boundary-layer flow characteristics on wind-turbine wakes: A wind-tunnel study. *Bound. Layer Meteorol.* **2010**, *136*, 515–533. [CrossRef]
19. Abkar, M.; Porté-Agel, F. Influence of atmospheric stability on wind-turbine wakes: A large-eddy simulation study. *Phys. Fluids* **2015**, *27*, 035104. [CrossRef]
20. Yang, X.; Howard, K.B.; Guala, M.; Sotiropoulos, F. Effects of a three-dimensional hill on the wake characteristics of a model wind turbine. *Phys. Fluids* **2015**, *27*, 025103. [CrossRef]
21. Yang, X.; Hong, J.; Barone, M.; Sotiropoulos, F. Coherent dynamics in the rotor tip shear layer of utility-scale wind turbines. *J. Fluid Mech.* **2016**, *804*, 90–115. [CrossRef]
22. Foti, D.; Yang, X.; Sotiropoulos, F. Similarity of wake meandering for different wind turbine designs for different scales. *J. Fluid Mech.* **2018**, *842*, 5–25. [CrossRef]

23. Calderer, A.; Yang, X.; Angelidis, D.; Khosronejad, A.; Le, T.; Kang, S.; Gilmanov, A.; Ge, L.; Borazjani, I. *Virtual Flow Simulator*; Technical Report; University of Minnesota: Minneapolis, MN, USA, 2015.
24. Yang, X.; Sotiropoulos, F.; Conzemius, R.J.; Wachtler, J.N.; Strong, M.B. Large-eddy simulation of turbulent flow past wind turbines/farms: The Virtual Wind Simulator (VWiS). *Wind Energy* **2015**, *18*, 2025–2045. [CrossRef]
25. Yang, X.; Sotiropoulos, F. A new class of actuator surface models for wind turbines. *Wind Energy* **2018**, *21*, 285–302. [CrossRef]
26. Germano, M.; Piomelli, U.; Moin, P.; Cabot, W.H. A dynamic subgrid-scale eddy viscosity model. *Phys. Fluids A Fluid Dyn.* **1991**, *3*, 1760–1765. [CrossRef]
27. Yang, X.; Sotiropoulos, F. Wake characteristics of a utility-scale wind turbine under coherent inflow structures and different operating conditions. *Phys. Rev. Fluids* **2019**, *4*, 024604. [CrossRef]
28. Foti, D.; Yang, X.; Shen, L.; Sotiropoulos, F. Effect of wind turbine nacelle on turbine wake dynamics in large wind farms. *J. Fluid Mech.* **2019**, *869*, 1–26. [CrossRef]
29. Yang, X.; Sotiropoulos, F. A Review on the Meandering of Wind Turbine Wakes. *Energies* **2019**, *12*, 4725. [CrossRef]
30. Kelley, C.L.; Maniaci, D.C.; Resor, B.R. Horizontal-axis wind turbine wake sensitivity to different blade load distributions. In Proceedings of the 33rd Wind Energy Symposium, Kissimmee, FL, USA, 5–9 January 2015; p. 0490.
31. Kelley, C.L.; White, J. *An Update to the SWiFT V27 Reference Model*; Technical Report SAND2018-11893; Sandia National Laboratories (SNL-NM): Albuquerque, NM, USA, 2018.
32. Kang, S.; Yang, X.; Sotiropoulos, F. On the onset of wake meandering for an axial flow turbine in a turbulent open channel flow. *J. Fluid Mech.* **2014**, *744*, 376–403. [CrossRef]
33. Blondel, F.; Cathelain, M. An alternative form of the super-Gaussian wind turbine wake model. *Wind. Energy Sci. Discuss.* **2020**, *2020*, 1–16, [CrossRef]
34. Doubrawa, P.; Debnath, M.; Moriarty, P.J.; Branlard, E.; Herges, T.G.; Maniaci, D.C.; Naughton, B. Benchmarks for Model Validation based on LiDAR Wake Measurements. *J. Phys. Conf. Ser.* **2019**, *1256*, 012024, [CrossRef]
35. He, G.; Jin, G.; Yang, Y. Space-time correlations and dynamic coupling in turbulent flows. *Annu. Rev. Fluid Mech.* **2017**, *49*, 51–70. [CrossRef]

energies

MDPI

Article

Fluid–Structure Interaction Numerical Analysis of a Small, Urban Wind Turbine Blade

Michal Lipian [1,*], Pawel Czapski [2] and Damian Obidowski [1]

1 Institute of Turbomachinery, Lodz University of Technology, 90-924 Lodz, Poland; damian.obidowski@p.lodz.pl
2 Department of Strength of Materials, Lodz University of Technology, 90-924 Lodz, Poland; pawel.czapski@p.lodz.pl
* Correspondence: michal.lipian@p.lodz.pl; Tel.: +48-42-631-2454

Received: 27 February 2020; Accepted: 8 April 2020; Published: 10 April 2020

Abstract: While the vast majority of the wind energy market is dominated by megawatt-size wind turbines, the increasing importance of distributed electricity generation gives way to small, personal-size installations. Due to their situation at relatively low heights and above-ground levels, they are forced to operate in a low energy-density environment, hence the important role of rotor optimization and flow studies. In addition, the small wind turbine operation close to human habitats emphasizes the need to ensure the maximum reliability of the system. The present article summarizes a case study of a small wind turbine (rated power 350 W @ 8.4 m/s) from the point of view of aerodynamic performance (efficiency, flow around blades). The structural strength analysis of the blades milled for the prototype was performed in the form of a one-way Fluid–Structure Interaction (FSI). Blade deformations and stresses were examined, showing that only minor deformations may be expected, with no significant influence on rotor aerodynamics. The study of an unorthodox material (PA66 MO polyamide) and application of FSI to examine both structural strength and blade deformation under different operating conditions are an approach rarely employed in small wind turbine design.

Keywords: small wind turbine (SWT); computational fluid dynamics (CFD); composites; fluid–structure interaction (FSI)

1. Introduction

Although the increasing importance of wind energy in the global energy market (see, e.g., [1]) is mainly due to large-scale installations, Small Wind Turbines (SWTs) are also an important player in this field. This includes prosumer applications (e.g., in hybrid installations [2]), isolated sites operation (for example, to power transmitter stations or remote islands [3]), and multi-rotor array applications (a substitute for the single, big-size rotor [4]). This article presents the outcomes of a case study of an SWT fit for prosumer suburban application in European wind conditions.

1.1. Small Wind Turbines—Interest and Research

Nordic Folkecenter for Renewable Energy regularly publishes *The Small Wind Turbine* catalogue, the 8th edition of which brought together 104 companies from 28 countries, a total of 302 models of the rated power below 50 kW [5]. Publications such as the *Small Wind Guidebook* by WINDExchange (supported by the US Department of Energy and National Renewable Energy Association (NREL)) give guidelines about how to estimate whether or not an SWT is fit for a particular location and application and how to choose a proper solution for particular demands and needs [6]. SWTs in urban applications are one of the key interests of programs like the Intelligent Energy–Europe and Horizon 2020 Energy Efficiency.

Stathopoulos et al. [7], in their review of urban SWT technologies, argue that, although Vertical-Axis Wind Turbines (VAWTs) tend to be quieter and visually pleasant, the Horizontal-Axis Wind Turbines (HAWTs) remain a preferred choice for urban applications. This is because VAWTs remain commercially less cost-efficient than HAWTs. The authors also stress the importance of reliable data on urban aerodynamics the more that these wind conditions tend to be more capricious and characterized by a high level of turbulence intensity due to obstructions. The latter is extensively discussed by Anup et al. [8], who stress the need to conceive particular standards pertaining to SWTs, as those referring to big-scale machines may not reproduce the adverse wind conditions correctly. The authors discuss the influence of stochastic flow phenomena on the power outcome and wind turbine loading, which leads to increased fatigue load and underline the need for the structural analysis of wind turbine rotors by means of numerical codes such as Fatigue, Aerodynamics, Structures and Turbulence (FAST). Mechanical analysis of wind turbine blades is also important from the point of view of their inertia. Pourrajabian et al. [9] optimized the wooden blade geometry using genetic algorithms, in order to maximize rotor efficiency while preserving blade loadings in a safe range and ensure low blade inertia for low cut-in wind speed. The authors concluded that not every timber may be successfully used over a wide spectrum of velocity and identified alder as a preferable choice for wooden SWT blades.

Contemporary computational methods offer the possibility to couple high-order simulation of fluid flow and structural response in the Fluid–Structure Interaction (FSI) approach. A one-way FSI is an operation of checking deformation once the whole fluid flow simulation is executed. In a two-way approach, in each coupling iteration of fluid flow simulation, the deformations of the structure are being calculated and according to it—fluid mesh is changing its shape [10,11]. In either case, the simulation requires a significant computational effort, hence its main interest is in case of large-scale wind turbines (see, e.g., [12]). Lee et al. [13] used a one-way FSI model in their NREL Phase VI [14] small HAWT structural studies. The authors claim that the deformation of the tested rotor blades is mainly due to operating conditions (stall, etc.) and not elevated wind speed. FSI also proved important input and validation data for simpler, Blade Element-Momentum theory-based computations. FSI computations are also crucial in the process of developing completely new SWT designs, such as VAWT with morphing blades by MacPhee and Beyene [15]. The authors claimed that the controllably deformable blades enabled an increase in efficiency by as much as 9.6% with respect to fully rigid ones.

SWTs, studied at Institute of Turbomachinery of Lodz University of Technology, incorporate both experimental [16] and numerical [17] research. The increasing use of new manufacturing technologies in SWT studies and prototype manufacturing [18] makes it essential to ensure rotor blade integrity and safe operation. In the current article, the authors summarize the outcomes of a one-way FSI case study of a Generative Urban Small Turbine (GUST) horizontal-axis SWT prototype (see Chapter 2). The research was performed in order to check the blade behavior and performance in different wind conditions, ranging from normal operation to extreme working and static loads, hoping to see if the resulting deformations (twisting, axial displacement) are a serious threat to blade performance. It is also important to find the blade regions most susceptible to load concentrations and compare them with the material strength parameters (see Table 1). To the knowledge of the authors, this kind of analysis is rarely performed for SWTs in general, and for the unorthodox selected material in particular.

Table 1. Blade and material properties.

Blade mass	m	0.586	kg
Density	ρ	1.15	kg·m^{-3}
Young modulus	E	3.40	GPa
Poisson's ratio	ν	0.39	-
Ultimate tensile strength	U_T	90	MPa
Flexural strength	U_F	109	MPa

1.2. Basic Mathematical Formulations

In this article, the dimensionless analysis of wind turbines is employed [19]. The non-dimensional rotational speed (Tip-Speed Ratio, *TSR*) is defined as:

$$TSR = \frac{\omega R}{V}, \tag{1}$$

with ω—wind turbine rotational speed, R—rotor radius, and V—the reference wind speed. In order to present the wind turbine efficiency, the power coefficient Cp is introduced, as follows:

$$Cp = \frac{P}{\frac{\rho \times \pi R^2 \times V^3}{2}} \tag{2}$$

with P—wind turbine power and ρ—fluid density. Lastly, pressure coefficient c_{press} relates static pressure p to the dynamic pressure, as:

$$c_{press} = \frac{p}{\frac{\rho V^2}{2}}. \tag{3}$$

2. Object of the Study

The rotor is a three-bladed GUST HAWT. This small wind turbine has a nominal power 350 W at a wind velocity of 8.4 m/s [20]. The blades are of variable chord and twist angle (Figure 1) and are constructed upon NREL S826 and S834 airfoils [21]. The examined rotor radius R is equal to 0.8 m, which is the same as that used in experimental investigation used as a reference. For wind speed V = 8.4 m/s, at optimum TSR = 5, the local Reynolds number (at each blade station, thus taking chord length as a characteristic dimension) is of an order of magnitude of 1.2×10^5.

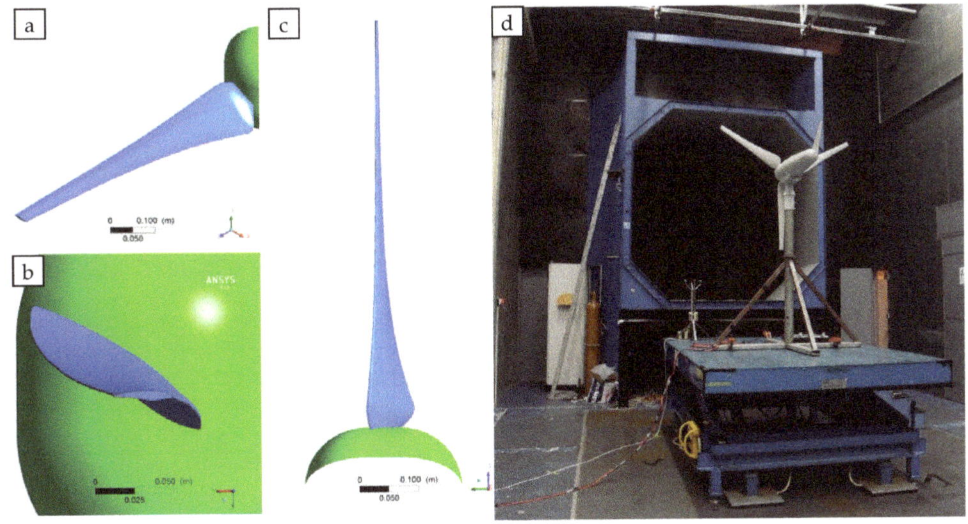

Figure 1. Blade geometry (left): (**a**) general view; (**b**) top view; (**c**) leading edge and suction side; and (**d**) wind turbine prototype during wind tunnel testing (own materials).

The material which was used to produce the prototype blades was Polyamide PA66 MO as an optimal choice between stiffness, allowable stresses, manufacturability, and price. The material was bought in sheets with dimensions 90 × 610 × 2000 mm and blades were milled on CNC machine. The material properties of this material, necessary for structural analysis, are grouped in Table 1. Since the blade was machined from a uniform piece of material, in the structural simulation, an isotropic material model is assumed.

3. Methodology

The simulation is performed using the commercial numerical simulation software ANSYS 18.0. Within the Workbench package ANSYS CFX was one-way coupled with ANSYS Structural (for mechanical analysis). The complete rotor geometry and domain encompassing it are discretized in the approach known as the Fully-resolved Rotor Model (FRM) [22]. The flow is assumed to be periodic, so only one-third of the rotor (one blade, 120° section) is considered.

3.1. Preprocessing

The overview of the considered geometry is visible in Figure 2. Two domains were distinguished in order to represent the fluid flow: rotor domain, in the form of a flat cylinder in which blade and hub are modelled, and stationary domain, which encompasses the former. In between the two domains, an interface is placed to permit the exchange of data. There is no relative movement between domains since the problem is considered as steady state. Instead, the frozen rotor model is employed, where the relative orientation of the components across the interface remains fixed along with the calculations and the flow from one component to the next changes the frame of reference [23]. The steady-state approach was chosen, as the simulation is performed under uniform, time-independent inflow conditions and one-way FSI. The transient analysis would require numerical and time resources prohibitively large for this—initial—phase of computations. The rotational velocity ω is imposed on the particles flowing through the rotor domain—analyzed cases are seen in Figure 4. The values of ω mimic those set in the benchmark experiment, performed independently by team GUST at TU Delft Open Jet Facility wind tunnel [24]. The experimental investigations were not a part of this study, and their results were made available as reference values at the courtesy of the GUST team. The authors estimate that the relative error of the experimental results is in range of 5%.

Figure 2. Side and front view of the analyzed problem geometry (elements are in scale).

Both domains were discretized together using tetrahedral, unstructured mesh (Figure 3) in ANSYS Mesher software. Refinement was performed in the rotor vicinity in order to better model the expected high gradients within the flow. The inflation layer was created around blades and hub surfaces to ensure a full resolution of boundary layer flow. The resulting mesh size is 10.8×10^6 nodes and 22.8×10^6 elements.

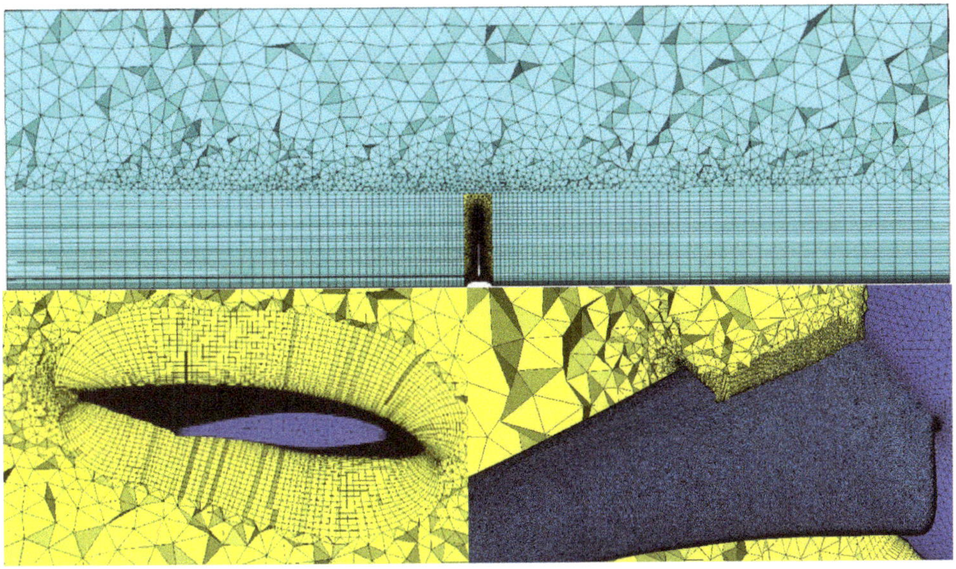

Figure 3. Mesh cross-section views: global (top) and around blade and hub (bottom).

Boundary conditions (see Figure 2 and Table 2) were set in order to mimic the outdoor wind turbine operation. The reference pressure was set to 1 atm (101,325 Pa). Chosen turbulence closure was k-ω SST with a standard set of coefficients, as this model proved trustworthy in previous wind turbine applications [22,23]. The considered flow medium (air) density $\rho = 1.185$ kg·m^{-3} and dynamic viscosity is 1.831×10^{-5} kg·m^{-1}·s^{-1}. The fluid is considered to be an incompressible ($Ma < 0.3$) ideal gas. Double precision, the fully coupled pressure-based solver was used, and the resolved equations involved flow continuity, momentum, and total energy conservation.

Table 2. Flow simulation boundary conditions.

Inlet	Wind velocity V, turbulence intensity $TI = 5\%$
Outlet	Static pressure $p = 0$ Pa
Side surfaces	Rotational periodicity
External surface	Opening (inflow and outflow), relative pressure 0 Pa
Blade, hub	No slip smooth wall

3.2. Solution and Convergence

A rudimentary numerical verification of the model was performed in order to check the correctness of the calculation. The tests were performed for $V = 12$ m/s and $TSR = 5$.

During the mesh independence study, the results obtained with the primary grid were compared with a reference grid, with approximately 75% more elements and nodes. Selected variables were compared between these two cases: torque in the axis of rotation (relative error $\delta \approx 0.5\%$) and flapwise bending moment ($\delta < 0.1\%$), as well as the axial force ($\delta \approx 0.6\%$). The relative error in the rotor power was of the order of magnitude 10^{-3}. The mesh quality was further assessed from the point of view of boundary layer flow solution correctness. The dimensionless distance $y+$ is the standard means for this verification. For turbulence closures incorporating the k-ω model for boundary layer flow, this parameter should be kept at level $y+ < 3$ [23], guaranteeing that the model will be able to correctly depict the velocity profile in the immediate vicinity of the wall. In the considered case, the mesh was inflated from the blade surface to ensure the proper transition of the element size. The differences in $y+$

values for the two meshes turned out to be more evident, with the average value on the blade surface being approximately 10.8 for the primary and 2.2 for reference mesh. For the latter grid, the highest values of $y+$ were identified near the blade tip, thus having only very little influence on the overall rotor performance. On the other hand, for the primary mesh, the local increase in $y+$ was observed in the flow separation region at both sides of the blade, which has a chief influence on the blade performance. This flaw of the primary mesh can significantly deteriorate the model fidelity, especially at lower wind speeds. Followed by this reasoning was the choice of the reference mesh for the actual studies.

The simulation convergence was evaluated based on the attained (normalized) level of residuals. For the momentum equations, this was approximately 3×10^{-5}, for the continuity equation 3×10^{-6}, for turbulent quantities transport equations 7×10^{-5} (k), and for 1×10^{-5} (ω). These values were deemed satisfactory to consider the solution as converged.

3.3. Fluid Flow Simulation Results

The first aspect of the performed results postprocessing was an analysis of the power output. In Figure 4 simulation (squares) and experiment (circles) results are compared for the same working points (V and TSR). A systemic error is noticeable, with the numerical results being approximately 120–140 W lower than experimental ones. While the current study is mostly dedicated to FSI investigation, these discrepancies may result in underestimation of actual aerodynamic loads and need addressing, for example, as a guideline for future works.

V, m/s	*TSR*, -	*Cpsim*, -	*Cpexp*, -
6.0	4.82	0.171	0.392
6.5	5.23	0.169	0.429
7.0	5.03	0.193	0.447
7.5	4.99	0.214	0.432
8.0	4.37	0.243	0.426
8.5	4.41	0.260	0.426
8.5	5.00	0.364	n/a
9.0	4.18	0.279	0.440
9.5	4.54	0.300	0.443
10.0	4.45	0.319	0.426
10.5	4.37	0.338	0.431
11.0	4.31	0.349	0.437
11.5	4.24	0.349	0.432
12.0	4.45	0.360	0.434
12.0	5.00	0.374	n/a

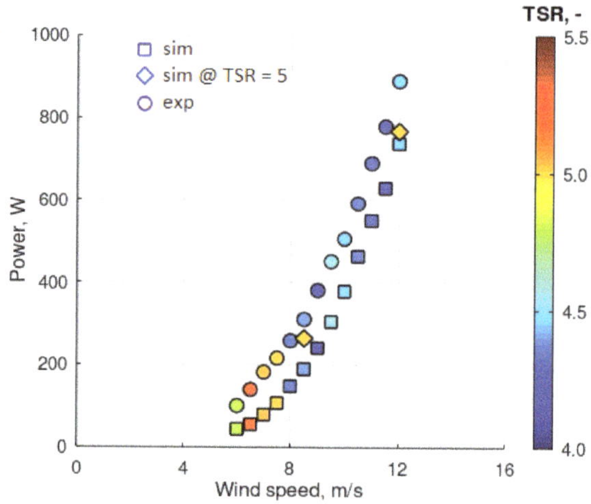

Figure 4. Analyzed cases and power obtained at different wind speeds and Tip-Speed Ratio *(TSR)*.

An important factor influencing the results are the simplifications of the numerical model. Firstly, the simulation performed in a steady-state mode may not entirely adequately depict the complex helical tip vortex wake structure, leading to modification in the induced velocities. This is further altered by the frozen rotor scheme–the most closely depicting the real flow, yet still simplifying it. For the full resolution of these flow phenomena, an unsteady model would be required. However, this was deemed prohibitively expensive in terms of computational resources and will be examined in the future, possibly with a two-way FSI. Another sources of error may be the size of the simulation domains (especially the one encompassing the rotor), and the fact that just 1/3 of the problem is being resolved. While this was done according to the standard procedures, these elements may influence the pressure fields, especially in the places where high gradients occur. The influence of this aspect was previously observed, also for rotor actuator models [22] and may be also connected with data transmission through interfaces.

In all, the obtained results may lead to a conclusion that the numerical model has the tendency to underestimate the aerodynamic loads. However, the performed structural assessment (Chapter 4) shows that the estimated stresses are well within the material flexural strength limit, and the deformations are very small, so even if a safety factor is taken into account, these observations will remain valid.

On a more general note, it is also visible how the flow is influenced by the relatively low Reynolds number phenomena: at lower wind speeds the rotor needs to turn at higher *TSR* in order to remain close to performance peaks. This becomes less evident as the wind speed increases, at which point, in experimental investigations, aeroelastic effects may start to play an important role.

In addition to cases mimicking experimental results, the simulation was also performed for design $TSR = 5$ at $V = 8.4$ m/s (rated wind speed) and 12 m/s (diamonds in Figure 4). This was done to assess the loads acting on the blade and provide data for further FSI simulation. The pressure fields on the blade are visible in Figure 5, providing input data for mechanical analysis. Special interest must be paid to the tip region, as in there the visibly high-pressure gradients between the two sides of the blade contribute to increased mechanical loads.

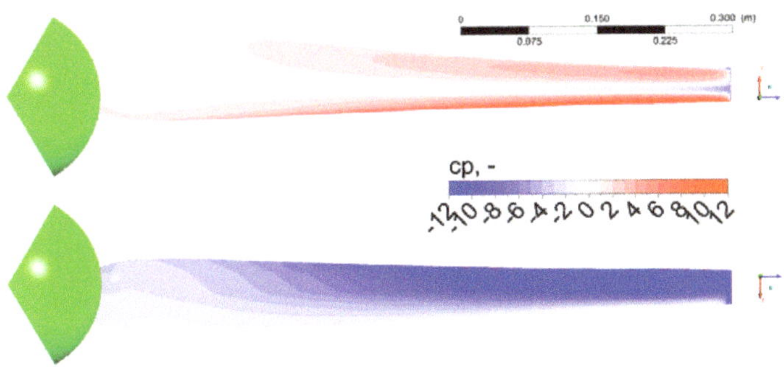

Figure 5. Pressure coefficient distribution at blade pressure (top) and suction (bottom) side ($V = 12$ m/s, $TSR = 5$).

Lastly, a general overview of the flow around the blade is visible in Figure 6. No boundary layer separation is visible whatsoever, suggesting that the wind turbine operates in pre-stall conditions. Quite visibly, the flow speed increases along with radius and it is fair to say that near the blade tip it is almost 3 times higher than at the bottom.

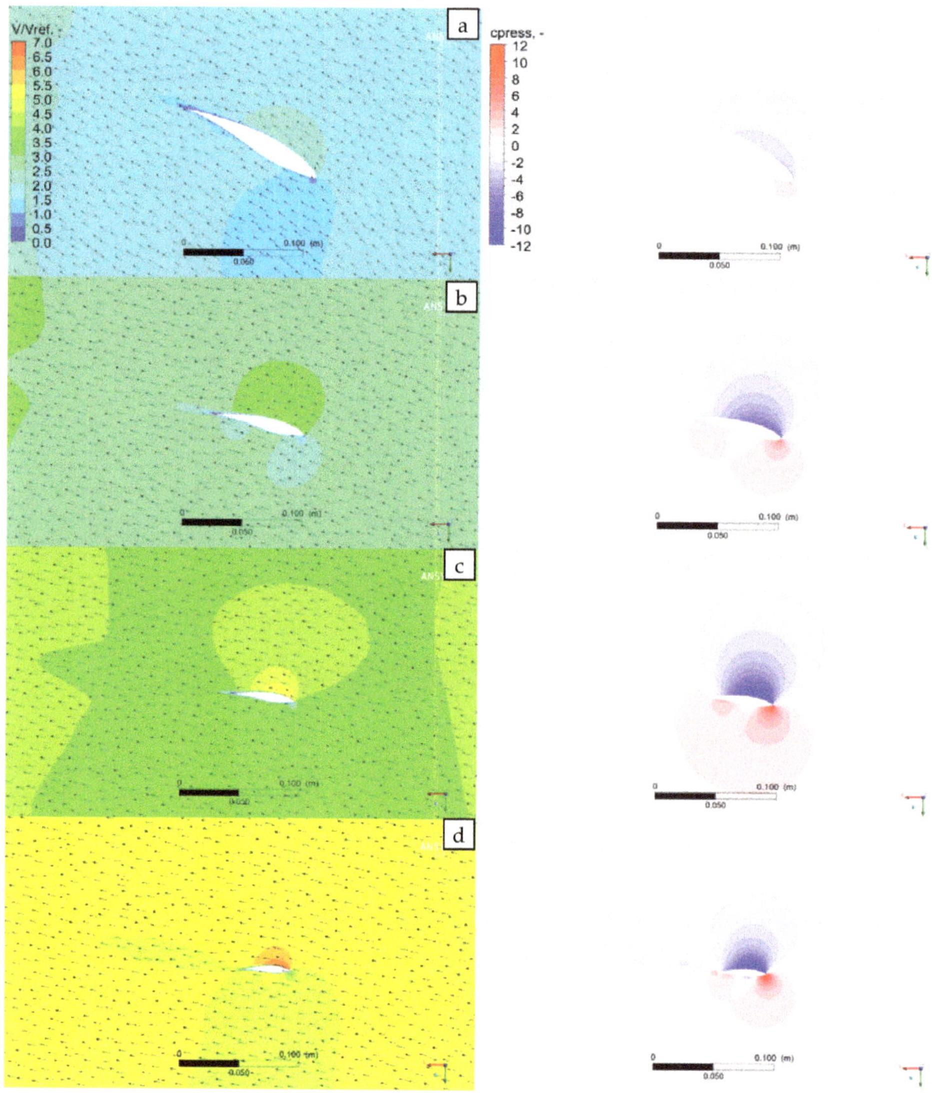

Figure 6. Normalized velocity (*Vref* = 12 m/s, *TSR* = 5) contours and vectors (left) and pressure coefficient contours (right) at *r*/*R* equal to (**a**) 0.25, (**b**) 0.5, (**c**) 0.75, and (**d**) 0.9875.

4. Fluid–Structure Interaction—Structural Assessment

The next step of the study covers the structural assessment of the blade and the entire wind turbine. Under inspection will be the taken stresses and deformations. The evaluation of stresses appearing in the material allows assessing the safety of the structure—whether the structure will not start to crack or creep. Deformations, meaning tip deflections and cross-sectional twist, provide information about potential changes of the airflow around the blade.

It is important to underline that the FSI simulations receive relatively little attention for SWTs of 1-kW power range since their blades are usually less prone to rupture than those of megawatt-size machines (due to, for instance, more uniform rotor disc loading). The current study shows that the

FSI successful application due to two chief aspects: to ensure that the applied, unorthodox material can withstand severe loads, and to try to estimate the influence of blade deformation due to the aerodynamic loads and check if this is a major concern for this size of machines.

4.1. Operational Cases and Boundary Conditions

The turbine is designed to operate in the wind speed range varying from 6 to 12 m/s. However, the numerical tests were also performed for velocities above this range, in extreme cases, for the structural strength evaluation. Four particular wind cases were investigated: the maximum thrust operation at normal operation, overload condition (twice the normal wind speed), and extreme loads ($V = 42$ m/s) for two rotor modes: working and stopped. Extreme working load corresponds to the point just before stopping the turbine, while static corresponds to when the turbine has stopped. Overload conditions correspond to the case between the upper limit of operational condition (max thrust) and extreme loads. Each of these cases corresponds to different flow parameters given in Table 3.

Table 3. Load cases under inspection in structural analysis.

	V, m/s	*TSR*, -	Rotational Velocity [rpm]	[rad/s]
Max thrust	12	5	716	74.98
Overload conditions	24	5	1432	149.96
Extreme working load	42	5	2500	261.80
Extreme static load	42	-	0	0

The blade of the wind turbine is subjected to two types of loads: pressure from flowing fluid and centrifugal forces. Therefore, in the case of the analyzed blade, following loads and boundaries are assumed (see Figure 7): fixed support at the beginning of the blade – in the place where the blade is in contact with the hub (A); the rotational velocity of the blade enabling to calculate deformations due to centrifugal forces (B); pressure fields imported from fluid flow simulation (C, see exemplary imported pressure in Figure 8).

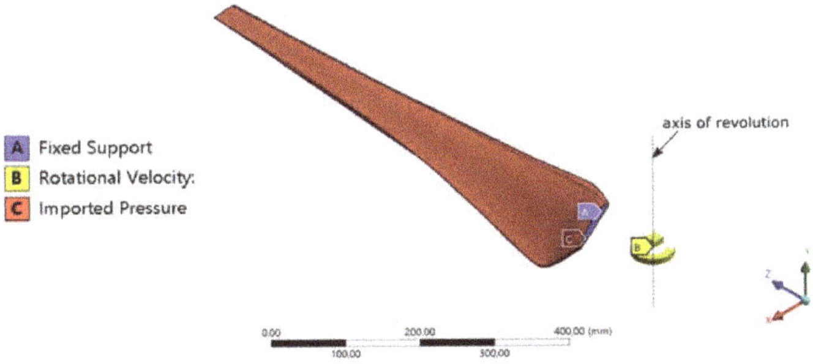

Figure 7. Applied loads and boundary conditions.

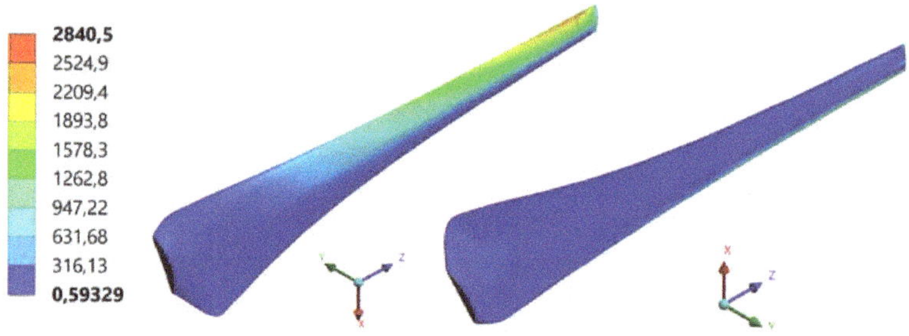

Figure 8. Map of imported pressure on the pressure and suction sides of the blade (results in [Pa]).

4.2. Mesh Creation and Choice

Wind turbine blades, under operation, are mainly subjected to bending and torsion. To capture these phenomena correctly, proper structural mesh must be prepared. The analyzed blade is a sweepable body, making it possible to only use solid hexahedron elements instead of tetrahedrons. The body was meshed using ANSYS Solid 185 elements with eight nodes with three degrees of freedom at each node (translations in all directions). Moreover, the body is going to be subjected to bending, which means that at least three elements through the thickness are necessary to capture the stresses properly. For the mesh convergence study, four meshes are prepared—the mesh operations and sizings are as follows (see Figure 9):

1. Sweep method—this mesh parameter forces the software to sweep the elements across the whole length of the blade;
2. Face sizing—face sizing of the cross-section enables to create required size element throughout the thickness of the blade;
3. Edge sizing of the trailing edge is set in order to obtain in this place at least three elements throughout the blade thickness.

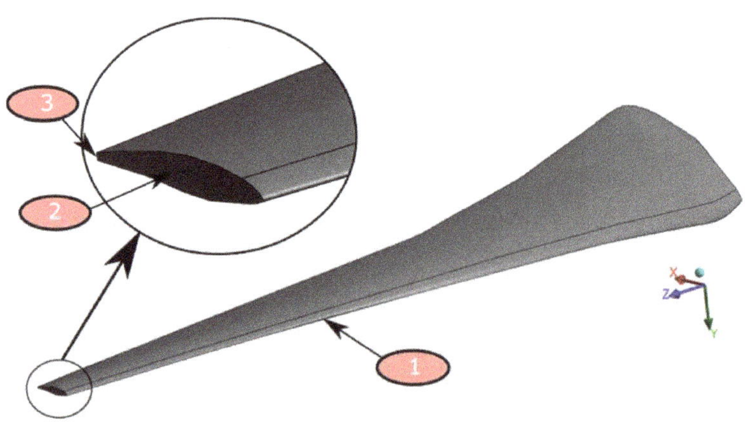

Figure 9. Mesh operations on the blade.

Four meshes under inspection are denoted as I, II, III, and IV. First, three meshes differ only by means of the sizing of elements. In mesh IV, additionally, the bias in the length of the blade direction is applied. The mesh parameters for these densities are presented in Table 4. The cross-sections of the blades for meshes II, III, and IV are the same. The comparison of these cross-sections with a cross-section of the mesh I is presented in Figure 10.

Table 4. Mesh sizing.

Mesh	Sweep El. Length, mm	Method Sizing of the Tip Cross-Section El. Size, mm	Edge Sizing N° of Divisions	Number of Elements
I	3	0.4	4	243,780
II	5	0.6	3	69,498
III	5	0.6	3	49,572
IV	5 [1]	0.6	3	69,498

[1] In mesh IV, a bias is applied in order to increase the number of elements near the fixing point.

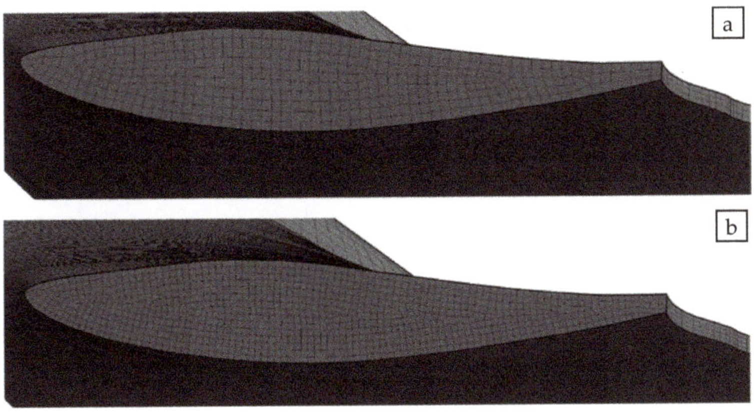

Figure 10. Comparison of blade grids in cross-sections in meshes (**a**) II, III, IV, and (**b**) I.

Mesh convergence study has been performed for maximal thrust case (wind speed equal to 12 m/s and $TSR = 5$). The criteria for investigating these three meshes are as follows:

1. Maximal deflection of the blade tip δ—see the map of vertical deflections in Figure 11;
2. Maximal equivalent stress in the blade σ_{max}—see the map of equivalent stresses in Figure 12;
3. The maximal angle of additional, deformation blade twist $\Delta\beta$—see Figure 13.

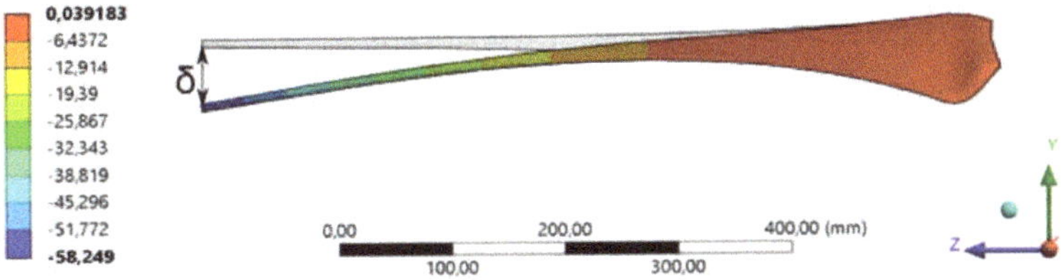

Figure 11. Blade tip deflection δ (in mm).

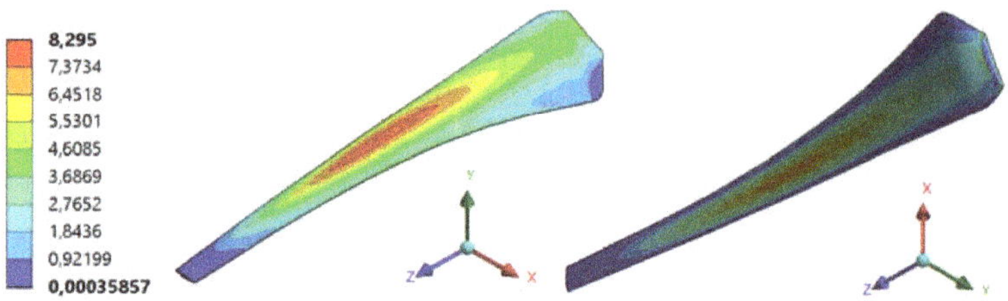

Figure 12. Map of equivalent (von Mises) stresses appearing in the blade σ_{max} (in MPa).

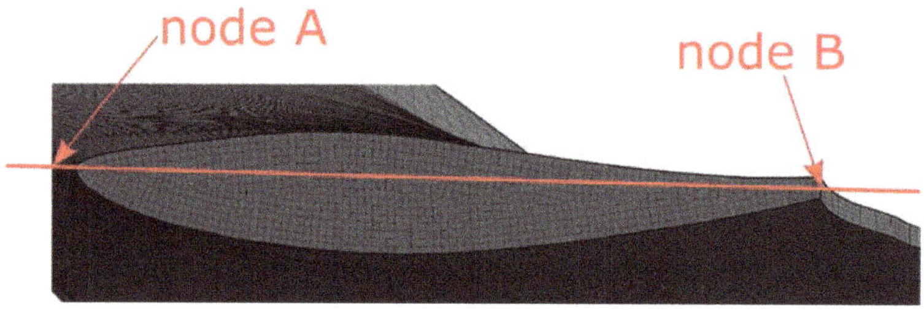

Figure 13. Control line passing through the leading and trailing edge of the blades for determining blade tip twist.

The last parameter, which is the twist of the blade due to deformation (in addition to the design section twist), must be quantified manually by means of comparing coordinates of the nodes before and after deformation. It is determined for the blade tip, as the highest angular deformations of the blade appear there. Coordinates of the leading edge node with coordinates of the mid-node at the trailing edge are being extracted from the software and twist in blade-tip cross-section plane is being calculated.

Based on the coordinates of points A and B, a twisting of the blade tip is determined. A line is traced through points at the blade tip leading edge (A) and trailing edge (B). The angle of inclination of this line with respect to the rotor plane of rotation may be computed using simple mathematical transformations, both before and after the deformation. A difference between these two values, $\Delta\beta$, is a measure of the deformation twist angle of the blade due to deformation.

A summary of all three examined parameters is shown in Table 5. It is visible that all mesh densities are providing the results, which are in agreement with each other in terms of the deformations—both total deflections and deformation twist angles are comparable to each other for all structural meshes (the relative difference is not higher than 2%). The situation is similar when talking about the convergence of the value of allowable stress—the relative differences between two diverging results are not exceeding the value of 2%.

Table 5. Mesh convergence study results.

Mesh	δ, mm	σ_{max}, MPa	$\Delta\beta$, deg
I	58.249	8.295	0.032
II	57.882	8.453	0.036
III	57.817	8.641	0.034
IV	57.511	8.815	0.033
average	57.864	8.551	0.034

4.3. Stresses and Deformations—Discussion

The structural assessment of the wind turbine has been performed on structural mesh no. III as the mesh with the smallest number of elements and fulfilling all convergence criteria. For all considered cases, tip deflection, maximal equivalent stresses and deformation twist have been calculated using one-way coupling FSI. The data are presented in Table 6. Moreover, tip deflections and maximal equivalent stresses appearing in the structure as the function of wind speed and TSR are presented in Table 6, respectively.

Table 6. Tip deflection, maximal equivalent stresses, and deformation twist for all considered cases (mesh III).

	δ, mm	σ_{max}, MPa	$\Delta\beta$, deg
Max thrust	58.249	8.551	0.034
Overload conditions	223.939	31.131	0.453
Extreme working load	621.616	90.141	1.403
Extreme static load	56.824	8.365	-

From the point of view of this study, a very significant parameter is the deformation twist of the blade. The value of this parameter strongly affects the efficiency of the turbine, as it results in operation of the airfoil in adverse, out-of-design conditions. If $\Delta\beta$ is too high, a stall may occur and provoke efficiency drop. In this case, however, the study proves that the angular deformation of the blade should not have a significant influence on the rotor efficiency. Numerical tests had shown that considered structure is not twisting more than 0.05°, which is a very good result, indicating that the consecutive blade sections are well in their operation regions. The total deflection of the blade is not playing a significant role because the total area of the blade is almost constant. Summing up, according to these results, it can be stated that deformations of the blade will not affect its operation.

The last parameters to be analyzed are equivalent stresses (Von Mises). In the case of the blades, the highest stresses are appearing in the middle of the length of the blade and near connection with the hub. The flexural strength of Polyamide PA66 MO is equal to 109 MPa. This value is not exceeded for any of the considered cases, meaning that the turbine is structurally safe.

5. Conclusions

The article summarizes the observations made in frames of a numerical study of a small HAWT performance. The simulation was based on the reference experiment, performed on a prototype in wind tunnel conditions. The results of fluid flow assessment prove to underestimate the HAWT performance, which was attributed to the simplifications of the model, but a satisfactory qualitative agreement is found between simulation and experiment nonetheless.

Furthermore, thanks to one-way coupling FSI, it was possible to perform a structural assessment of the turbine. Firstly, stresses occurring in the blade during different operational load cases were inspected. The study proved that the structural integrity of CNC-milled blades is not threatened in its operating regions, and there are still high safety margins. Next, the study of deformations delivered an overview of potential blade twisting and bending due to the experienced loads. Under normal operating conditions of the assessed wind turbine, these additional deformations turned out to be negligible, with very little to no harm to blade aerodynamics.

The study proved that—in the context of small wind turbines—the FSI analysis may be an interesting choice not only for the strength analysis of non-orthodox materials but may also try to help estimate the influence of the blade deformation on rotor performance. Future plans include extending the simulation towards transient investigations in order to examine the fatigue influence on the blade endurance.

Author Contributions: Data curation, M.L. and P.C.; Investigation, P.C.; Supervision, D.O.; Writing – original draft, M.L., P.C. and D.O. All authors have read and agreed to the published version of the manuscript.

Funding: The investigations have been performed under the research project POWR.03.02.00-00-I042/16-00 of the National Centre for Research and Development.

Acknowledgments: The authors would like to thank the GUST team for sharing the experience and data collected during the TU Delft Open Jet Facility wind turbine prototype testing.

Conflicts of Interest: The authors declare no conflict of interest.

Nomenclature

FRM		Fully-resolver Rotor Model
FSI		Fluid–Structure Interaction
GUST		Generative Urban Small Turbine
HAWT		Horizontal-Axis Wind Turbine
NREL		National Renewable Energy Laboratory
SST		Shear Stress Transport (turbulence model)
SWT		Small Wind Turbine
VAWT		Vertical-Axis Wind turbine
Cp	-	Power coefficient
E	GPa	Young modulus
Ma	-	Mach number
P	W	Wind turbine power
R	m	Rotor radius
TI	%	Turbulence intensity
TSR	-	Tip-Speed Ratio
U	MPa	Tensile strength
V	m/s	(Reference) wind speed
c_{press}	-	Pressure coefficient
m	kg	(Blade) mass
p	Pa	Pressure
$y+$	-	Dimensionless distance
$\Delta\beta$	deg	Maximal angle of blade twist due to deformation
δ	mm	Maximal deflection of the blade tip
ν	-	Poisson's ratio
ρ	kg/m^3	Density
σ_{max}	MPa	Maximal equivalent stress in the blade
ω	rad/s	Rotational speed

References

1. BP Plc. *BP Statistical Review of World Energy 2019*; BP Plc: London, UK, 2019.
2. Ma, T.; Javed, M.S. Integrated sizing of hybrid PV-wind-battery system for remote island considering the saturation of each renewable energy resource. *Energy Convers. Manag.* **2019**, *182*, 178–190. [CrossRef]
3. Neto, P.B.L.; Saavedra, O.R.; Oliveira, D.Q. The effect of complementarity between solar, wind and tidal energy in isolated hybrid microgrids. *Renew. Energy* **2020**, *147*, 339–355. [CrossRef]
4. Jamieson, P. Multi Rotor Solution for Large Scale Offshore Wind Power. In Proceedings of the EERA Deepwind 2017, Trondheim, Norway, 18–20 January 2017.
5. Folkecenter for Renewable Energy. *Catalogue of Small Wind Turbines (under 50 kW)*; Folkecenter Print; Folkecenter for Renewable Energy: Hurup Thy, Denmark, 2016.
6. WINDExchange. *WINDExchange, Small Wind Guidebook*; US Department of Energy, Office of Energy Efficiency and Renewable Energy: Washington, DC, USA, 2020.
7. Stathopoulos, T.; Alrawashdeh, H.; Al-Quraan, A.; Blocken, B.; Dilimulati, A.; Paraschivoiu, M.; Pilay, P. Urban wind energy: Some views on potential and challenges. *J. Wind Eng. Ind. Aerodyn.* **2018**, *179*, 146–157. [CrossRef]

8. Anup, K.C.; Whale, J.; Urmee, T. Urban wind conditions and small wind turbines in the built environment: A review. *Renew. Energy* **2019**, *131*, 268–283.
9. Pourrajabian, A.; Dehghan, M.; Javed, A.; Wood, D. Choosing an appropriate timber for a small wind turbine blade: A comparative study. *Renew. Sustain. Energy Rev.* **2019**, *100*, 1–8. [CrossRef]
10. Glück, M.; Breuer, M.; Durst, F.; Halfmann, A.; Rank, E. Computation of fluid–structure interaction on lightweight structures. *J. Wind Eng. Ind. Aerodyn.* **2001**, *89*, 1351–1368. [CrossRef]
11. Löhner, R.; Haug, E.; Michalski, A.; Muhammahd, B.; Drego, A.; Nanjundaiah, R.; Zarfam, R. Recent advances in computational wind engineering and fluid–structure interaction. *J. Wind Eng. Ind. Aerodyn.* **2015**, *144*, 14–23. [CrossRef]
12. Wang, L.; Quant, R.; Kolios, A. Fluid structure interaction modelling of horizontal-axis wind turbine blades based on CFD and FEA. *J. Wind Eng. Ind. Aerodyn.* **2016**, *158*, 11–25. [CrossRef]
13. Lee, K.; Huque, Z.; Kommalapati, R.; Han, S.-E. Fluid-structure interaction analysis of NREL phase VI wind turbine: Aerodynamic force evaluation and structural analysis using FSI analysis. *Renew. Energy* **2017**, *113*, 512–531. [CrossRef]
14. Hand, M.M.; Simms, D.A.; Fingersh, L.J.; Jager, D.W.; Cotrell, J.R.; Schreck, S.; Larwood, S.M. *Unsteady Aerodynamics Experiment Phase VI: Wind Tunnel Test Configurations and Available Data Campaigns*; National Renewable Energy Laboratory: Golden, CO, USA, 2001.
15. MacPhee, D.W.; Beyene, A. Fluid–structure interaction analysis of a morphing vertical axis wind turbine. *J. Fluid Struct.* **2016**, *60*, 143–159. [CrossRef]
16. Lipian, M.; Dobrev, I.; Karczewski, M.; Massouh, F.; Jozwik, K. Small wind turbine augmentation: Experimental investigations of shrouded- and twin-rotor wind turbine systems. *Energy* **2019**, *186*, 115855. [CrossRef]
17. Kulak, M.; Lipian, M.; Zawadzki, K. Investigation of performance of Small Wind Turbine blades with winglets. *Int. J. Numer. Method Heat* **2020**, in press. [CrossRef]
18. Lipian, M.; Kulak, M.; Stepien, M. Fast Track Integration of Computational Methods with Experiments in Small Wind Turbine Development. *Energies* **2019**, *12*, 1625. [CrossRef]
19. Burton, T.; Jenkins, N.; Sharpe, D.; Bossanyi, E. *Wind Energy Handbook*, 2nd ed.; John Wiley & Sons, Ltd.: Chichester, West Sussex, UK, 2011.
20. Grapow, F.; Raszewska, D.; Skalski, R.; Czarnecki, J.; Telega, K.; Miller, M.; Rogowski, P.; Prociow, M. Small wind, big potential: HAWT design case study. *MATEC Web Conf.* **2018**, *234*, 01005. [CrossRef]
21. Kadrowski, D.; Kulak, M.; Lipian, M.; Stepien, M.; Baszczynski, P.; Zawadzki, K.; Karczewski, M. Challenging low Reynolds—SWT blade aerodynamics. *MATEC Web Conf.* **2018**, *234*, 01004. [CrossRef]
22. Lipian, M.; Karczewski, M.; Jozwik, K. Analysis and comparison of numerical methods for design and development of small Diffuser-Augmented Wind Turbine (DAWT). In Proceedings of the IECON 2016—42nd Annual Conference of the IEEE Industrial Electronics Society, Florence, Italy, 23–26 October 2016.
23. *Ansys CFX, Release 18.0, Help System, Book: CFX*; ANSYS, Inc.: Canonsburg, PA, USA, 2017.
24. Open Jet Facility, TU Delft. Available online: https://www.tudelft.nl/lr/organisatie/afdelingen/aerodynamics-wind-energy-flight-performance-and-propulsion/facilities/low-speed-wind-tunnels/open-jet-facility/ (accessed on 23 February 2020).

energies

MDPI

Article

Development and Validation of CFD 2D Models for the Simulation of Micro H-Darrieus Turbines Subjected to High Boundary Layer Instabilities †

Rosario Lanzafame [1], Stefano Mauro [1,*], Michele Messina [1] and Sebastian Brusca [2]

1 Department of Civil Engineering and Architecture (DICAR), University of Catania, Via Santa Sofia 64, 95125 Catania, Italy; rosario.lanzafame@unict.it (R.L.); mmessina@dii.unict.it (M.M.)
2 Department of Engineering, University of Messina, Contrada Di Dio, 98166 Messina, Italy; sbrusca@unime.it
* Correspondence: mstefano@diim.unict.it; Tel.: +390957382455

† This paper is an extended version of our paper published in December 2019 on AIP Conference Proceedings 2191(1):020109, doi:10.1063/1.5138842.

Received: 30 August 2020; Accepted: 20 October 2020; Published: 23 October 2020

check for updates

Abstract: The simulation of very small vertical axis wind turbines is often a complex task due to the very low Reynolds number effects and the strong unsteadiness related to the rotor operation. Moreover, the high boundary layer instabilities, which affect these turbines, strongly limits their efficiency compared to micro horizontal axis wind turbines. However, as the scientific interest toward micro wind turbine power generation is growing for powering small stand-alone devices, Vertical Axis Wind Turbines (VAWTs)might be very suitable for this kind of application as well. Furthermore, micro wind turbines are widely used for wind tunnel testing, as the wind tunnel dimensions are usually quite limited. In order to obtain a better comprehension of the fluid dynamics of such micro rotors, in the present paper the authors demonstrate how to develop an accurate CFD 2D model of a micro H-Darrieus wind turbine, inherently characterized by highly unstable operating conditions. The rotor was tested in the subsonic wind tunnel, owned by the University of Catania, in order to obtain the experimental validation of the numerical model. The modeling methodology was developed by means of an accurate grid and time step sensitivity study and by comparing different approaches for the turbulence closure. The hybrid LES/RANS Delayed Detached Eddy Simulation, coupled to a transition model, demonstrated superior accuracy compared to the most advanced unsteady RANS models. Therefore, the CFD 2D model developed in this work allowed for a thorough insight into the unstable fluid dynamic operating conditions of micro VAWTs, leading the way for the performance improvement of such rotors.

Keywords: CFD; Delayed DES; H-Darrieus; VAWT; micro wind power generation

1. Introduction

The interest for micro wind power generation is growing, since micro wind turbines appear to be very suitable for powering low-power devices such as wireless sensors, actuators, controllers, and small lightning systems. Usually, these devices are powered by means of chemical batteries, which however must be periodically replaced, therefore representing a challenge for a reliable power supply. For this reason, small-size energy harvesters are increasingly used in order to realize reliable self-powered systems. Small-scale wind turbines represent an attractive solution for the electricity generation in such small-size energy harvesters.

Previous studies about miniature wind turbines mainly focused on Horizontal Axis Wind Turbine (HAWT) due to the fact they had acceptable efficiency despite the small scale. Xu et al. [1] developed an experimental test and a numerical predictive model on a miniature HAWT with a diameter of

15 cm. They obtained a maximum efficiency of approximately 27%. Kishore et al. [2,3] designed and characterized a small-scale portable wind turbine of approximately 40 cm that showed very low cut-in velocity and a maximum efficiency of 30% with a rated power of 1.4 W. Zakaria et al. [4] performed an experimental investigation of a centimeter-scale wind turbine, thus showing the strong effect of the very low Reynolds number on the rotor performance as they obtained an efficiency of only 3–4%. An interesting application of miniature HAWT was proposed again by Xu et al. [5]. They developed a physical-based model for the prediction of the optimal load resistance and the experimental characterization of the micro turbine. The efficiency was found to be lower than 10%. Ionescu et al. [6] studied the possibility to optimize small VAWTs through the use of different techniques such as specific low Re airfoils, blade shapes, and passive and active flow control. A special design optimization of a cost-effective micro turbine was implemented by Leung et al. [7]. The multi-bladed rotor, with a radius of approximately 12 cm, reached a maximum power coefficient (C_p) of 12%. Howey et al. [8] designed a miniature shrouded multi-bladed wind turbine by means of the BEM theory. The experimental study result demonstrated a maximum efficiency just over 10%. Park et al. [9] made a feasibility study about the use of micro wind turbines to power wireless sensors on a bridge. They were able to demonstrate how a micro turbine, with a diameter of 14 cm, can generate sufficient energy for this specific application.

Micro wind turbines are often used for wind tunnel experiments as well. As the wind tunnel dimensions are usually limited, only small size rotors can be tested. In this regard, Bastankhah et al. [10] designed and analyzed a miniature wind turbine with a rotor diameter of 15 cm. They demonstrated that an accurate fluid dynamic design for specific low Reynolds number was of utmost importance for reaching high efficiency. In this case they obtained a maximum C_p of approximately 40%. The authors of the present paper presented a numerical and experimental study regarding a three bladed micro HAWT for wind tunnel applications [11]. An efficiency of about 30% was found.

The studies presented above demonstrate the scientific interest toward micro wind rotors and highlight the fact that small rotors need a very accurate fluid dynamic design in order to obtain high efficiency. This is mainly due to two factors. Very low Reynolds numbers drastically reduce the airfoil performances and the small dimensions emphasize the unsteadiness and the instabilities, as will be demonstrated hereinafter. It is no coincidence that VAWTs have not been taken into consideration in the aforementioned studies. Indeed, in these rotors, the unsteady phenomena affect the performance much more than in HAWTs and the scale reduction drastically augments negative effects such as dynamic stall, blade-wake interaction as well as low Reynolds number influence. However, the advantages of VAWTs, such as constructive simplicity, omnidirectionality with respect to the flow, and positioning of the generation unit on the ground, make these turbines deserve further investigation in the aforementioned small-scale applications. For instance, Mutlu et al. [12] evaluated the performance of in-pipe VAWT for turbo solenoid valve system, finding interesting results.

In light of the above, in the present work the authors developed a 2D CFD model of a H-Darrieus VAWT with a diameter of 20 cm. The CFD model was validated by means of wind tunnel experiments carried out in the subsonic wind tunnel at the University of Catania. This micro rotor operated at very low tip speed ratios and very low Re, which caused strong and sudden boundary layer instability (separation and unsteady vortex shedding) leading to early dynamic stall development and large lift losses on the blade. This involved that most of the CFD procedures, proposed in the literature for largest rotors, may not be suitable in this case.

The experimental H-Darrieus rotor had 4 NACA 0012 blades and it was designed and constructed with a 3D printer. Further details about the experimental set up are presented in the next section. In order to develop an accurate and reliable CFD model of such micro rotor, a thorough sensitivity study was carried out. The study analyzed the spatial and temporal discretization sensitivity and the influences of three different turbulence models. The turbulence models evaluated were the widely used RANS fully turbulent SST k-ω model, the transition SST model by Menter, and the hybrid RANS/LES Delayed Detached Eddy Simulation model (DDES) coupled to the transition SST model by Menter for the RANS region. Furthermore, the results were compared to those obtained through the use of a double multiple stream-tube 1D model (DMSTM) using the commercial software Qblade with the suitable aerodynamic polar for an average rotor Reynolds number of approximately 40,000. The global comparison with the experimental data demonstrated that the DDES turbulence model with a very fine spatial and temporal discretization was the one able to provide accurate predictions of the rotor performance in terms of average power coefficient (C_P). The other turbulence models highly underestimated the average C_P despite the very fine spatial and temporal discretization. On the other hand, Qblade DMSTM results demonstrated a surprising proximity to the C_P experimental data, at least in a portion of the curve.

The results obtained through the post-processing of the CFD 2D model allowed the authors to gain an important insight into to the fluid dynamics of micro Darrieus rotors and to clarify the physical reasons for the very poor performance of these turbines when the geometrical scale is reduced. The boundary layer instability, accurately detected through the use of DDES, seemed to be the most important reason for the poor performance of such small rotors. Furthermore, this instability was strongly related to the reduced geometrical scale, which led to very low operating Reynolds numbers. At these very low Re, the boundary layer is mostly laminar and, as is widely known, a laminar boundary layer is more sensitive to adverse pressure gradients that trigger laminar bubbles, separation, and sudden transition.

2. Numerical Model and Experimental Setup

2.1. Literature Background

The CFD 2D numerical model presented in this paper was based on a wide literature background. The literature dealt mainly with larger rotors that in general operated in more stable conditions and higher Reynolds numbers, therefore their 3modelling3 usually led to accurate results. On the other hand, the present study focused on the development of a numerical 3model3 specifically dedicated to micro H-Darrieus rotor in which highly unstable conditions like boundary layer instability, dynamic stall, and blade-wake interaction affected the rotor operation more than in large rotors. However, numerous advices were found in the literature as reported hereinafter. Balduzzi et al. [13–15] provided essential guidelines for the development of accurate CFD 2D models of VAWTs, particularly concerning the spatial and temporal independency study. Rezaeiha et al. [16] provided the guidelines for the development of accurate CFD simulations for H-Darrieus rotors, focusing on the domain dimension and the azimuthal increment, which imposed the temporal discretization size. In this case, the authors found a good compromise using a smaller domain than Balduzzi et al. Actually, there is no accordance concerning the computational domain size, which, however, must be sufficiently large to guarantee that the flow around the rotor is not affected by the domain dimensions. Through wind tunnel experiments, Raciti Castelli et al. [17] validated a CFD 3modelling3 strategy, which regarded the near blade spatial discretization depending on the turbulence models. Recently, Rogowski et al. [18] developed a 2D CFD model of a two bladed H-Darrieus rotor. The model was validated using experimental data. Bangaa et al. [19] performed a numerical study on a single bladed VAWT under dynamic stall conditions using CFD. Additionally, the authors of the present paper previously developed a 2D CFD 3model for moderately large H-Darrieus [20], demonstrating the good accuracy of the transition turbulence model by Menter in cases in which the laminar to turbulent boundary layer transition played an essential role.

The above showed that, beyond the domain size, there was no strong agreement about the spatial and temporal discretization as well. All the reviewed papers agreed with the fact that the SST k-ω based turbulence models demonstrated best accuracy among the RANS turbulence models when used with sufficiently refined spatial and temporal discretization. For example, a y^+ less than one near the blades was universally considered an essential constraint. A time step size, which limited the azimuthal increment to less than 0.5 degree, was equally important. When boundary layer transition phenomena are detected, a transition turbulence model strongly increased the accuracy of the CFD models [20,21]. However, other authors evidenced the superior accuracy of advanced turbulence 4models4 like hybrid RANS-LES formulations, specifically when the rotors are subjected to unstable conditions. Strong dynamic stall, high blade-wake interaction, and vortices related to the flow separation were certainly more accurately predicted through the use of hybrid RANS-LES models like DES and Delayed DES. For example, Lei et al. [22] demonstrated that improved DDES simulation was able to capture real flow characteristics, like those generated by vortices related to dynamic stall phenomena, that were not predicted by the SST k- ω model. Li et al. [23] optimized the blade pitch in a two-blade H-Darrieus turbine using a 2D CFD model based on DDES simulation, therefore evidencing its superior predictive capability with respect to the RANS models. Thè et al. [24], in their thorough review, showed that DDES simulations were the best compromise between high accuracy and computation requirements for the simulation of VAWTs under unstable conditions. Simão Ferreira et al. [25,26] performed a complete numerical-experimental comparison between RANS, DES, and LES turbulence models using accurate Particle Image Velocimetry (PIV) data. They proved that "the DES model is not only able to predict the generation and shedding of vorticity and its convection, it also shows an acceptable sensitivity to grid refinement (both space and time), thus making it suitable for simulations where validation data are limited or non-existent. URANS models proved insufficient because of their inability to correctly model the large eddies, and the influence of this in the development of forces in the downwind passage of the rotor. The LES performed worse than the DES model, probably because of a less accurate 4modelling4 of the wall region." Again, DDES techniques demonstrated an ability to accurately predict massively separated turbulent flow structures in an Orthopter-type VAWT [27] in which the operating conditions are inherently unstable. Abdellah et al. [28] 4analyzed the effect of the spatial discretization when DDES simulations were implemented for VAWT rotors. Wang et al. [29] also demonstrated the high accuracy of DDES turbulence models for deep dynamic stall simulations at low Reynolds numbers of the NACA 0012 airfoil. The paper fixed grid and time step requirements such that the LES region was able to capture at least 80% of the turbulent kinetic energy of the flow. Furthermore, they evidenced the superior accuracy of the SST transition model for the RANS region among the RANS models at low Reynolds numbers. The above suggested that a DDES, based on a transition formulation for the RANS region, would be the optimal choice for the simulation of airfoil subjected to deep dynamic stall at low Reynolds number. This was certainly the typical operating condition of the H-Darrieus rotor in the present work. This option was studied by Sa et al. [30] on the flow past an Eppler 387 wing. The results indicated that the hybrid DES/Transition model predicted both strong laminar/turbulent transition phenomena, including the laminar separation bubble, and flow separation at high angles of attack. Therefore, the DDES turbulence model with the transition turbulence model by Menter [31–33], recently implemented in ANSYS Fluent solver, appeared to be very suitable for the scope of the present work. In order to support this assumption a comparison between the RANS fully turbulent SST k-ω model, the SST transition model and the DDES, coupled to the SST transition model, was carried out in the present paper.

2.2. Computational Domain and Experimental Setup

The experimental rotor was a four bladed H-Darrieus rotor with a ratio between height and diameter equal to one. Two endplates were used for the reduction of the tip losses in such a way to make the experimental results, as much as possible, consistent with the 2D simulations. Even though a 3D simulation would have been very interesting, the computation time would have been unaffordable, therefore a 2D simulation was the only way to gain a thorough insight into the blade aerodynamics. The blades were built in a resin 3D printer, which allowed for very high accuracy of the details and a very fine surface roughness. The endplates were 3D printed in PLA material. A steel threaded bar was used as shaft. Table 1 reports the geometrical and experimental rotor features while Figure 1 shows an image of the rotor assembled on the experimental setup.

Table 1. Experimental rotor features.

Geometrical and Experimental Details	
Diameter (m)	0.2
Height (m)	0.2
Cord (m)	0.03
Number of blades	4
Solidity	0.19
Blade airfoil	NACA 0012
Pitch (deg)	0
Shaft diameter (m)	0.01
Maximum Re	~50,000
Bearings	2 Needle Roller Bearings
Flow speed range (m/s)	11–21
Rotational speed range (r/min)	27–580
Tip speed ratio range	0.025–0.29
Blade attachment point	$\frac{1}{4}$ chord length from LE

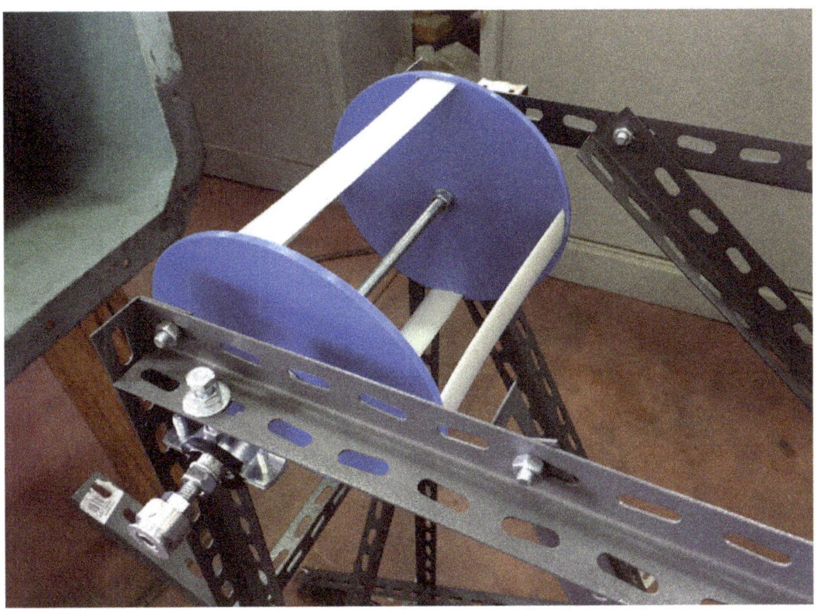

Figure 1. Experimental rotor.

The rotor was widely tested in the subsonic wind tunnel owned by the University of Catania. The wind tunnel was a closed circuit wind tunnel with a test section of 0.5 × 0.5 m, a maximum achievable flow speed of 31 m/s, and a maximum measured turbulent intensity of 0.4%. The test was carried out in open test section configuration. More details about the wind tunnel are reported in [11,34]. The flow speed was measured through the use of a Pitot probe, placed at the center of the test section inlet and at half diameter from it. The rotational speed was measured by means of a digital laser tachometer. The torque was evaluated by using a specifically designed braking system based on the principle of the belt brake. The instantaneous brake force was measured through a load cell and the torque was obtained. The rotor was anchored at the support structure through two needle roller bearings and the torque losses were experimentally evaluated as a function of the rotational speed. To measure the torque, the wind tunnel was set in such a way to obtain the maximum achievable flow speed in open test chamber. The rotor was free to rotate, without braking load, until an equilibrium rotational speed was reached. A constant braking load was then applied, slowing the rotor down until a new equilibrium rotational speed was reached. The net instantaneous breaking force was measured through the load cell, and the average torque for the operating point was calculated. Then, keeping fixed the braking load, the flow speed was reduced by 2 m/s and the new torque was measured each time together with the new equilibrium rotational speed. This process was repeated in steps of 2 m/s until the rotor stopped. The experiment was with three different braking loads, five times for each. In order to obtain the fluid dynamic power, to be compared to the numerical simulation results for the validation, the power losses in the bearings were added at each measured operating point. The flow and rotational speed range is reported in Table 1, while in Figure 2 a sketch of the experimental setup is shown.

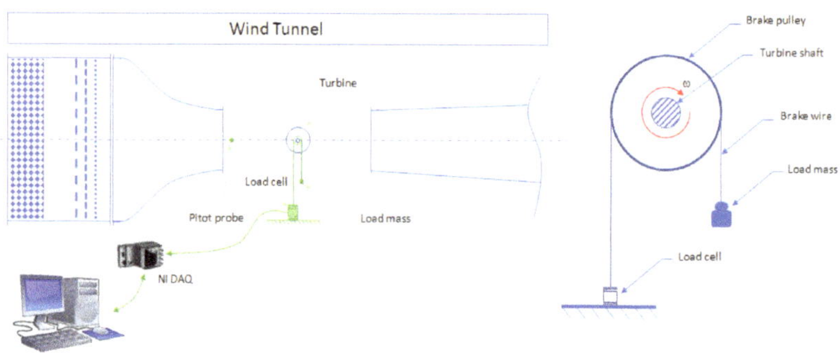

Figure 2. Experimental setup and detail of the mechanical brake system.

The CFD simulations were implemented for five different couples of flow and rotational speed to cover the entire range of measured tip speed ratio. Since there was no agreement about the adequate domain dimensions to have independent results, the first step in the development of the CFD 2D model was the definition of the suitable computational domain. In this case, the best compromise is shown in Figure 3. It was verified that larger domains did not affect the solution further in terms of torque prediction. The use of the symmetry condition for the lateral boundaries reduced the possible influences of the domain dimensions on the flow-field.

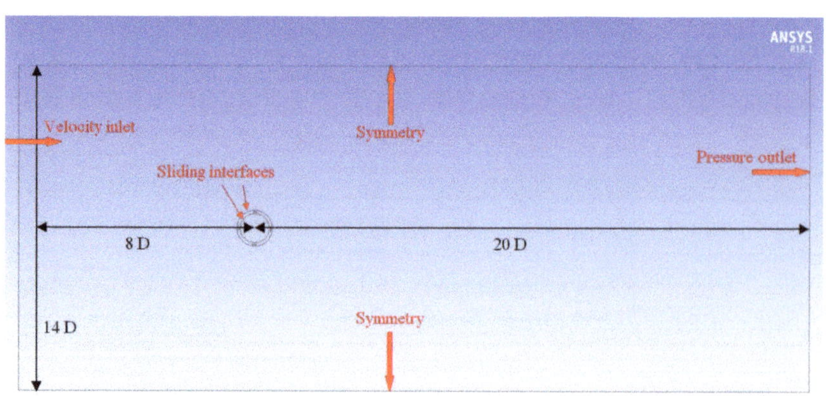

Figure 3. Computational domain dimensions and boundary conditions.

The boundary conditions used in the present work are shown in Figure 3. A velocity inlet condition was used for the inlet and a pressure outlet condition for the outlet. For the rotating zone, the domain was divided into three sub-domains in order to implement the unsteady sliding mesh model (SMM) in a rotating ring, which contained the four blades as in [20]. The internal circle and the outer domain remained stationary. The external and internal circumferences of the ring were thus set as sliding interfaces as highlighted in Figure 3.

2.3. Spatial and Temporal Sensitivity Study

In order to obtain an adequate level of spatial and temporal discretization, a thorough sensitivity study was made in this work, based on the papers of Balduzzi et al. [13–15]. This was done to ensure the best compromise between accuracy and computation time. Specifically, it was suggested in [13–15] that, as the tip speed ratio was reduced, the grid refinement must be increased in order to adequately capture the higher vorticity gradients related to the increasing unstable operating conditions [13–15]. Furthermore, the temporal discretization must be reduced accordingly to the grid refinement so as the Courant number was below 10, thus providing the optimal error damping properties in the implicit scheme solver used in the present work. Therefore, due to the highly unstable condition related to the very low operating tip speed ratios of the micro rotor analyzed (between 0.025 and 0.29 in Table 1), a specific spatial and temporal sensitivity study was made. The study was carried out for each of the three turbulence models. In this way, the sensitivity of the turbulence models to the grid refinements was evaluated as well. Nine sensitivity tests for each rotor operating condition were made. Three meshing levels and three time steps were tested for each of the three turbulence models. The sensitivity was tested for the maximum ($\lambda = 0.29$) and the minimum ($\lambda = 0.025$) tip speed ratio in order to find the best compromise for all the intermediate simulations. A global amount of 54 simulations were carried out in this sensitivity study. In this way, an optimal spatial and temporal discretization level was found to be valid for all the simulated operating conditions.

The three grid refinement levels used are reported in Table 2, where M1 is the coarsest mesh and M3 is the finest one. To refine the mesh, the number of nodes on the airfoil was increased and the cell sizing of the rotating domain was reduced accordingly. The same cell sizing of the rotating ring was imposed to the inner circle to have a very fine discretization leading to an accurate transport of the vortices detached from the upwind blades. This constant fine mesh reduced the numerical diffusion on the upwind blade wakes, thus improving the accuracy of the downwind blade-vortex interactions. Exactly the same refinement was imposed to both the sliding interfaces so as to reduce the interpolation errors at the non-conformal sliding mesh. The boundary layer of the blades was discretized using quadrangular layers of elements. The quadrangular elements allowed for a more accurate resolution of the boundary layer compared to the triangular elements used for the rest of the domain.

Table 2. Grid independence study meshing characteristics.

Grid Features	Mesh M1	Mesh M2	Mesh M3
Number of elements	320,000	800,000	2,170,000
Nodes on airfoil	1000	2000	4000
Max rotating domain sizing	0.75 mm	0.5 mm	0.25 mm
Max inner circle sizing	0.75 mm	0.5 mm	0.25 mm
Max outer domain sizing	10 mm	10 mm	10 mm
Global growth rate	1.2	1.1	1.05
Inflation layers	20	40	60
First layer height	10^{-3} mm	10^{-3} mm	10^{-3} mm
Skewness max	0.77	0.72	0.79
Y^+ max	0.25	0.23	0.22

The three meshes were developed with a different number of quadrangular layers and growth rates as reported in Table 2. The first layer height was fixed in such a way to guarantee a $y^+ < 1$ for all the simulations, as required by the turbulence models [29,31–33]. A sizing function, which limited the maximum dimension of the elements, was used in the stationary outer domain as well. The same growth rate as that used for the inflation layers was applied to the whole domain, thus allowing for a gradual mesh coarsening from the rotating interfaces to the boundaries.

In Figure 4, some details of the Mesh M2 are shown. It is worth remarking that Mesh M2 and Mesh M3 did satisfy the Grid Reduced Vorticity (GRV) criterion proposed by Balduzzi et al. [14]. GRV is a quantitative parameter for a qualitative a priori evaluation of the spatial discretization accuracy. GRV is defined as a vorticity normalized with respect to the local grid size, and thus gives an estimate of the velocity variation within a single element. Therefore, it represents the capability of the mesh itself of correctly computing the gradient flow features. The mesh sensitivity analysis and the evaluation of GRV are therefore strictly related: grid independent results are obtained when the discretization error becomes irrelevant, i.e., when GRV is sufficiently small [14]. In the present paper, it was verified that Mesh M2 and Mesh M3 had GRV < 1%, as recommended by [14], while Mesh M1 presented slightly larger values. Furthermore, all three grids satisfied the LES filter constraint imposed by the use of the DDES model, which required to have cell dimension lower than 1/30 the cord length [35].

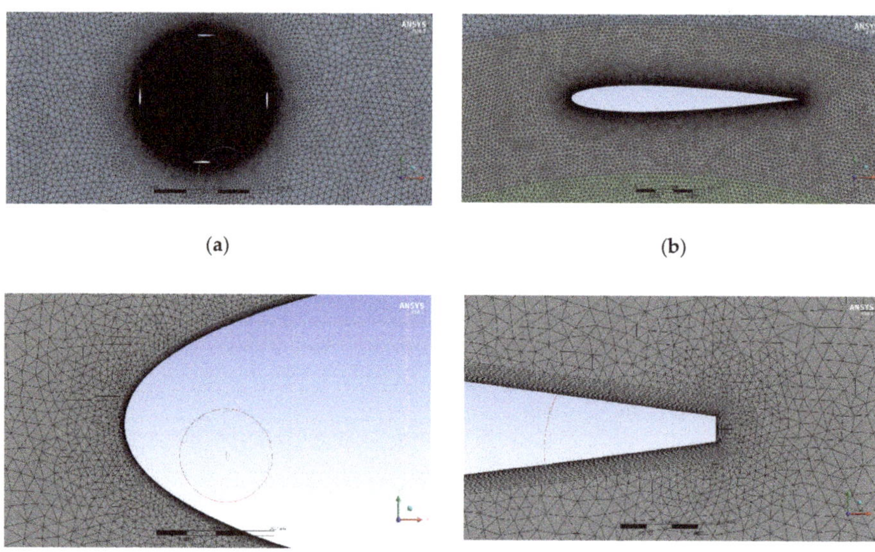

(a) (b) (c) (d)

Figure 4. Details of the rotor region (**a**), airfoil region (**b**) leading edge (**c**) and trailing edge (**d**) for mesh M2.

The temporal discretization must be defined so that all the relevant time scales of the flow were resolved. For this purpose, the time step dimension was chosen in such a way to try a wide set of Courant numbers. The Courant number was defined as:

$$Co = V\ \Delta t/\Delta x \tag{1}$$

As reported in [14], for VAWTs, V is the peripheral velocity of the airfoil, Δx is the average distance between two cell centroids on the airfoil wall and Δt is the time step. The Courant Number expresses the ratio between the temporal time step (Δt) and the time required by a fluid particle, moving with V velocity, to be convected throughout a cell of dimension Δx.

Since V was the tangential velocity, in order to obtain similar Courant numbers for the different grids and rotational speeds, the time step must be varied together with the tip speed ratio. This obviously involved the angular step being kept constant by reducing the time step with increasing tip speed ratios. In Table 3, the Courant numbers, obtained for different combination of grids and time steps, are shown. Only the finest grids with the largest time step presented slightly high Courant numbers while, in all the other combinations, the Courant numbers were well below 10. This was in order to obtain the best error damping properties, as recommended in [14]. The adaption of the time step was made specifically for all the simulated operating conditions, thus ensuring the same angular step and approximately the same Courant number. In this way, the spatial and temporal discretization was suitable for the specific spatial and temporal scales in each simulation.

The results of the sensitivity study are shown in Figure 5. The charts present the trend of the time-averaged torque coefficient for the various combinations of grids, angular steps and turbulence models. The time-averaged torque coefficient is plotted as a function of the angular steps corresponding to the time steps in Table 3. The number of grid elements is reported on the horizontal axis. Figure 5a,b refers to the DDES model results at maximum (a) and minimum (b) tip speed ratio, respectively. Figure 5c,d refers to the RANS SST transition model while Figure 5e,f shows the RANS SST k-ω model results. The sensitivity analysis demonstrates that the grids M2 and M3 with angular steps $\Delta\theta = 0.035°$ and $\Delta\theta = 0.01°$ lead to results which are in very close proximity to each other, for both λ_{max} and λ_{min}. For the DDES turbulence model, the difference is approximately 1%. Therefore, mesh M2 with an angular step equal to $\Delta\theta = 0.035°$ was the best compromise for all the simulated range of tip speed ratio. Specifically, for all the other simulated operating conditions, the time step was adapted to obtain $\Delta\theta = 0.035°$.

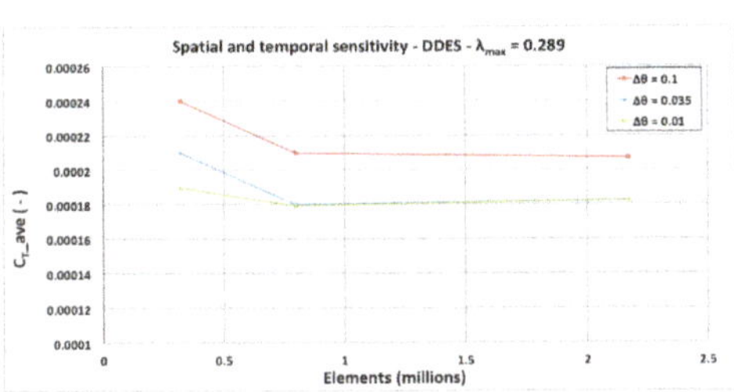

(**a**)

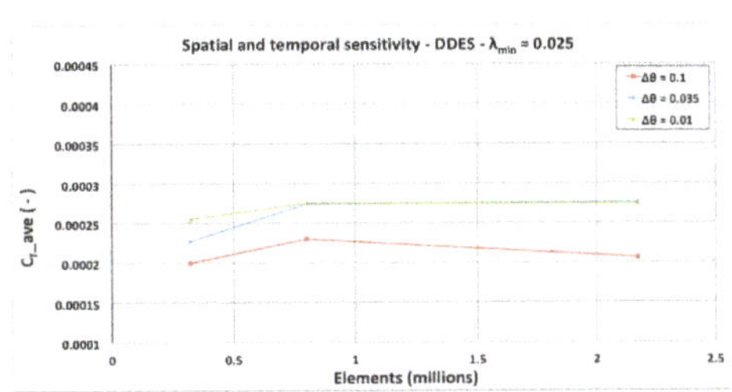

(**b**)

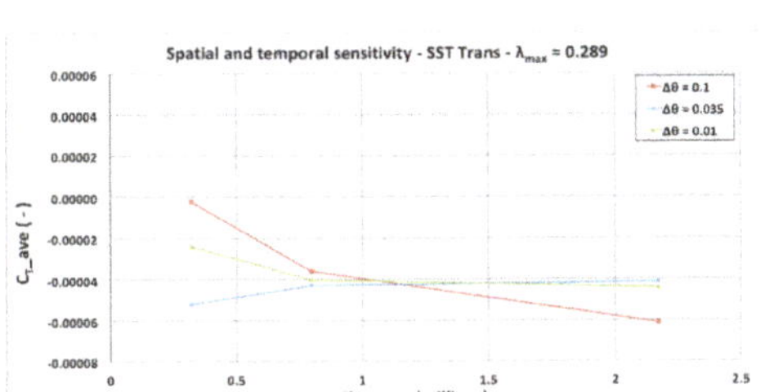

(**c**)

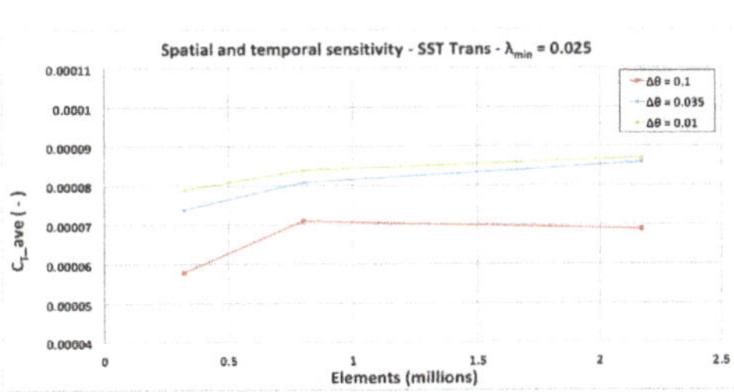

(**d**)

Figure 5. *Cont.*

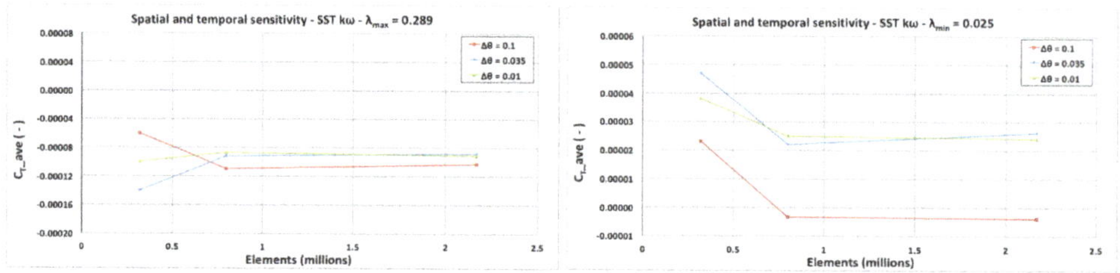

(e) (f)

Figure 5. Average torque coefficient sensitivity analysis at λ_{max} for DDES (**a**), SST Trans (**c**), SST kω (**e**) and at λ_{min} for DDES (**b**), SST Trans (**d**), SST kω (**f**).

Table 3. Courant numbers for different grids and time steps at maximum and minimum λ.

Δθ (deg)	Δt (s)	Δx (mm) 0.03 (M1)	0.015 (M2)	0.0075 (M3)	λ (-)
		Co (-)			
0.1	0.00003	6.07	12.15	24.29	0.29
0.035	0.00001	2.02	4.05	8.10	
0.01	0.000003	0.61	1.21	2.43	
0.1	0.00062	5.84	11.69	23.37	0.025
0.035	0.00021	1.98	3.96	7.92	
0.01	0.000062	0.58	1.17	2.34	

It is already evident that both RANS turbulence models highly underestimate the average torque coefficient with respect to the DDES model. Even negative time-averaged torque coefficients are predicted at λ_{max}. This would suggest that in highly unstable boundary layer conditions, the RANS turbulence models lead to wrong physical predictions of the flow-field. On the contrary, the DDES model results agreed very well with the experiments as demonstrated in the following sections.

2.4. CFD SOLVER Settings

The CFD solver setup is reported in Table 4. The ANSYS Fluent transient solver was used with a coupled algorithm for pressure-velocity coupling. The three aforementioned turbulence models were tested. Optimized local correlation parameters were used for the SST transition formulation both in URANS and in DDES. This optimization was carried out according to a previous work by the authors [36]. These local correlation parameters triggered and controlled the onset of the laminar to turbulent boundary layer transition within the transport equation for intermittency and momentum thickness Reynolds number. The number of iterations within each time step was set in order to ensure that all the residuals, within each sub-iteration, were well below 10^{-4}. The turbulence boundary conditions were set according to wind tunnel data and literature suggestions [20]. The convergence criterion was to have a time-averaged torque coefficient variation lower than 0.1% between two consecutive revolutions [14]. This was achieved in about five to ten complete rotor revolutions. Once the convergence was reached, the data were sampled for two consecutive revolutions for the torque time-averaging. The simulated operating conditions are shown in Table 4. The simulations were carried out on a HP Z820 workstation with 24 available cores for parallel calculation and 128 GB of RAM memory. The calculation time per revolution was approximately 58 h with the SST k-ω model, 65 h with the Transition SST model and 71 h with the DDES model.

Table 4. Main CFD solver settings.

<table>
<tr><th>Solver</th><th colspan="3">ANSYS Fluent—Transient—Coupled</th></tr>
<tr><td>Turbulence models</td><td colspan="3">URANS Transition SST
URANS SST k-ω
Delayed DES with Transition SST</td></tr>
<tr><td>Numerical schemes</td><td colspan="3">Least squares cell based for gradients
Second order upwind for all the equations
Bounded central differencing for momentum in DDES
Second order implicit for time differencing in URANS
Bounded second order implicit for time differencing in DDES</td></tr>
<tr><td>Rotation model</td><td colspan="3">Sliding Mesh Model</td></tr>
<tr><td>Iterations per time step</td><td colspan="3">60</td></tr>
<tr><td>Turbulence boundary conditions</td><td colspan="3">Inlet: TI = 0.1%, TVR = 10
Outlet: TI = 5%, TVR = 10</td></tr>
<tr><td>Convergence criterion</td><td colspan="3">Average torque coefficient variation lower than 0.1% between two subsequent revolutions</td></tr>
<tr><td>Simulated operating conditions</td><td>V_w = 11.1 m/s
V_w = 15.2 m/s
V_w = 17 m/s
V_w = 19.1 m/s
V_w = 21 m/s</td><td>n = 27 r/min
n = 133 r/min
n = 227 r/min
n = 362 r/min
n = 580 r/min</td><td>λ = 0.025
λ = 0.092
λ = 0.14
λ = 0.198
λ = 0.29</td></tr>
</table>

3. Model Validation and Analysis of the Results

3.1. Experimental Validation of the CFD Model

The experimental tests, carried out in the subsonic wind tunnel owned by the University of Catania, demonstrated that the small VAWT rotor object of this study is much less efficient than a HAWT of comparable size. The maximum measured power of the micro rotor presented here was approximately only 0.55 W at a flow speed of 21 m/s. A micro HAWT with comparable dimensions, tested by the authors in a previous work [11], showed a power of approximately 10 W at the same flow speed.

Concerning the experimental validation of the CFD model, in Figure 6 the comparison between the measured and the calculated power coefficients is shown. The experimental data in Figure 6 refer to the three different braking loads equal to 25 g, 50 g, and 75 g in mass. Due to the low torque generated by the rotor, the torque dissipated by the bearings limited the range of measurement approximately at λ = 0.3. Owing to the unphysical prediction obtained with the RANS turbulence models, only the DDES simulation results are reported in Figure 6. Both the URANS SST transition and SST k-ω model predicted unrealistic negative power coefficients for almost the entire operating range of the rotor. Furthermore, in Figure 6, the Qblade Double Multiple Stream Tube Model (DMSTM) power coefficient prediction is plotted. Qblade is an open source 1D code [37] that required the use of suitable polars for the airfoils. In this case, the experimental data of the NACA 0012 were taken from the literature [38], for an average cord Reynolds number of approximately 40,000. The DMSTM implemented within Qblade uses advanced models and corrections for tip losses, dynamic stall and virtual camber, similar to those used in the BEM theory for HAWTs [39].

The close proximity between experimental measurements and simulation results proved an excellent accuracy of the CFD 2D model based on DDES [40]. Considering the unphysical predictions obtained with the RANS turbulence models despite the very fine spatial and temporal discretization level, this result is very meaningful. In the simulation of strong boundary layer instabilities in VAWTs, the use of an advanced turbulence model like DDES appears to be necessary.

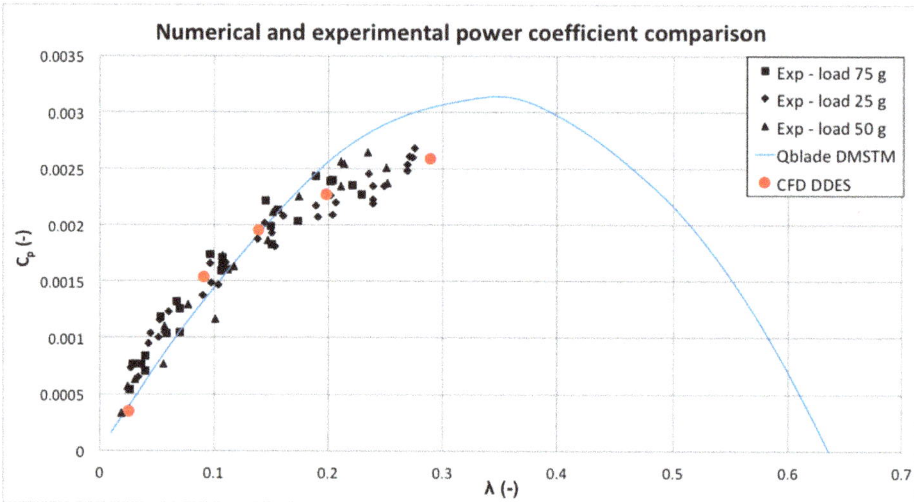

Figure 6. Comparison between numerical and experimental power coefficient.

The 1D Qblade results in Figure 6 showed surprising accuracy as well. At least the first part of the C_p curve is well predicted in light of the extreme simplicity and rapidness of the Qblade model. This certainly deserves further investigations on different geometries and operating conditions, in order to verify whether it is a mere chance or a generalizable result.

Nevertheless, CFD still remains the only way to get a thorough insight into the rotor aerodynamics, which is of utmost importance for the comprehension of the causes of the poor efficiency of such micro VAWTs. For this purpose, the present work demonstrated that only an advanced turbulence modeling like the hybrid RANS-LES formulation, implemented in DDES, was able to provide accurate and physically reliable results. Moreover, the availability of an accurate CFD model will allow the authors to identify an optimization strategy for these rotors in order to increase their efficiency. The use of more suitable airfoils, specific pitch angles, and vortex generators are just some of the simplest and cheapest techniques, whose effectiveness will be evaluated thanks to the CFD model developed in this work.

3.2. Post-Processing and Analysis of the Results

A thorough post-processing analysis of the flow-field in terms of velocity, turbulence production, and vorticity was carried out for two specific azimuthal rotor positions evidenced in Figure 7. In blue, the instantaneous position named as "position 0°" in which the blades are located at θ = 0°, 90°, 180°, and 270°. In red, the instantaneous position named as "position 45°" in which the blades are located at θ = 45°, 135°, 225°, and 315°. In this way the analysis is quite representative of the rotor behavior within a revolution. Only the λ_{max} condition results are presented as the other operating conditions lead to very similar considerations. The images showed hereinafter explain both the reasons for the poor accuracy of the RANS turbulence models and the low efficiency of these rotors.

In Figure 8, the contours of velocity magnitude for the three turbulence models demonstrate the strong differences in the flow-field prediction. The DDES results (Figure 8a) demonstrate the capability to accurately capture structures which are smaller and more defined than those obtained from the RANS simulations. Indeed, the RANS models seem to diffuse the swirling structures more than the DDES model. This result was expectable since the LES modeling inside the DDES was inherently more accurate. Furthermore, the flow separation dynamic of the blades at azimuthal positions θ = 90° and 270° appears to be different, also between the transition and the SST k-ω models. In this condition, indeed, the transition model predicted larger scale vortices compared to the SST k-ω model. The above suggests that the RANS models overestimate the dimensions of the flow structures and their diffusion within the grid. Therefore, the negative effects of unsteady phenomena like dynamic stall and blade-vortex interaction, in the down-wind sector, would be probably overestimated as well. For example, the blade at $\theta = 0°$ shows a recirculation area near the trailing edge in Figure 8b,c, which is larger than that in Figure 8a. This seems to denote that the RANS models in this case of micro rotor, tend to predict an earlier flow separation, which in turn results in higher turbulence production and finally lower torque. Considering the fact that both the RANS models predicted negative torque at λ_{max}, while the DDES showed excellent accuracy, it can be supposed that the RANS models led to unphysical flow-field predictions. A possible explanation may be related to the fact that the small dimensions of the rotor generate flow structures that the RANS averaging is not able to capture despite the very high spatial and temporal discretization. Thus, the more advanced physical solution provided by the DDES model appears to be more suitable in the case of such micro rotors.

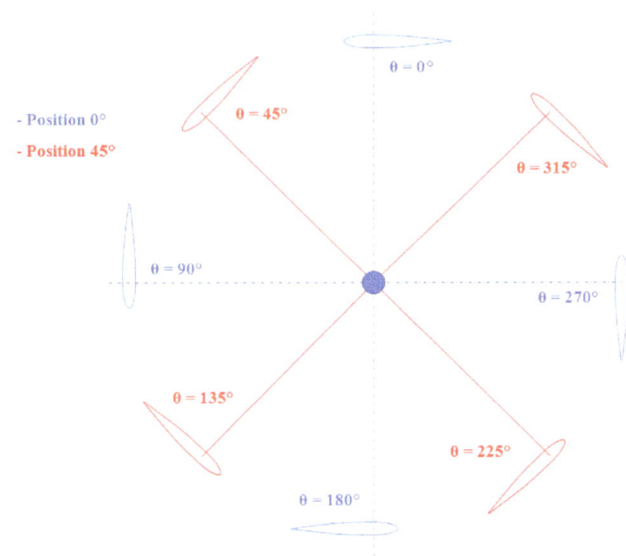

Figure 7. Azimuthal rotor position reference for the post-processing analysis.

The contours of turbulent intensity for the three turbulence models in Figure 9 further confirm this assumption. The RANS models predict a massive and smoothed turbulence production. The DDES model instead predicts much less turbulence production in smaller and more defined structures. The massive turbulence production of the RANS models results in high rotor energy dissipation, which is not realistic. Moreover, the large turbulence structures in Figure 9b,c produce much more influences on the downwind blades than those in the DDES model. This further confirms that the physics beyond the RANS models does not seem to be suitable for the simulation of the highly unstable conditions related to such small rotors.

The vorticity field presented in Figure 10 shows very different results between the three turbulence models. Again, the DDES (Figure 10a) demonstrate the capability to predict more defined and less smoothed structures but with much higher vorticity within the cores of the shed vortices, compared to the RANS model results. In light of the widely known capability of the LES to accurately predict the eddies behavior, and thanks to the excellent numerical-experimental agreement evidenced in Figure 6, the DDES model appears to be more physically realistic. In cases like that in the present work, in which the operating conditions are massively dominated by unstable shedding of vortices, the RANS modeling clearly leads to wrong physical predictions. This issue is reduced on larger rotors in which more stable flow-field conditions lead to less shedding of vortices, and therefore to a better physical agreement of the RANS models.

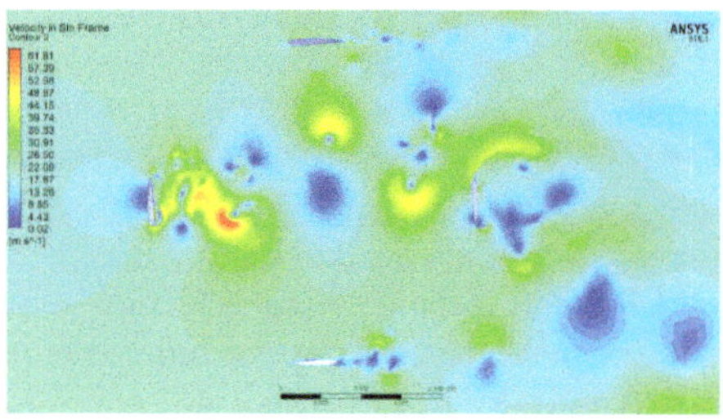

(**a**)

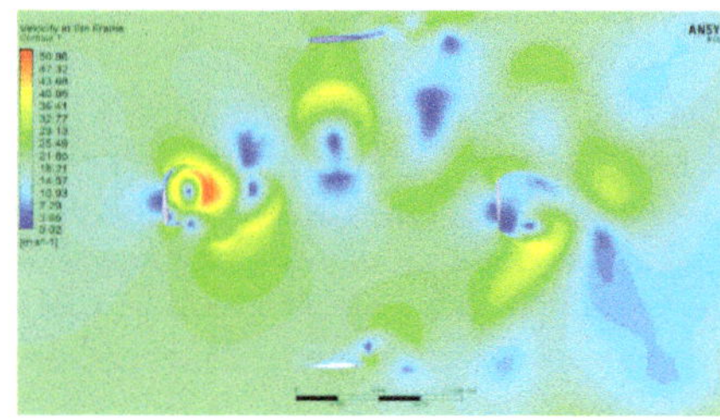

(**b**)

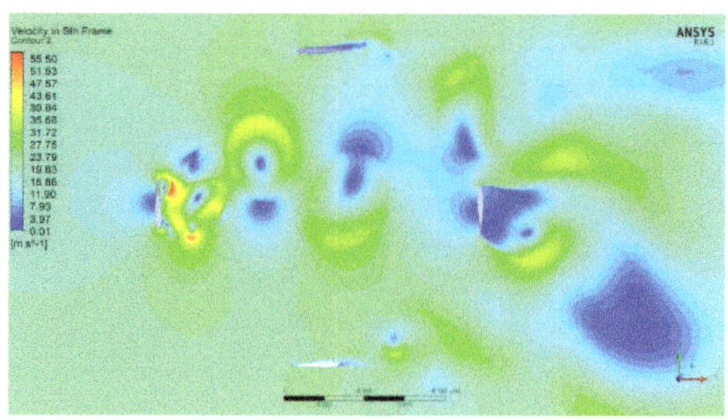

(**c**)

Figure 8. Contours of velocity magnitude for DDES (**a**), RANS Transition (**b**) and RANS SST k-ω (**c**) at 0° azimuthal position at $\lambda_{max} = 0.29$.

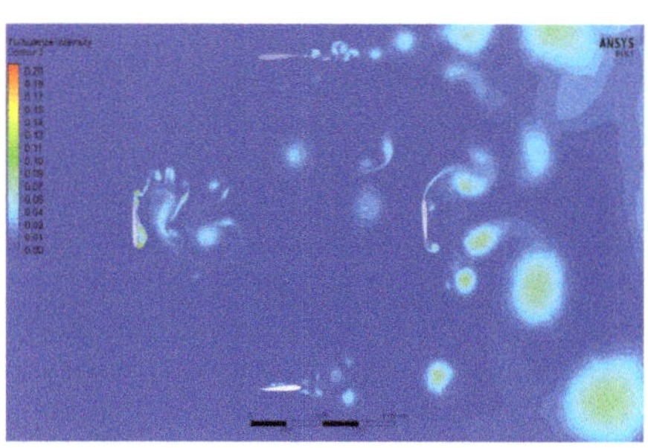

(**a**)

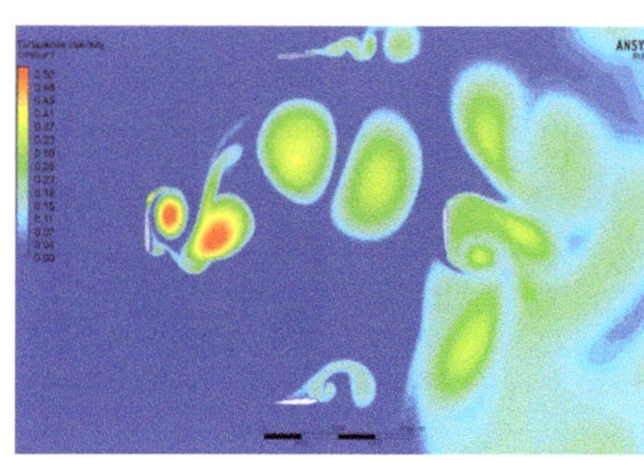

(**b**)

Figure 9. *Cont.*

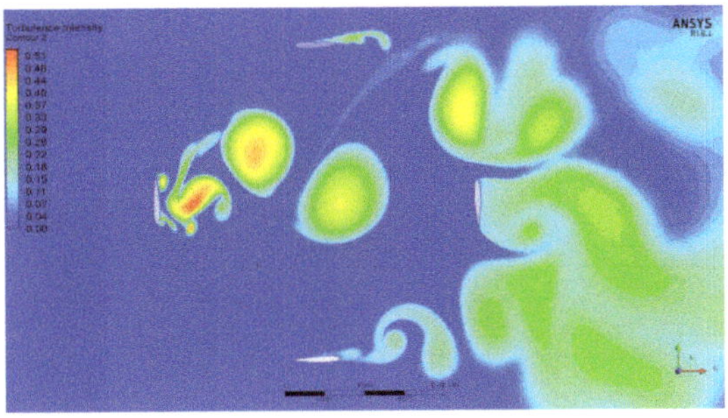

(c)

Figure 9. Contours of turbulent intensity for DDES (**a**), RANS Transition (**b**) and RANS SST k-ω (**c**) at 0° azimuthal position at $\lambda_{max} = 0.29$.

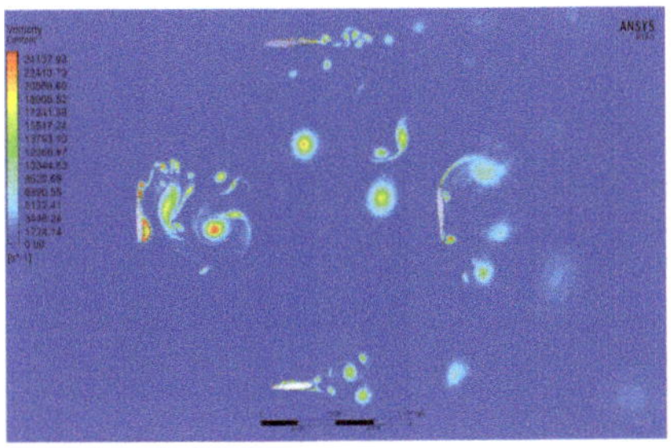

(**a**)

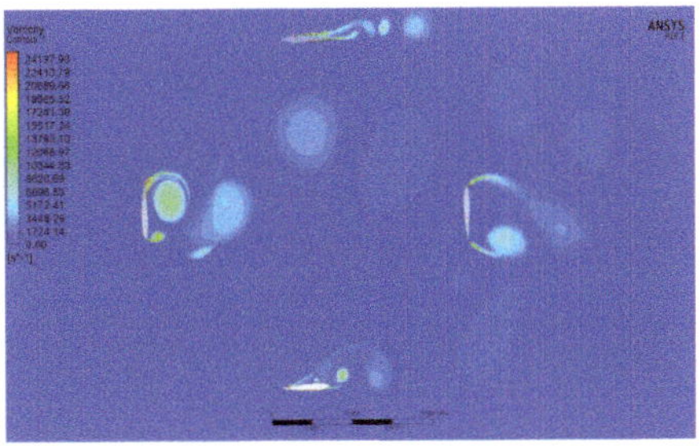

(**b**)

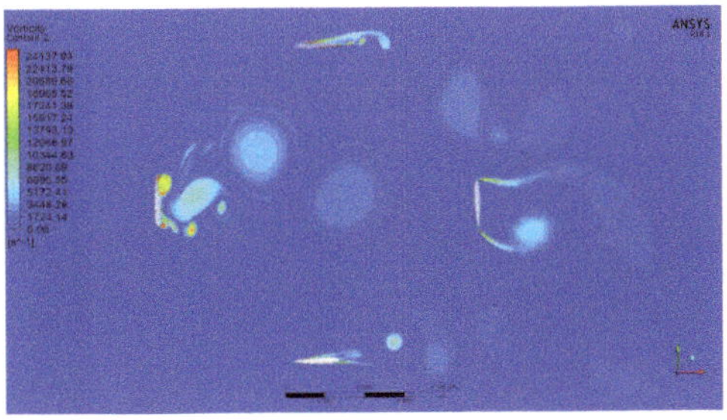

(c)

Figure 10. Contours of vorticity for DDES (**a**), RANS Transition (**b**) and RANS SST k-ω (**c**) at 0° azimuthal position at $\lambda_{max} = 0.29$.

Similar considerations can be made for the position 45°. In Figures 11–13 the contours of velocity magnitude, the contours of turbulence intensity, and the contours of vorticity are shown, respectively. In this case, all the blades are subjected to angles of attack such that the flow is fully separated. The differences between the DDES and the RANS results are even more evident. The velocity field appears smoother in the RANS compared to the DDES. In Figure 12, the massive unrealistic turbulence production, predicted by the RANS models, is even higher than in Figure 9, with respect to the DDES results. The contours of Figure 13 confirm the complexity of the vorticity field predicted by the DDES simulation, compared to the smoothed and less complex structures obtained thorough the RANS simulations.

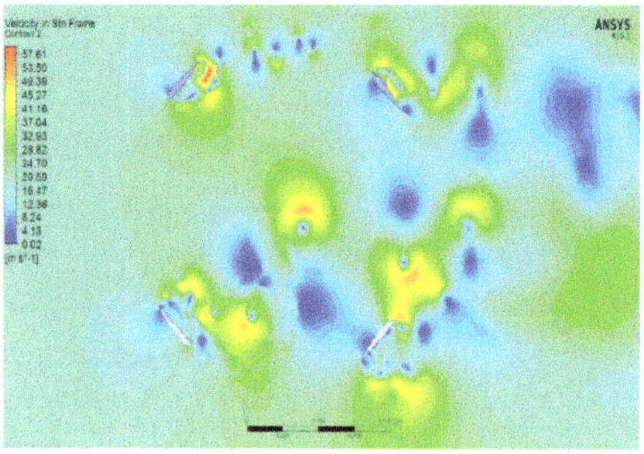

(a)

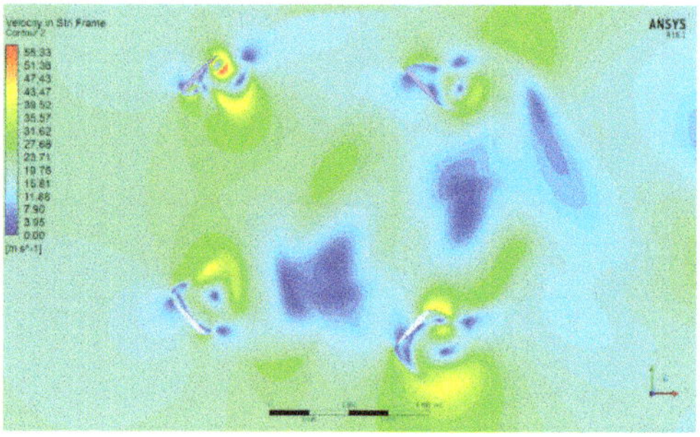

(b)

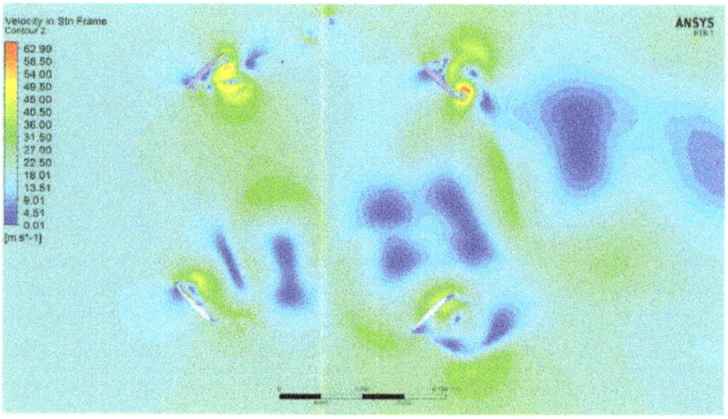

(c)

Figure 11. Contours of velocity magnitude for DDES (**a**), RANS Transition (**b**) and RANS SST k-ω (**c**) at 45° azimuthal position at $\lambda_{max} = 0.29$.

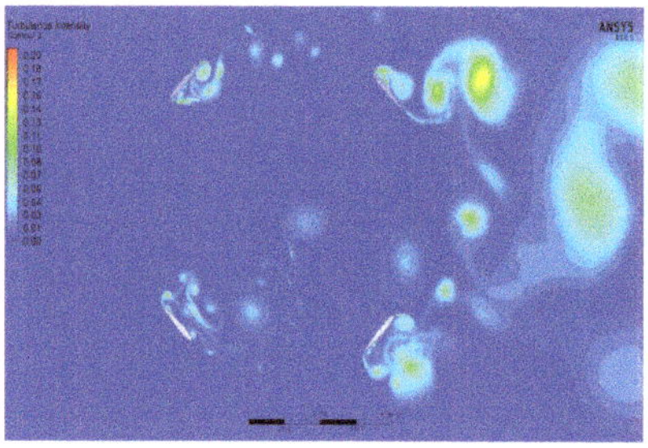

(a)

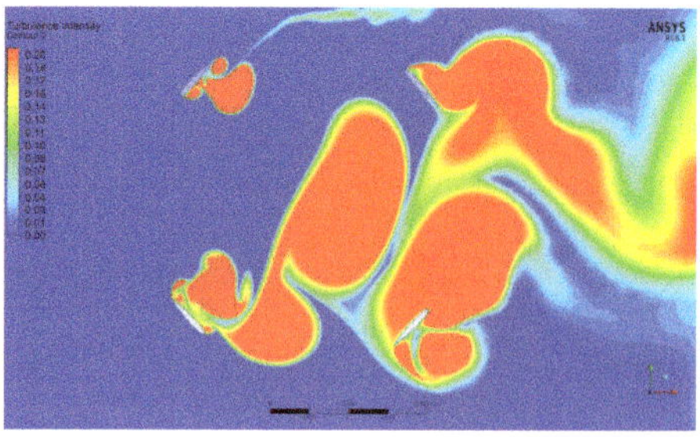

(b)

Figure 12. *Cont.*

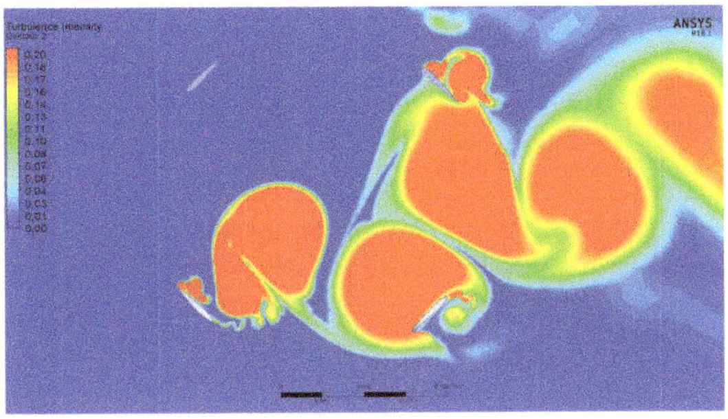

(c)

Figure 12. Contours of turbulent intensity for DDES (**a**), RANS Transition (**b**) and RANS SST k-ω (**c**) at 45° azimuthal position at $\lambda_{max} = 0.29$.

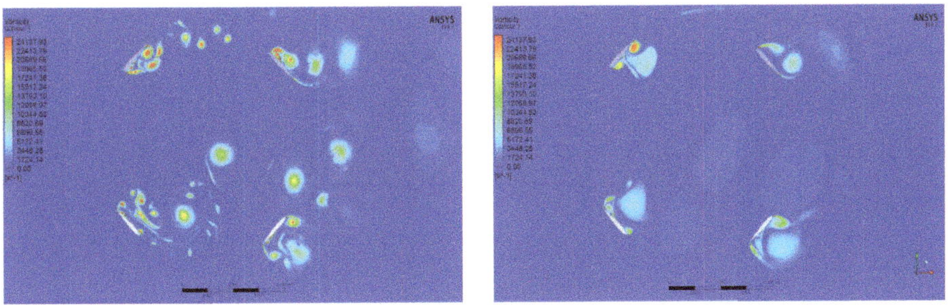

(**a**) (**b**)

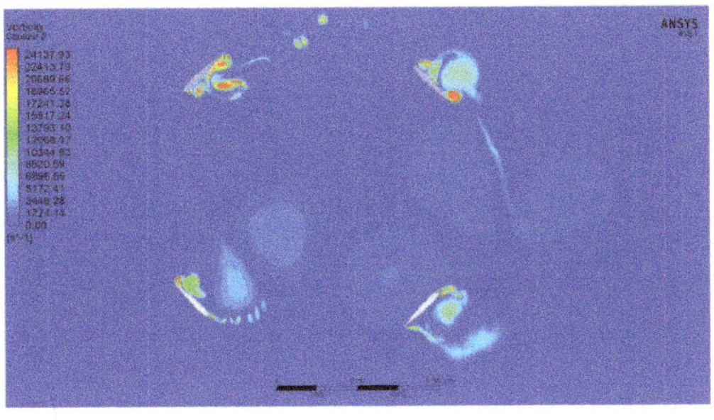

(c)

Figure 13. Contours of vorticity for DDES (**a**), RANS Transition (**b**) and RANS SST k-ω (**c**) at 45° azimuthal position at $\lambda_{max} = 0.29$.

The accuracy demonstrated by the DDES results allows for an insight into the poor efficiency that characterizes such micro rotors. Figures 8, 9 and 10a show that the blade at the azimuthal position θ = 0° is just affected by an incipient boundary layer instability even though the local AoA is approximately 0°. A shedding of vortices is already clearly evident in the airfoil wake.

In this regard, the details reported in Figure 14a,b clarify this assumption. The velocity vectors show boundary layer instabilities, which generate a consistent turbulent kinetic energy production. This is probably due to the low Reynolds number and the related strong sensitivity of the laminar boundary layer to the adverse pressure gradients. This earlier instability leads also to earlier flow separation, which can emphasize the negative effects of the dynamic stall as the blade moves towards higher azimuthal positions. Since the instability limits the attached flow condition phase, the blade azimuthal angle phases in which high lift to drag ratios are achievable are very limited as well. This means that the rotor operates under stall conditions for almost the entire blade rotation. As the blades are subjected to separation for most of the rotation, the dynamic stall, which inherently affects VAWTs, develops more than in large rotors, thus causing higher cyclic losses. The chart in Figure 15 reports the trend of the local blade AoA as a function of the azimuthal blade position, calculated by means of Qblade code at different tip speed ratios. Despite the fact that Qblade results are not as accurate, they provide a fast estimation of the local blade AoA during a complete rotor revolution. The rapid increase of the local AoA is related to the low peripheral speed, which in turn is due to the low torque produced, caused by the aforementioned instability.

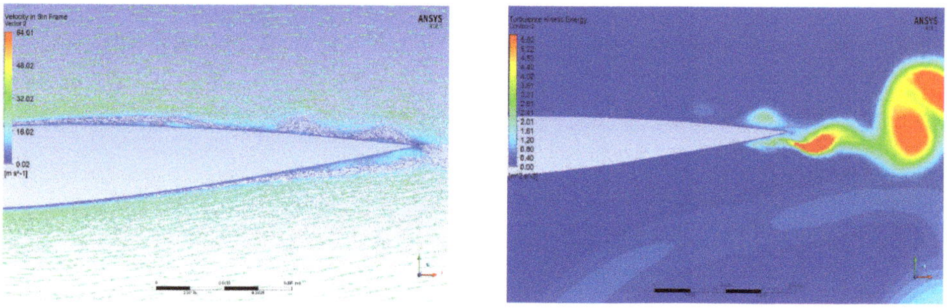

(**a**) (**b**)

Figure 14. Velocity vectors (**a**) and contours of turbulence kinetic energy (**b**) for a blade at $\theta = 0°$ and $\lambda_{max} = 0.29$.

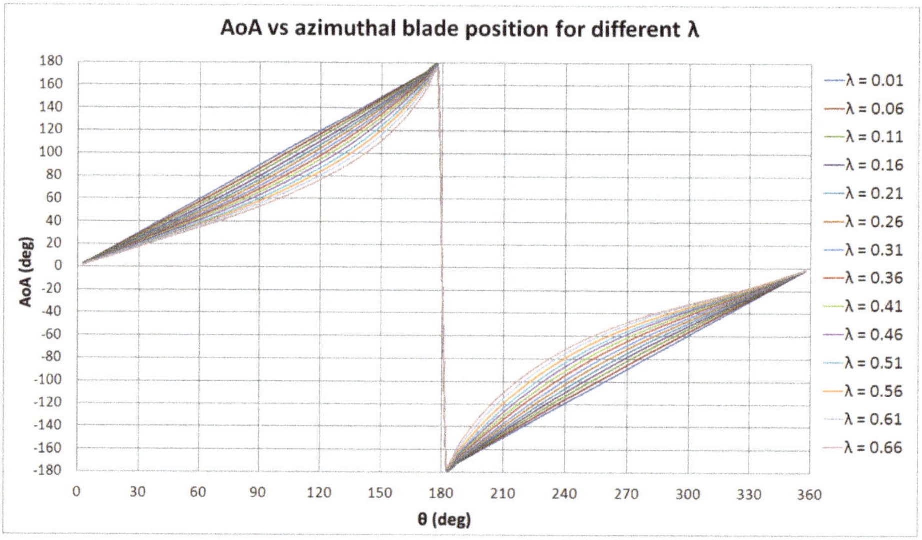

Figure 15. Qblade calculated local AoA as a function of the azimuthal blade position for different tip speed ratios.

Figure 16 further confirms the above considerations. The velocity vectors (a), the contours of turbulent kinetic energy (b) and the pressure coefficient (c) for a blade at azimuthal position θ = 15° are shown. In this specific condition, at λ_{max} = 0.29, the local blade AoA is approximately 8°. Despite the low AoA, the velocity vectors demonstrate an extensive area of instability, in the form of a kind of laminar bubble, near the trailing edge of the suction side of the blade. This instability produces high turbulent kinetic energy (Figure 16b), which increases the energy dissipation. The pressure coefficient in Figure 16c, calculated with respect to the flow velocity at inlet, shows the effects of these bubbles on the pressure distribution over the suction side of the blade. This in turn impacts negatively on the lift generated by the blade. As the blade moves toward higher azimuthal positions, the instability rapidly moves toward the leading edge, thus influencing the entire suction side of the blade with large flow separation and massive production of swirling structures. The subsequent onset of the dynamic stall further worsens the aerodynamic performance of the blade.

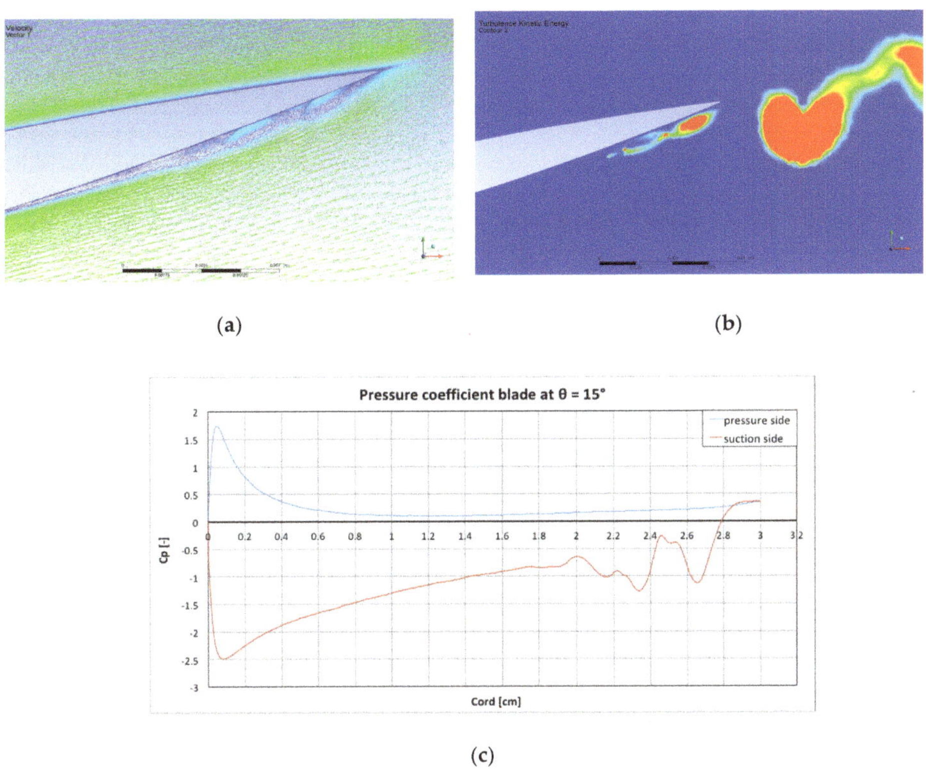

Figure 16. Velocity vectors (**a**), contours of turbulence kinetic energy (**b**) and pressure coefficient (**c**) for a blade at azimuthal position θ = 15° and λ_{max} = 0.29.

Thus, summarizing, the most important macroscopic evidence for the poor performance of micro H-Darrieus rotors is the fact that the turbine cannot accelerate sufficiently even at high flow speed. The cause of this is essentially due to the precocity with which boundary layer instability and separation occur in terms of AoAs. The very low Reynolds numbers, due to the small dimensions, seem to be the main responsible factor for this early instability, which manifests itself through massive production of vortices and turbulence. Furthermore, the early separation affects the development of the dynamic stall, since the range of AoAs in which it can be triggered is certainly wider. All these effects strongly limit the lift generation as they represent a source of cyclic losses. Moreover, the small dimensions of the radius do not allow for the generation of sufficient torque so that the rotor is not able accelerate at tangential speeds such to overcome the instabilities.

4. Summary and Conclusions

Although one may think micro wind turbines are of little scientific and industrial interest, the literature analysis in the present paper demonstrates that micro rotors represent an attractive solution for small size energy harvesters in many industrial applications. However, at such small scales, the advantages of the vertical axis solution are reduced by the poor rotor efficiency due to the onset of highly unstable boundary layer conditions as the rotor size decreases. The possibility to thoroughly analyze the fluid dynamic causes of the poor efficiency certainly represents a key step for the improvement of these micro rotors. Furthermore, in general, the simulation of the unstable conditions of VAWTs, which arise at low tip speed ratio, is still a very complex problem.

For these reasons, in this work, the authors present the implementation of a 2D CFD model for the simulation of a micro VAWT designed, constructed and tested in the subsonic wind tunnel of the University of Catania. For this purpose, three different turbulence models were evaluated: the SST Transition and SST k-ω RANS models and the hybrid RANS/LES Delayed DES. The thorough temporal and spatial discretization sensitivity study demonstrated that a very fine spatial and temporal discretization was necessary in order to capture the high vorticity gradients related to the unstable conditions, and thus to obtain independent results. While the RANS SST Transition model and SST k-ω model both led to unphysical torque predictions, the Delayed Detached Eddie Simulation, coupled to a transition formulation for the RANS region, was able to accurately reproduce the measured power coefficient.

The thorough post-processing comparison between the RANS and the DDES model results allowed for the identification of the reasons of the poor accuracy of the RANS models. In the highly unstable conditions related to micro rotors, both the RANS models predicted a massive turbulence production which was smoothed in large structures according to the inherent characteristics of the Reynolds averaging. The DDES model instead predicted a higher vorticity but in smaller and more defined swirling structures, leading to much lower turbulence production, and hence lower energy dissipation. In light of the excellent agreement between the DDES and the experimental results, it could be concluded that, in general, for a reliable simulation of the unstable condition of VAWTs the advanced DDES model appear to be the most suitable. However, a fine spatial and temporal discretization is needed to resolve the high vorticity gradients near the blades.

However, the most important result of this work was that the DDES, with transitional treatment for the RANS region, allowed for a thorough insight into the causes of the poor performance of micro H-Darrieus rotors. The small radius of these rotors together with the low cord Reynolds numbers of the blades involved the onset of early separation and boundary layer instabilities. These phenomena constantly affected the micro rotor operating conditions also because of a wider development of the dynamic stall for large part of the azimuthal blade position. All this involved a significant limitation of the torque generated since the rotor operated under high cyclic losses. The negative impact of the reduction of the geometrical scale was evidently much more significant in Darrieus VAWT than in HAWT.

The results of the present work provide some guidelines for the improvement of the efficiency of micro VAWTs as well. First of all, the use of specific low Reynolds number airfoils could reduce the onset of the instabilities, therefore allowing for higher torque generation and, therefore, higher rotational speeds. Different blade pitch angles may be more efficient in these small rotors than in the larger rotors. Simple variable blade pitch systems could be easily and cheaply applied to these rotors thanks to 3D printers. All this will be investigated in detail in future works.

Author Contributions: Conceptualization, S.M., S.B., R.L., M.M.; methodology, S.M. and S.B.; software, S.M.; validation, S.M. and S.B.; formal analysis, S.M., S.B., R.L., M.M.; investigation, S.M., S.B., M.M.; resources, R.L.; data curation, S.M. and S.B.; writing—original draft preparation, S.M.; writing—review and editing, S.M.; visualization, S.M.; supervision, R.L. and M.M.; project administration, R.L. and M.M. All authors have read and agreed to the published version of the manuscript.

Funding: This research received no external funding.

Acknowledgments: This research is supported by "Programma ricerca di Ateneo UNICT 2020-22 linea 2" of the University of Catania.

Conflicts of Interest: The authors declare no conflict of interest.

References

1. Xu, F.J.; Yuan, F.-G.; Liu, L.; Hu, J.Z.; Qiu, Y.P. Performance Prediction and Demonstration of a Miniature Horizontal Axis Wind Turbine. *J. Energy Eng.* **2013**, *139*, 143–152. [CrossRef]
2. Kishore, R.A.; Coudron, T.; Priya, S. Small-scale wind energy portable wind turbine (SWEPT). *J. Wind Eng. Ind. Aerodyn.* **2013**, *116*, 21–31. [CrossRef]
3. Kishore, R.A.; Priya, S. Design and eperimental verification of a high efficiencysmall wind energy portable turbine (SWEPT). *J. Wind Eng. Ind. Aerodyn.* **2013**, *118*, 12–19. [CrossRef]
4. Zakaria, M.Y.; Pereira, D.A.; Hajj, M.R. Experimental investigation and performance modeling of centimeter-scale micro-wind turbine energy harvesters. *J. Wind Eng. Ind. Aerodyn.* **2015**, *147*, 58–65. [CrossRef]
5. Xu, F.J.; Yuan, F.G.; Hu, J.Z.; Qiu, Y.P. Miniature horizontal axis wind turbine system for multipurpose application. *Energy* **2014**, *75*, 216–224. [CrossRef]
6. Ionescu, R.D.; Szava, I.; Vlase, S.; Ivanoiu, M.; Munteanu, R. Innovative solutions for portable wind turbines, used on ships. *Procedia Technol.* **2015**, *19*, 722–729. [CrossRef]
7. Leung, D.Y.C.; Deng, Y.; Leung, M.K.H. Design Optimization of a Cost-Effective Micro Wind Turbine. In Proceedings of the World Congress on Engineering (WCE 2010), London, UK, 30 June–2 July 2010; Volume 2, pp. 988–993. [CrossRef]
8. Howey, D.A.; Bansal, A.; Holmes, A.S. Design and performance of a cm-scale shrouded wind turbine for energy harvesting. *Smart Mater. Struct.* **2011**, *20*, 85021. [CrossRef]
9. Park, J.W.; Jung, H.-J.; Jo, H.; Spencer, B.F., Jr. Feasibility Study of Micro-Wind Turbines for Powering Wireless Sensors on a Cable-Stayed Bridge. *Energies* **2012**, *5*, 3450–3464. [CrossRef]
10. Bastankhah, M.; Porté-Agel, F. A New Miniature Wind Turbine for Wind Tunnel Experiments. Part I: Design and Performance. *Energies* **2017**, *10*, 908. [CrossRef]
11. Lanzafame, R.; Mauro, S.; Messina, M. Numerical and experimental analysis of micro HAWTs designed for wind tunnel applications. *Int. J. Energy Environ. Eng.* **2016**, *7*, 199. [CrossRef]
12. Mutlu, Y.; Çakan, M. Evaluation of in-pipe turbine performance for turbo solenoid valve system. *Eng. Appl. Comput. Fluid Mech.* **2018**, *12*, 625–634. [CrossRef]
13. Balduzzi, F.; Bianchini, A.; Maleci, R.; Ferrera, G.; Ferrari, L. Critical issues in the CFD simulation of Darrieus wind turbines. *Renew. Energy* **2016**, *85*, 419–435. [CrossRef]
14. Balduzzi, F.; Bianchini, A.; Ferrara, G.; Ferrari, L. Dimensionless numbers for the assessment of the mesh and timestep requirements in CFD simulations of Darrieus wind turbines. *Energy* **2016**, *97*, 246–261. [CrossRef]
15. Bianchini, A.; Balduzzi, F.; Bachant, P.; Ferrara, G.; Ferrari, L. Effectiveness of two-dimensional CFD simulations for Darrius VAWTs: A combined numerical and experimental assessment. *Energy Convers. Manag.* **2017**, *136*, 318–328. [CrossRef]
16. Rezaeiha, A.; Kalkman, I.; Blocken, B. CFD simulation of a vertical axis wind turbine operating at a moderate tip speed ratio: Guidelines for minimum domain size and azimuthal increment. *Renew. Energy* **2017**, *107*, 373–385. [CrossRef]
17. Castelli, M.R.; Ardizzon, G.; Battisti, L.; Benini, E.; Pavesi, G. Modeling strategy for a Darrieus vertical axis micro-wind turbine. In Proceedings of the ASME 2010 International Mechanical Engineering Congress & Exposition IMECE2010-39548, Vancouver, BC, Canada, 12–18 November 2010; pp. 409–418. [CrossRef]
18. Rogowski, K.; Hansen, M.O.L.; Lichota, P. 2-D CFD Computations of the Two-Bladed Darrius-Type Wind Turbine. *J. Appl. Fluid Mech.* **2018**, *11*, 835–845. [CrossRef]

19. Bangga, G.; Huotomo, G.; Wiranegara, R.; Sasongko, H. Numerical study on a single bladed vertical axis wind turbine under dynamic stall. *J. Mech. Sci. Technol.* **2017**, *31*, 261–267. [CrossRef]
20. Lanzafame, R.; Mauro, S.; Messina, M. 2D CFD modeling of H-Darrieus wind turbines using a transition turbulence model. *Energy Procedia* **2014**, *45*, 131–140. [CrossRef]
21. Mannion, B.; Leen, S.B.; Nash, S. A two and three-dimensional CFD investigation into performance prediction and wake characterization of a vertical axis turbine. *J. Renew. Sustain. Energy* **2018**, *10*, 34503. [CrossRef]
22. Lei, H.; Zhou, D.; Bao, Y.; Li, Y.; Han, Z. Three-dimensional Improved Delayed Detached Eddy Simulation of a two-bladed vertical axis wind turbine. *Energy Convers. Manag.* **2017**, *133*, 235–248. [CrossRef]
23. Li, C.; Xiao, Y.; Xu, Y.; Peng, Y.; Hu, G.; Zhu, S. Optimization of blade pitch in H-rotor vertical axis wind turbines through computational fluid dynamics simulations. *Appl. Energy* **2018**, *212*, 1107–1125. [CrossRef]
24. Thé, J.; Yu, H. A critical review on the simulations of wind turbine aerodynamics focusing on hybrid RANS-LES methods. *Energy* **2017**, *138*, 257–289. [CrossRef]
25. Ferreira, C.J.S.; Bijl, H.; van Bussel, G.; van Kuik, G. Simulating Dynamic Stall in a 2D VAWT: Modeling strategy, verification and validation with Particle Image Velocimetry data. *J. Phys. Conf. Ser.* **2007**, *75*, 12023. [CrossRef]
26. Ferreira, C.J.S.; van Zuijlen, A.; Bijl, H.; van Bussel, G.; van Kuik, G. Simulating dynamic stall in a two-dimensional vertical-axis wind turbine: Verification and validation with particle image velocimetry data. *Wind Energ.* **2010**, *13*, 1–17. [CrossRef]
27. Elkhoury, M.; Kiwata, T.; Nagao, K.; Kono, T.; Elhajj, F. Wind tunnel experiments and Delayed Detached Eddy Simulation of a three-bladed micro vertical axis wind turbine. *Renew. Energy* **2018**, *129*, 63–74. [CrossRef]
28. Abdellah, B.; Belkheir, N.; Khelladi, S. Effect of Computational Grid on Prediction of a Vertical axis Wind turbine Rotor Using Delayed Detached-Eddy Simulations. In Proceedings of the 2018 International Conference on Wind Energy and Applications in Algeria (ICWEAA), Algiers, Algeria, 6–7 November 2018; pp. 1–5. [CrossRef]
29. Wang, S.; Ingham, D.B.; Ma, L.; Pourkashanian, M.; Tao, Z. Turbulence modeling of deep dynamic stall at relatively low Reynolds number. *J. Fluids Struct.* **2012**, *33*, 191–209. [CrossRef]
30. Sa, J.H.; Park, S.H.; Kim, C.J.; Park, J.K. Low-Reynolds number flow computation for eppler 387 wing using hybrid DES/transition model. *J. Mech. Sci. Technol.* **2015**, *29*, 1837–1847. [CrossRef]
31. Menter, F.R.; Langtry, R.; Volker, S. Transition modeling for general purpose CFD codes. *Flow Turb. Combust.* **2006**, *77*, 277–303. [CrossRef]
32. Langtry, R.B.; Menter, F.R.; Likki, S.R.; Suzen, Y.B.; Huang, P.G.; Volker, S. A Correlation-Based Transition Model Using Local Variables—Part I: Model Formulation. In Proceedings of the ASME Turbo Expo 2004: Power for Land, Sea, and Air, Vienna, Austria, 14–17 June 2004.
33. Langtry, R.B.; Menter, F.R.; Likki, S.R.; Suzen, Y.B.; Huang, P.G.; Volker, S. A correlation-based transition model using local variables—Part II: Test cases and industrial applications. *ASME J. Turbomach.* **2006**, *128*, 423–434. [CrossRef]
34. Brusca, S.; Lanzafame, R.; Messina, M. Low speed wind tunnel: Design and build. In *Wind Tunnels: Aerodynamics, Models and Experiments*; Pereira, J.D., Ed.; Nova Science Publishers, Inc.: Hauppauge, NY, USA, 2011; ISBN 978-1-61209-204-1.
35. Spalart, P.R. *Young-Person's Guide to Detached-Eddy Simulation Grids*; NASA/CR-2001-211032; NASA Langley Research Center: Hampton, VA, USA, 2001.
36. Mauro, S.; Lanzafame, R.; Messina, M.; Pirrello, D. Transition turbulence model calibration for wind turbine airfoil characterization through the use of a Micro-Genetic Algorithm. *Int. J. Energy Environ. Eng.* **2017**, *8*, 359–374. [CrossRef]
37. Marten, D.; Wendler, J.; Pechlivanoglou, G.; Nayeri, C.N.; Paschereit, C.O. Qblade: An open source tool for design and simulation of horizontal and vertical axis wind turbines. *Int. J. Emerg. Technol. Adv. Eng.* **2013**, *3*, 264–269.
38. Shedahl, R.E.; Klimas, P.C. *Aerodynamic Characteristics of Seven Symmetrical Airfoil Sections through 180-Degree Angle of Attack for Use in Aerodynamic Analysis of Vertical Axis Wind Turbines*; Technical Report No. SAND80-2114; Sandia National Laboratories: Albuquerque, NM, USA, 1981.

39. Lanzafame, R.; Mauro, S.; Messina, M. HAWT design and performance evaluation: Improving the BEM therory mathematical models. *Energy Procedia* **2015**, *82*, 172–179. [CrossRef]
40. Mauro, S.; Brusca, S.; Lanzafame, R.; Messina, M. Micro H-Darrieus wind turbines: CFD modeling and experimental validation. *AIP Conf. Proc.* **2019**, *2191*. [CrossRef]

Publisher's Note: MDPI stays neutral with regard to jurisdictional claims in published maps and institutional affiliations.

energies

Article

Understanding the Aerodynamic Behavior and Energy Conversion Capability of Small Darrieus Vertical Axis Wind Turbines in Turbulent Flows

Francesco Balduzzi [1], Marco Zini [1], Andreu Carbó Molina [2,3], Gianni Bartoli [2], Tim De Troyer [4], Mark C. Runacres [4], Giovanni Ferrara [1] and Alessandro Bianchini [1,*]

1 Department of Industrial Engineering (DIEF), Università degli Studi di Firenze, via di Santa Marta 3, 50139 Firenze, Italy; francesco.balduzzi@unifi.it (F.B.); marco.zini@unifi.it (M.Z.); giovanni.ferrara@unifi.it (G.F.)

2 Department of Civil and Environmental Engineering (DICEA), Università degli Studi di Firenze, via di Santa Marta 3, 50139 Firenze, Italy; andreu.carbo.molina@dicea.unifi.it (A.C.M.); gianni.bartoli@unifi.it (G.B.)

3 Aerospace Systems and Transport Research Group, Universidad Rey Juan Carlos, Camino del Molino S/N, Fuenlabrada, 28943 Madrid, Spain

4 Thermo and Fluid Dynamics (FLOW), Vrije Universiteit Brussel, Pleinlaan 2, 1050 Brussels, Belgium; tim.de.troyer@vub.be (T.D.T.); mark.runacres@vub.be (M.C.R.)

* Correspondence: alessandro.bianchini@unifi.it; Tel.: +39-055-275-8773; Fax: +39-055-275-8755

Received: 19 May 2020; Accepted: 5 June 2020; Published: 8 June 2020

Abstract: Small Darrieus vertical-axis wind turbines (VAWTs) have recently been proposed as a possible solution for adoption in the built environment as their performance degrades less in complex and highly-turbulent flows. Some recent analyses have even shown an increase of the power coefficient for the large turbulence intensities and length scales typical of such environments. Starting from these insights, this study presents a combined numerical and experimental analysis aimed at assessing the physical phenomena that take place during the operation of a Darrieus VAWT in turbulent flows. Wind tunnel experiments provided a quantification of the performance variation of a two-blade VAWT rotor for different levels of turbulence intensity and length scale. Furthermore, detailed experiments on an individual airfoil provided an estimation of the aerodynamics at high turbulence levels and low Reynolds numbers. Computational fluid dynamics (CFD) simulations were used to extend the experimental results and to quantify the variation in the energy content of turbulent wind. Finally, the numerical and experimental inputs were synthetized into an engineering simulation tool, which can nicely predict the performance of a VAWT rotor under turbulent conditions.

Keywords: VAWT; Darrieus; turbulence; experiments; CFD

1. Introduction

1.1. Impact of Turbulence in New Wind Energy Applications

The installation of small wind turbines in built environments is being studied by the scientific community to complement wind energy conversion in large wind farms with a distributed generation. Smaller rotors positioned at the top of tall buildings could theoretically exploit a higher zone of the wind profile (more energetic) without the need for tall towers. The delocalized generation would also allow the production of energy where it is needed, saving transportation costs and contributing to the sustainable design of new buildings [1]. The real feasibility of this scenario, however, still needs to be proved in terms of energy conversion efficiency and social acceptance [2].

From a more technical point of view, the main challenge is the complexity of urban flows. The terrain presents high roughness, with the displacement height often at the level of the building heights themselves [3]. In addition, the flow reaching the rotors is often modified by the interaction with multiple obstacles of different shapes and permeability, e.g., upstream buildings or street furniture. All these effects result in mean wind speeds significantly lower than those available in the countryside areas [4], and poorer flow quality in terms of skew angles [4], fluctuations and, in the end, of high values of turbulence [5], which is one of the preeminent characteristics of urban flows. Turbulence in these contexts is characterized not only by high intensities, but also by large length scales [6]. This kind of turbulence is then quite different with that usually considered both in the experimental testing of airfoils or turbine prototypes and in the design phase using either engineering models or computational fluid dynamics (CFD) [7]. Experimental evidence suggests, however, that it can play a relevant role in the effective turbine operation in terms of increase of fatigue, unpredictability of energy production, or influence on stall conditions [8].

1.2. Evidence of VAWT Behaviour in Turbulent Flows

Vertical-axis wind turbines (VAWTs), especially with the Darrieus configuration, are often suggested as a possible alternative to conventional horizontal-axis wind turbines (HAWTs) for power production in small-sized applications, and especially in the aforementioned peculiar conditions found in the urban environment [2]. Despite the lower nominal efficiency with respect to HAWTs, this research trend is connected to the inherent advantages of the Darrieus concept: the omnidirectionality with respect to the wind direction (so that no yaw system is needed [9]), the possibility of putting the generator on the ground, the low noise emissions [10], and the good performance in case of misaligned (skewed) flows thanks to an increased virtual swept area [11]. In addition, Darrieus VAWTs are often preferred to other turbine layouts in view of an integration with the landscape of populated areas, since they are commonly perceived as aesthetically more pleasant by people [12]. Up to now, the study of turbulence effects on Darrieus VAWTs has been based on little field data correlating the performance of installed turbines with the on-site wind turbulence measured by a meteorological station. These studies often draw unclear and contradictory conclusions, since the effects of turbulence are said to be positive [13,14], negative [15,16], or velocity-dependent [17]. The difficulty of properly acquiring experimentally the inflow condition in a real urban environment suggests that detailed laboratory measurements are needed in order to reproduce in an accurate and repeatable way the turbine power curves and wake characteristics in turbulence. Nevertheless, replicating the urban flow characteristics inside a wind tunnel is not an easy task. In fact, most of existing experimental facilities are designed for aeronautical purposes and, thus, have a very low background turbulence intensity ($I_u < 1\%$). This value is far from the turbulence intensity typically found within a built environment, which is typically higher than 10% [5]. In addition, the characteristic integral length scales (L_{ux}, indication of the size of the most energetic eddies) are quite large, with values of L_{ux} in the order of 1 m [6]. As one may argue, these values are barely replicable in wind tunnels of limited dimensions: large ones are therefore needed. This latter characteristic is also motivated by the need of limiting the blockage effect inside the tunnel, which could be responsible of creating an interference that may deform the expected wind conditions during turbine operation [18].

As a result, only few studies on the influence of turbulence on VAWT performance in wind tunnels have been carried out so far; in all cases, physical grids were used to increase turbulence in the test section. Regarding power generation, [19] measured a slight increase of the power coefficient for a five-bladed Darrieus turbine, while [20] performed a full set of experiments regarding the interaction between turbulent flows and an H-Darrieus, and despite detecting a very large increase of the power produced, they could not retrieve the complete power curves. The influence of turbulence in the wake was studied by [21] for a five-bladed VAWT, recording better self-starting capabilities and a lower speed deficit in the far-wake; [22] also recorded faster wake recovery for a three3-bladed VAWT. The combination of the two features (turbine performance and wake recovery) was measured by [23,24] for a two-bladed Darrieus, also concluding that turbulence enhanced turbine production and wake recovery. The impact of turbulence was also studied in the past at a structural level [25], to understand the dynamic response of the rotors to gusts. Numerical studies were also presented, using different levels of fidelity ranging from engineering models [26] to U-RANS CFD [27]. However, very often the studies seem to lack extensive validation or a deep understanding of the involved physics.

1.3. Research Outline

The present study is aimed at understanding the different physical phenomena stimulating the enhancement of the aerodynamic power output of Darrieus VAWTs in turbulent flows. In particular, focus is given to the case of medium-small machines, where the Reynolds numbers are sufficiently low to induce the presence of transition on the airfoils.

The starting point of the analysis is represented by the experimental tests of [23], which demonstrated a strong enhancement of the power output as the turbulence intensity of the wind tunnel flow is increased. In the study, some conjectures were made on the reasons underlying this evidence, but no demonstration was available at that time. Later, in [7] the same case study was tested with a dedicated CFD approach (discussed in the following Section 2), showing that one of the contributions to the power enhancement is indeed represented by the higher energy content in the turbulent wind in comparison to a uniform one. Starting from this background, in the present work we quantify the aerodynamic efficiency improvement taking place on the airfoils working in cycloidal conditions at low Reynolds numbers and high turbulence. This represents the second key element contributing to the overall physics. To this end, both dedicated wind tunnel measurements and CFD calculations were performed in the framework of the activity. Finally, efforts were made in trying to synthetize the results into proper corrections and best practices to be included in a state-of-the-art engineering model for the simulation of Darrieus turbines based on the blade element momentum (BEM) theory in order to better estimate the actual performance in turbulent conditions.

2. Energy Content of Turbulent Wind: Evidence from Recent CFD Simulations

An important effort has been devoted in recent years to increase the accuracy of CFD simulations for the prediction of the performance of Darrieus wind turbines. To this end, a variety of approaches have been proposed, ranging from detailed unsteady Reynolds-averaged Navier–Stokes (U-RANS) approaches (e.g., [28–31]) to more advanced Detached Eddy Simulation (DES) [32] or Large Eddy Simulation (LES) approaches [33]. These latter approaches however, often require enormous calculation costs, which are often prohibitive, leading the researchers to compromises that often do not ensure fully reliable results. A recent study on the effects of turbulence on Darrieus VAWTs is reported in [7]. In this work, a new method to reproduce with a U-RANS approach the turbulence conditions typically found in wind energy applications (in terms of realistic scales of turbulence intensity and length scale) was developed. An innovative user-defined function was developed to assign a randomly variable velocity inlet boundary conditions able to create flow structures having a prescribed length scale; additionally, numerical settings have been tuned so as to allow the correct inflow turbulence level, controlling the numerical dissipation due to the U-RANS approach.

The authors then applied the method to the case study of [23] (see Figure 1), which also represents the one used in the present work, showing that an increase of the turbine performance was indeed achieved. In particular, the impact of the flow macrostructures on the turbine was analyzed as well as the contraction effect on the wake apparent in Figure 1.

Upon examination of the simulation data, the turbine performance increase was connected to two main phenomena:

- An improved stall resistance of the airfoils thanks to the turbulent flow, particularly relevant for small Reynolds numbers;
- An increase of the kinetic energy of the flow due to the macro-turbulence and to the exponential relationship that exists between wind energy content and wind speed.

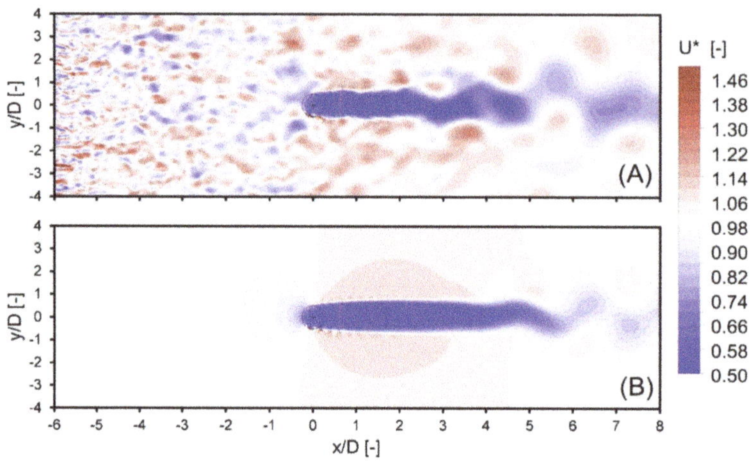

Figure 1. Computed contours for the dimensionless streamwise velocity: (**A**) turbine simulation with turbulence, and (**B**) turbine simulation without turbulence.

The first phenomenon was not addressed in [7], and it will represent the key element of the present study. The second one was instead somehow quantified, and it is briefly reported here since it is pivotal for the correct understanding of the present approach. In particular, if a turbulent velocity profile can be reasonably characterized as a randomly fluctuating one around the mean flow speed (see Figure 2), its energy content has to be scaled with the cube of the speed itself, thus leading in general to higher energy levels in turbulent wind with respect to a uniform one. In detail, an *equivalent velocity* to be considered to reflect the energy potential can be defined as in Equation (1), where E is the wind specific energy content of the turbulent flow, ρ is the air density:

$$U_{eq} = \sqrt[3]{\frac{\frac{1}{T}\int_{o}^{T} P(t)dt}{\frac{1}{2}\rho}} \tag{1}$$

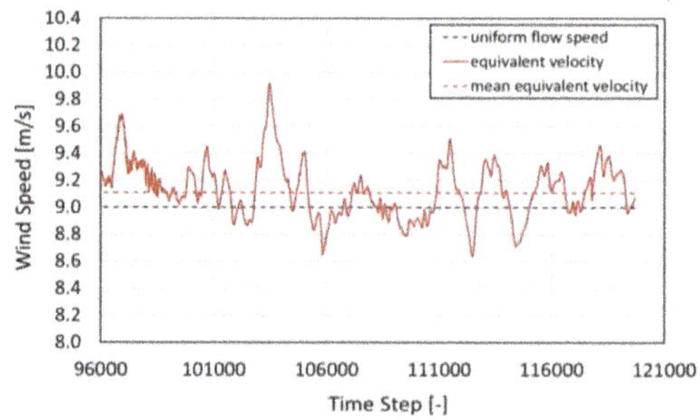

Figure 2. Comparison between a uniform wind speed, a real fluctuating one and its equivalent speed as in Equation (1).

The same authors, however, also pointed out that the flow macrostructures corresponding to the turbulence length scale typical of wind energy applications generate a non-periodic fluctuation of the relative speed oncoming on the airfoils in cycloidal motion onboard a Darrieus turbine. This, in turn, generates a variation of the local tip-speed ratio (TSR), which is defined as in Equation (2):

$$TSR = \frac{\omega R}{U} \tag{2}$$

It has to be stressed that these fluctuations happen in very short times, which are generally much lower than those required by the regulation system to act. The turbine revolution speed ω can then reasonably be considered as constant. The effect of the above is that it is indeed true that energy content is increased, but it is also true that it is extracted with a lower efficiency, as the optimal TSR is never maintained, as illustrated in Figure 3.

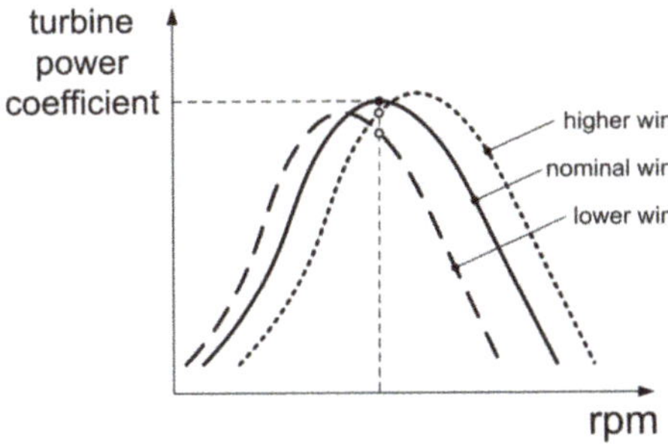

Figure 3. Schematic representation of the variation of turbine power coefficient due to a local variation of the incoming wind velocity at a fixed revolution speed.

As a result, in [7] a performance increase lower than that theoretically predictable based on Equation (1) was found. In any case, both targets were much lower than the experimental evidence on [23]. This was connected to the possible increase of low Reynolds airfoils' aerodynamics in turbulent wind, which are addressed instead in the following of the present study.

3. Experimental Tests

In the framework of the present work, two types of experimental tests were carried out. The most important ones were focused on reproducing correctly the turbulence levels of interest inside a wind tunnel [33] and on assessing the impact of turbulence on a two-blade small Darrieus VAWT [23]. These tests have been already presented in relevant technical papers: they will be then only briefly recalled here to give the reader a complete overview on the broad results of the entire research program. In addition, however, detailed measurements of isolated airfoils in turbulent conditions were repeated recently. The scope of these additional tests was to quantify the effect of turbulence on stall delay and maximum lift-to-drag ratio enhancement, in order to obtain a new set of aerodynamic polars for use in engineering simulation models.

3.1. Wind Tunnel Tests of the Turbine

The research on the turbine performance was carried out in the wind tunnel of CRIACIV, which is an Inter-University Research Consortium for Building Aerodynamics and Wind Engineering clustering eight Italian Universities. The laboratory hosting the open-circuit, boundary layer wind tunnel (Figure 4a) is located in Prato (Italy). The tunnel has a total length of about 22 m, including a nozzle at the inlet (characterized by a contraction ratio of 4.2 after the honeycomb) and a T-diffuser at the outlet. The test section has a rectangular shape and it is 2.42 m wide and 1.60 m high. Thanks to the possibility of changing both the speed of the motor driving the fan and the pitch of its blades, the flow velocity inside the section can be varied almost continuously up to approximately 30 m/s. The normal free-stream turbulence intensity in the tunnel is around 0.7% [34]. In preparation to the tests carried out for the present study, the growth of the freestream turbulence was obtained by means of squared-mesh wooden grids placed in development zone of the wind tunnel (see Figure 4a). To verify the flow conditions during the acquisitions with the turbine installed in the tunnel (Figure 4b), a single-component hot-wire probe was used to determine the spectral properties of the turbulence induced by the wooden grid and the homogeneity of the flow. The model used for the study is a two-blade, H-type Darrieus turbine having a diameter D = 0.5 m and a frontal area A = 0.4 m^2. Blades are made of a single NACA0018 airfoil and feature a constant chord c = 0.05 m; they are connected to the rotating shaft by two inclined struts per blade (see Figure 4b). Due to the small size of the rotor, revolution speeds (and thus centrifugal loads) were high (around 1200 rpm) in order to achieve significant Reynolds numbers.

(a) (b)

Figure 4. (**a**) Wind tunnel with a wooden grid for turbulence generation installed; (**b**) analyzed wind turbine.

As a consequence, the rotor was manufactured using carbon-epoxy composite [35] to ensure high-quality mechanical properties. The chosen turbulent conditions for the experimental campaign are presented in Table 1. These values represent a compromise between the levels found in urban environments [4,5] and the size limitations of the wind tunnel, as explained in [34]. A uniform wind speed profile with a mean of 9 m/s was chosen as a benchmark as it allowed the widest range of rotational speeds in the turbine prototype available and a clear comparison between the different levels of turbulence.

Table 1. Tests conditions reproduced during experiments.

Condition	Grid Distance x	Mean I_u	Mean L_{ux}
Smooth flow	-	0.7%	-
Low I_u	7.6 m	5.4%	18 cm
High I_u	3.75 m	9.2%	15 cm

3.2. Experimental Airfoil Polars

In order to emphasize the impact of turbulence on the polar curves of the airfoil an additional experimental campaign was performed in the wind tunnel of the FLOW (Fluid Dynamic and Thermodynamics) research group at the Vrije Universiteit Brussel (Belgium). This boundary layer wind tunnel has a length of 11 m and a test section with 1.04 m of height and 2 m width. Its maximum wind speed is 20 m/s and its base turbulence level is 0.5%. As in the previous experiments, the turbulence was increased by using wooden grids and the wind speed was set in order to match the mean Reynolds numbers experienced by the VAWT blades over a rotation. In this case only two flow conditions were tested: smooth flow (I_u = 0.5%) and high turbulence (I_u = 9.5%).

The blade tested was a NACA0018 adapted to the virtual camber effect that the blade experiences while rotating in VAWTs of high *c/R* ratio (equal to 0.2 in the present case), as it is described in [36]. A 20 cm chord model was built by laser cutting and gluing together wooden aerofoil sections. The airfoil was mounted on an aerodynamic balance to record the forces acting on it, and endplates where added to avoid 3-D effects. The setup can be observed in Figure 5.

Figure 5. Airfoil model mounted inside the FLOW wind tunnel, with the turbulence grid installed in the background.

3.3. Main Results

The main experimental results for the turbine have been presented in [23]. Measured power curves are also reported here for completeness to help the validation of the numerical approach developed in this work. In particular, Figure 6 reports the power coefficient variation of the turbine as a function of the TSR and the turbulence characteristics in terms of intensity and length scale. As discussed, a clear improvement of the power coefficient has been noted when increasing the turbulence, with a slight (and constant) shift of the curve peak at a lower tip-speed ratio.

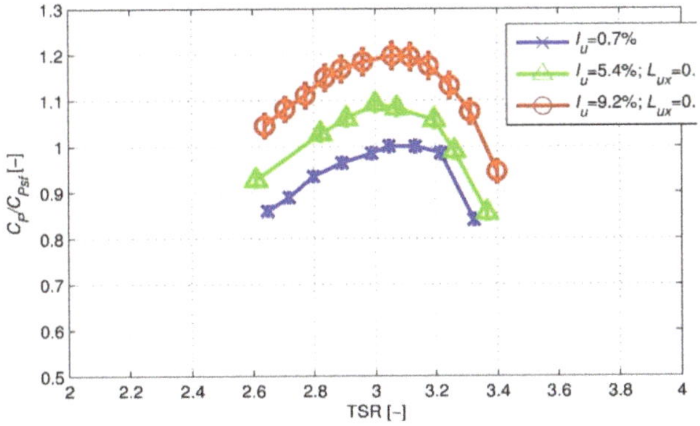

Figure 6. Power curves obtained with the VAWT under different turbulence conditions, normalized with the optimum C_P in smooth flow.

Focusing instead on the results of the measurements of the airfoil, which are an innovative contribution of the present study, Figure 7 reports the measured lift and drag polars for the virtually cambered airfoil at a Reynolds number of 80 k. Upon examination of the results, it is apparent that—for this *low-Re* condition—the attended performance increase was achieved, obtaining a constantly lower drag, a maximum lift coefficient increase of about 13% and a delay of the static stall angle of attack of approximately 6°. It is also worth noticing that the slope of the lift curve decreases somewhat. This is probably due to the fact that the laminar flow present along the blade in smooth flow conditions ensures a more intense pressure gradient in comparison to what happens in turbulent flow, where transition takes place earlier. The error bars in Figure 7 (note that only a few were included not to compromise readability) represent the uncertainty related to the calibration error of the balance (at 95% confidence level).

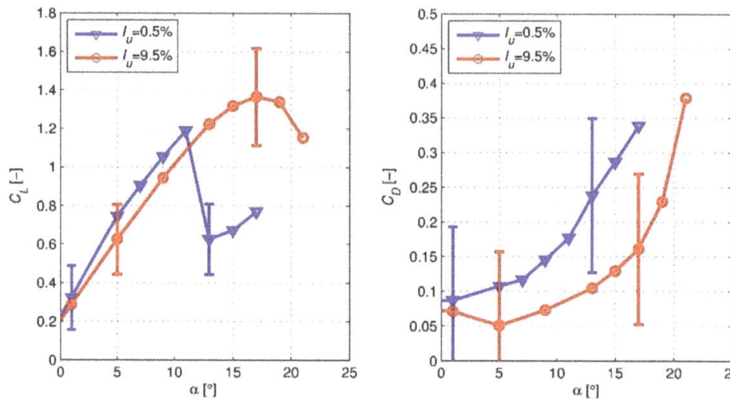

Figure 7. Experimental polars for the virtually cambered airfoil at Re = 80 k.

The error bars are especially large for the drag coefficient. Therefore, the authors double-checked the results by calculating the drag curve also by means of a measurement of the velocity deficit in the wake as in [37]: these measurements showed good agreement with those taken with the balance. The data in Figure 7 is sufficiently robust and significant for the scope of the work.

4. CFD Modeling

Since the experimental measurements under high turbulence inflow conditions were carried out for a single wind tunnel velocity, the lift and drag polars were "extended" by means of a dedicated numerical campaign. A wider range of Reynolds numbers is indeed necessary when using lift and drag polars in a low-order simulation tool for predicting the turbine performance in case of a turbulent inflow. To this end, a high-fidelity CFD simulation model was first calibrated against measurements obtained in wind tunnel laboratory experiments at a Reynolds number of Re = 80 k. Then, the calibrated model was exploited to assess the expected polars at the Reynolds number relevant for the present application.

4.1. Numerical Setup and Validation

The numerical simulations presented in this paper were performed using the commercial CFD software ANSYS® FLUENT® [38]. The solver-setting process for the numerical solutions involved the use of a two-dimensional unsteady Reynolds-averaged Navier–Stokes (U-RANS) approach with a pressure-based formulation of the solver and a *Coupled* algorithm for the pressure-velocity coupling. The second order upwind discretization scheme was used for momentum, energy, and turbulence parameters, and the pressure interpolation was second order. As far as the turbulence settings are concerned, the Transitional κ-ω SST (Shear Stress Transport) model was adopted for predicting the transition between laminar to turbulent flow of the boundary layer. The choice was related to three different requirements:

1. The need of a *Low-Re* wall treatment in order to resolve the boundary layer down to the viscous sub-layer;
2. The need of capturing the laminar-turbulent transition region due to working conditions characterized by low Reynolds number and high inflow turbulence levels;
3. The higher suitability of ω-based models for boundary layer flows with adverse pressure gradient and separation.

It has to be noted that in case of a three-equation U-RANS simulation characterized by an elevated level of turbulence intensity, the dominant turbulent parameter is the turbulence intensity itself.

Therefore, the length scale ultimately affects the rate of decay of the turbulence intensity, while having a negligible impact on the turbulent shear stress [7].

The computational domain is an open-field type having an overall extent of 60 chords and a width of 40 chords to avoid any blockage effect on the blade profile. A Dirichlet boundary condition was employed to prescribe a uniform velocity profile at the inlet boundary. The operating pressure input is set at the outlet boundary as the static pressure of the environment. The blade surface is modeled using the standard smooth no-slip wall. To replicate the conditions of the experiments, the turbulent intensity was specified at the inlet boundary in order to obtain a turbulence level (after the intrinsic decay along the domain) of roughly 9.5% at the blade location. Among the different levels tested for the turbine (Figure 6), the selection of the highest one was motivated by the fact that effects are expected to be more visible and less affected by the possible uncertainty of the CFD approach. The application of the transitional turbulence model used in the URANS approach is in fact critical in case of very low turbulence levels and Reynolds numbers.

A grid independence test was carried out for different mesh sizes in order to define the optimum mesh for the best tradeoff between accuracy and fast computation. The convergence study was performed considering three different grids. Due to the *Low-Re* number wall treatment, the size of the wall-adjacent cell for all meshes was defined such as to satisfy the requirement of the dimensionless wall distance (y^+) lower than ~1. The baseline coarse mesh was defined by adopting a discretization of the airfoil surface with 750 nodes, thus obtaining a size of 1.6×10^5 cells in the whole domain. The medium and fine meshes were defined by progressively doubling the overall elements count. In particular, the fine mesh was featuring 1500 nodes on the blade and 6.8×10^5 domain cells. The error in the estimation of lift and drag coefficients at Re = 80 k between the medium and fine mesh was lower than 0.2%. Therefore, the medium mesh featuring 1100 nodes on the blade and 3.3×10^5 domain cells, whose details are shown in Figure 8, was adopted for all of the computations presented in the paper. The core region of the flow is discretized by means of an unstructured triangular mesh, with a clustering of the mesh elements in airfoil walls. An O-grid of quadrilateral elements was used around the blade, with an extrusion of 40 layers off the wall to guarantee a sufficient boundary layer resolution.

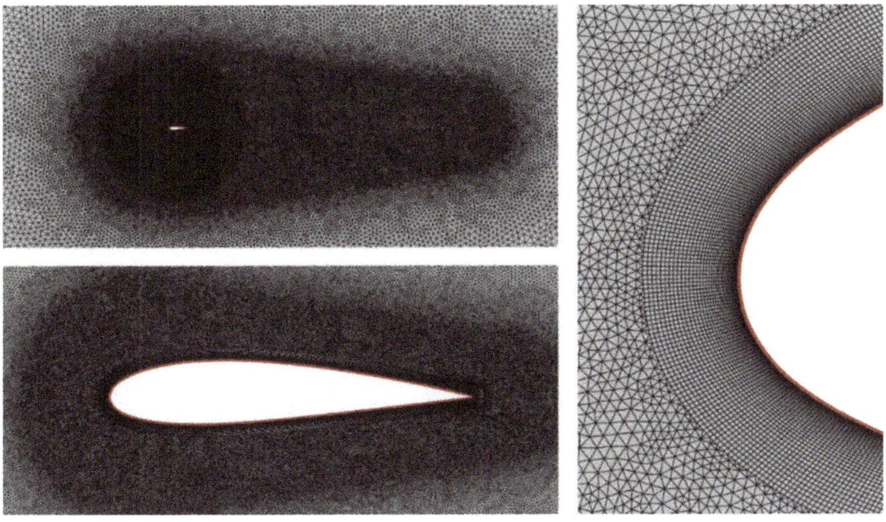

Figure 8. Computational grid used for the airfoil polars calculations.

As a result, Figure 9 shows the comparison between the experimental polars for the airfoil tested at Re = 80 k with a high turbulence flow and those resulting from the CFD simulations of the current study in the same flow conditions. The validity of the CFD approach is clearly confirmed, since measurements and numerical results are in good agreement. The CFD model is able to properly capture the trends of lift and drag coefficients as well as the maximum values, although a 1 degree shift of the static stall location between the experimental and the simulated curves can be observed.

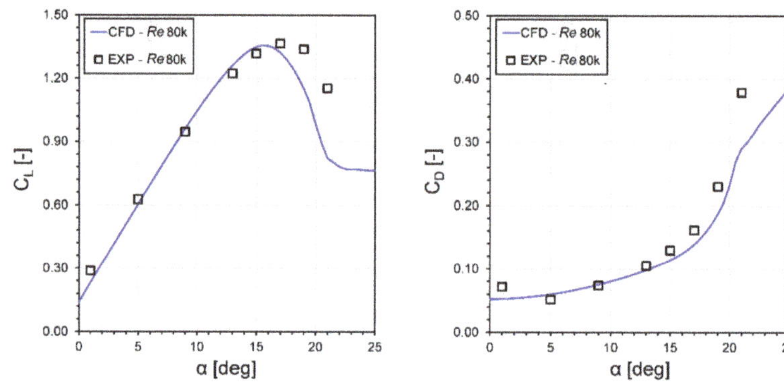

Figure 9. Comparison between CFD and experimental polars with a high turbulence flow.

4.2. Numerical Polars

The calibrated CFD model was then used to assess the influence of the Reynolds number in case of a high turbulence flow. The tested *Re* values were selected in order to cover the operating conditions of the study case VAWT for all of the analyzed TSRs. Figures 10 and 11 report the lift and drag polars for both positive and negative incidence angles starting from the lowest simulated Reynolds number (i.e., 40 k) up to the highest value of 160 k, with a step of 20 k. The lift increase, delay in stall angle and drag reduction is clearly visible from the CFD results, due to the higher resistance to flow separation as the Reynolds number is increased. Such behavior can be noticed by analyzing the flow field around the blade at 15° angle of attack for different Reynolds values, as shown in Figure 12: as the freestream velocity increases, the high vorticity region on the suction side due to separation is reduced and the flow tends to be more attached to the blade.

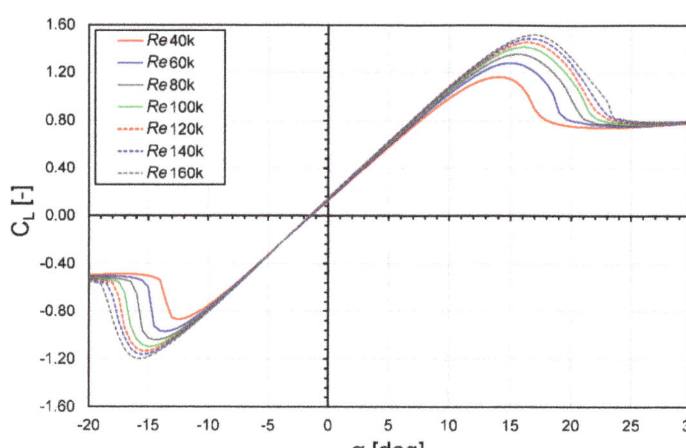

Figure 10. Lift polars for different Reynolds values with a high turbulence flow.

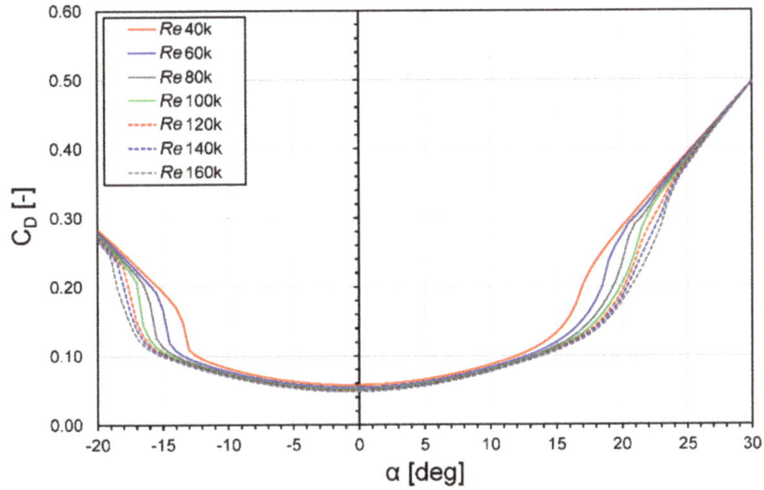

Figure 11. Drag polars for different Reynolds values with a high turbulence flow.

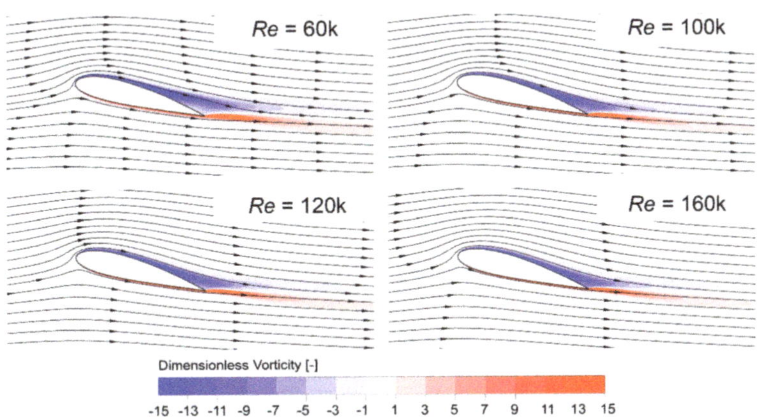

Figure 12. Vorticity contours and velocity streamlines around the blade at 15° incidence for different Reynolds.

5. Synthesis of the Effects into the BEM Modeling

Finally, to prove that the two investigated key phenomena, i.e., the increased energy content and the improved airfoil performance at low *Re* in the turbulent wind, are indeed responsible for the increase of the power coefficient under turbulent conditions, they have been combined into an engineering Blade Element Momentum (BEM) model. It is well known that this theory is, on the one hand, able to deliver sufficiently reliable results in terms of overall performance, while providing, on the other hand, scarce definition of the torque profile during the revolution and of the flow field past the turbine [39]. In the perspective of the present study, however, the use of a simple BEM model (although making use of the most advanced features presently available for this theory) was thought to be of particular interest to test the impact of the discussed phenomena. More specifically, the claimed result is that a proper combination of corrections for the energy content in the flow and for the airfoil polars can accurately predict the turbine performance variation in turbulent flows even with a very simple theory.

5.1. Setup

The VARDAR code of the Università degli Studi di Firenze (Italy) [39] has been used for the analysis. The prediction capabilities of this research code have been validated over the last ten years on a variety of small H-Darrieus turbines, proving its high accuracy in comparison to other existing codes. The BEM formulation inside the VARDAR code is based on an improved version of a *Double Multiple Streamtubes Approach with Variable Interference Factors* originally proposed by Prof. Paraschivoiu in [9]. With respect to the "standard" formulation, the Glauert's correction for high-induction cases based on recent experimental data has been implemented. To increase the accuracy in predicting VAWT aerodynamics, several sub-models have been embedded within the code, including the corrections to account for the finite aspect ratio of the blades using the Lanchester–Prandtl model, the parasitic torque of the struts, and the streamtube expansion model presented in [9], although the impact of this on the simulation of small turbines like the one investigated in this work is negligible, as discussed in [34]. Furthermore, based on recent findings about the aerodynamics of airfoils in cycloidal motion, specific corrections are included to correct airfoil polars in order to account for flow curvature effects, i.e., the *virtual camber* effect [36] and the *virtual incidence* [40]. Another important additional feature recently included in the code is the polar smoothing procedure discussed in [41]. Finally, several dynamic stall models are included, i.e., those proposed by Gormont, Berg, Strickland, and Paraschivoiu [9]; in the present study, Berg's one with a calibration factor of 30 has been used.

According to the discussed hypotheses, the CFD-based airfoil polars presented in Section 4 (obtained for a turbulence intensity of 9.5%) have been used in the present study, along with an average wind speed corrected by Equation (1) in comparison to experiments, equal to 9.2 m/s.

5.2. Results

Figure 13 displays the comparison between the measured increase of the power curve of the turbine (i.e., the power coefficient divided by the optimum one in smooth flow) and that obtained by means of the BEM approach.

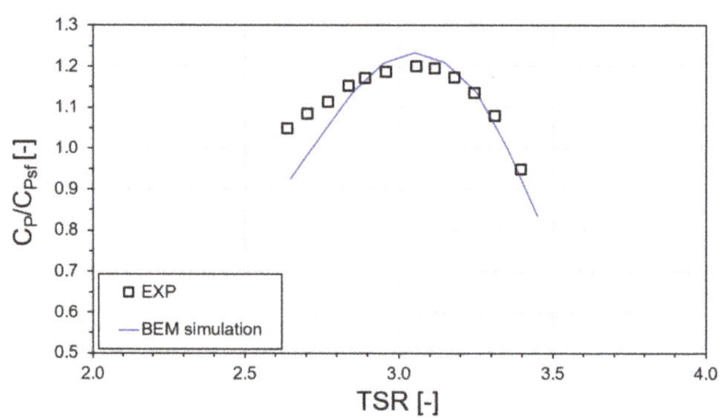

Figure 13. Comparison between the experimental turbine power curve (in dimensionless form) in turbulence and the predictions using the VARDAR code: the BEM code is run with the airfoil polars obtained for a 9.5% turbulence condition and at the equivalent wind speed of 9.2 m/s.

Upon examination of the figure, it is apparent that even a very simple model like the BEM one, if properly accounting for the analyzed phenomena, was able to nicely predict the overall tendency of the performance variation of the turbine, especially in terms of the maximum power coefficient increase. The numerical trend slightly anticipates the curve peak, probably due to the fact that the performance of real airfoils in motion could be slightly lower than that predicted via CFD. Additionally, the numerical curve is a slightly steeper than the experimental one. It also has to be remembered that the left-hand side of the curve is notably affected by dynamic stall, which could be further affected by turbulence, being barely reproducible with the engineering models embedded in the BEM code.

Overall, however, it is worth remarking that being able to estimate the power coefficient variation increase in turbulence with a simple engineering tool has to be considered like a very promising result. This also proves that, as hypothesized, the two discussed phenomena (more energized flow and better airfoil performance) impact consistently on the overall physics.

6. Conclusions

In this study, different techniques and data types were combined together to prove that two main physical phenomena concur in defining the performance increase (testified by unique experimental wind tunnel data reproducing realistic turbulence features) of small Darrieus turbines in turbulent flows.

The first one, assessed by means of detailed CFD simulations of the rotor, is the higher energy content in the turbulent wind, which induces a slight power surplus. This is, however, limited by the fact that the turbine constantly operates at a non-optimal tip-speed ratio, depending on the flow macrostructures that enter the turbine instant by instant. Overall, an equivalent wind speed can be defined.

The second one—which is thought of major relevance in case of small turbines—is the improved response of the airfoils in terms of delayed stall angle and increased lift-to-drag ratio. This second phenomenon has been verified with dedicated experimental tests in the wind tunnel, which also allowed for the calibration of a CFD tool to virtually replicate the polars.

Finally, the two elements have been combined into a state-of-the-art BEM code, which was able —despite its simplicity—to nicely predict the turbine behavior, thus suggesting that the two highlighted phenomena are really playing a major role in defining the aerodynamic behavior and energy conversion capability of small Darrieus vertical axis wind turbines in turbulent flows.

Future work will be devoted to providing an on-field validation to prove the feasibility of small Darrieus VAWTs in turbulent sites. In particular, gaining a better understanding of the discussed phenomena could lead in the near future to design strategies for small rotors specifically tailored to maximize the performance in the turbulent flows that are typical, for example, of the urban environment.

Author Contributions: Conceptualization: A.B. and F.B.; Methodology: A.B., F.B. and M.Z.; Software: A.B. and F.B.; Experiments: A.C.M., T.D.T. and G.B.; Validation: A.C.M.; Formal Analysis: A.B.; Investigation: A.B., F.B. and M.Z.; Resources: M.C.R., G.B. and G.F.; Data Curation: A.C.M., M.Z. and F.B.; Writing—Original Draft Preparation: M.Z., F.B. and A.B.; Writing—Review and Editing: T.D.T.; Visualization: F.B.; Supervision: T.D.T., M.C.R., G.B., G.F. and A.B.; Project Administration: M.C.R., G.B. and G.F.; Funding Acquisition: T.D.T., M.C.R. and G.B. All authors have read and agreed to the published version of the manuscript.

Funding: The experimental campaigns are part of the European Innovative Training Network (ITN) AEOLUS4FUTURE "Efficient Harvesting of the Wind Energy". The project is funded by the Horizon 2020 Research and Innovation program under the Marie Skłodowska-Curie grant agreement no. 643167.

Acknowledgments: Thanks are due to all the staff of the CRIACIV wind tunnel in Prato for the support during the wind tunnel tests of the turbine. The authors would like also to acknowledge the VUB Bachelor student Julen Echeverria for the help in the construction of the airfoil model for polars in turbulence.

Conflicts of Interest: The authors declare no conflict of interest.

References

1. Mertens, S. *Wind Energy in the Built Environment Concentrator Effects of Buildings*; Multi Science: Essex, UK, 2006.
2. Balduzzi, F.; Bianchini, A.; Carnevale, E.; Ferrari, L.; Magnani, S. Feasibility analysis of a Darrieus vertical-axis wind turbine installation in the rooftop of a building. *Appl. Energy* **2012**, *97*, 921–929. [CrossRef]
3. Balduzzi, F.; Bianchini, A.; Ferrari, L. Microeolic turbines in the built environment: Influence of the installation site on the potential energy yield. *Renew. Energy* **2012**, *45*, 163–174. [CrossRef]
4. Drew, D.; Barlow, J.; Cockerill, T. Estimating the potential yield of small wind turbines in urban areas: A case study for Greater London, UK. *J. Wind. Eng. Ind. Aerodyn.* **2013**, *115*, 104–111. [CrossRef]

5. Janajreh, I.; Su, L.; Alan, F. Wind energy assessment: Masdar City case study. *Renew. Energy* **2013**, *52*, 8–15. [CrossRef]
6. Dallman, A.R. Flow and Turbulence in Urban Areas. Ph.D. Thesis, University of Notre Dame, Notre Dame, IN, USA, 2013.
7. Balduzzi, F.; Zini, M.; Ferrara, G.; Bianchini, A. Development of a Computational Fluid Dynamics Methodology to Reproduce the Effects of Macroturbulence on Wind Turbines and Its Application to the Particular Case of a VAWT. *J. Eng. Gas Turbines Power* **2019**, *141*, 111010–111022. [CrossRef]
8. Lubitz, W.D. Impact of ambient turbulence on performance of a small wind turbine. *Renew. Energy* **2014**, *61*, 69–73. [CrossRef]
9. Paraschivoiu, I. *Wind Turbine Design: With Emphasis on Darrieus Concept*; Presses inter Polytechnique: Montréal, QC, Canada, 2002.
10. Mohamed, A. Aero-acoustics noise evaluation of H-rotor Darrieus wind turbines. *Energy* **2014**, *65*, 596–604. [CrossRef]
11. Bianchini, A.; Ferrara, G.; Ferrari, L.; Magnani, S. An Improved Model for the Performance Estimation of an H-Darrieus Wind Turbine in Skewed Flow. *Wind. Eng.* **2012**, *36*, 667–686. [CrossRef]
12. Cooper, P. *Development and Analysis of Vertical-Axis Wind Turbines*; WIT Press Ltd.: Billerica, MA, USA, 2010; Volume 1, pp. 277–302.
13. Möllerström, E.; Ottermo, F.; Goude, A.; Eriksson, S.; Hylander, J.; Bernhoff, H. Turbulence influence on wind energy extraction for a medium size vertical axis wind turbine. *Wind. Energy* **2016**, *19*, 1963–1973. [CrossRef]
14. Bertenyi, T.; Wickins, C.; McIntosh, S. Enhanced Energy Capture Through Gust-Tracking in the Urban Wind Environment. In Proceedings of the 48th AIAA Aerospace Sciences Meeting Including the New Horizons Forum and Aerospace Exposition, Orlando, FL, USA, 4–7 January 2010. [CrossRef]
15. Pagnini, L.C.; Burlando, M.; Repetto, M.P. Experimental power curve of small-size wind turbines in turbulent urban environment. *Appl. Energy* **2015**, *154*, 112–121. [CrossRef]
16. Kooiman, S.J.; Tullis, S. Response of a Vertical Axis Wind Turbine to Time Varying Wind Conditions found within the Urban Environment. *Wind. Eng.* **2010**, *34*, 389–402. [CrossRef]
17. Lee, K.-Y.; Tsao, S.-H.; Tzeng, C.-W.; Lin, H.-J. Influence of the vertical wind and wind direction on the power output of a small vertical-axis wind turbine installed on the rooftop of a building. *Appl. Energy* **2018**, *209*, 383–391. [CrossRef]
18. Dossena, V.; Persico, G.; Paradiso, B.; Battisti, L.; Dell'Anna, S.; Brighenti, A.; Benini, E. An Experimental Study of the Aerodynamics and Performance of a Vertical Axis Wind Turbine in a Confined and Unconfined Environment. *J. Energy Resour. Technol.* **2015**, *137*, 051207. [CrossRef]
19. Miau, J.J. *Wind Tunnel Study on Aerodynamic Performance of Small Vertical-Axis Wind Turbines*; Cheng Kung University: Tainan, Taiwan, 2012.
20. Ahmadi-Baloutaki, M.; Carriveau, R.; Ting, D.S.-K. Performance of a vertical axis wind turbine in grid generated turbulence. *Sustain. Energy Technol. Assess.* **2015**, *11*, 178–185. [CrossRef]
21. Peng, H.; Lam, H. Turbulence effects on the wake characteristics and aerodynamic performance of a straight-bladed vertical axis wind turbine by wind tunnel tests and large eddy simulations. *Energy* **2016**, *109*, 557–568. [CrossRef]
22. Hohman, T.; Martinelli, L.; Smits, A.J. The effects of inflow conditions on vertical axis wind turbine wake structure and performance. *J. Wind. Eng. Ind. Aerodyn.* **2018**, *183*, 1–18. [CrossRef]
23. Carbò Molina, A.; Massai, T.; Balduzzi, F.; Bianchini, A.; Ferrara, G.; De Troyer, T.; Bartoli, G. Combined experimental and numerical study on the near wake of a Darrieus VAWT under turbulent flows. *J. Phys. Conf. Ser.* **2018**, *1037*, 072052. [CrossRef]
24. Carbò Molina, A.; De Troyer, T.; Massai, T.; Vergaerde, A.; Runacres, M.C.; Bartoli, G. Effect of turbulence on the performance of VAWTs: An experimental study in two different wind tunnels. *J. Wind. Eng. Ind. Aerodyn.* **2019**, *193*, 103969. [CrossRef]
25. Malcolm, D. Dynamic response of a darrieus rotor wind turbine subject to turbulent flow. *Eng. Struct.* **1988**, *10*, 125–134. [CrossRef]
26. Brahimi, T.; Paraschivoiu, I. Darrieus Rotor Aerodynamics in Turbulent Wind. *J. Sol. Energy Eng.* **1995**, *117*, 128–136. [CrossRef]
27. Ahmedov, A.; Ebrahimi, K.M. Numerical Modelling of an H-type Darrieus Wind Turbine Performance under Turbulent Wind. *Am. J. Energy Res.* **2017**, *5*, 63–78. [CrossRef]

28. Balduzzi, F.; Bianchini, A.; Maleci, R.; Ferrara, G.; Ferrari, L. Critical issues in the CFD simulation of Darrieus wind turbines. *Renew. Energy* **2016**, *85*, 419–435. [CrossRef]
29. Balduzzi, F.; Bianchini, A.; Ferrara, G.; Ferrari, L. Dimensionless numbers for the assessment of mesh and timestep requirements in CFD simulations of Darrieus wind turbines. *Energy* **2016**, *97*, 246–261. [CrossRef]
30. Bianchini, A.; Balduzzi, F.; Bachant, P.; Ferrara, G.; Ferrari, L. Effectiveness of two-dimensional CFD simulations for Darrieus VAWTs: A combined numerical and experimental assessment. *Energy Convers. Manag.* **2017**, *136*, 318–328. [CrossRef]
31. Rezaeiha, A.; Montazeri, H.; Blocken, B. Characterization of aerodynamic performance of vertical axis wind turbines: Impact of operational parameters. *Energy Convers. Manag.* **2018**, *169*, 45–77. [CrossRef]
32. Dessoky, A.; Bangga, G.; Lutz, T.; Krämer, E. Computational Study Using DDES with Higher Order Scheme Modeling to Predict Darrieus VAWT Noise Mechanisms. In *New Results in Numerical and Experimental Fluid Mechanics XII*; Springer Science and Business Media LLC: Berlin/Heidelberg, Germany, 2019; Volume 142, pp. 807–818.
33. Patil, R.; Daróczy, L.; Janiga, G.; Thévenin, D. Large eddy simulation of an H-Darrieus rotor. *Energy* **2018**, *160*, 388–398. [CrossRef]
34. Carbó Molina, A.; Bartoli, G.; De Troyer, T. Generation of Uniform Turbulence Profiles in the Wind Tunnel for Urban VAWT Testing. In *Wind Energy Exploitation in Urban Environment*; Battisti, L., Ricci, M., Eds.; Springer International Publishing: Cham, Switzerland, 2018. [CrossRef]
35. Vergaerde, A.; De Troyer, T.; Carbò Molina, A.; Standaert, L.; Runacres, M. Design, manufacturing and validation of a vertical-axis wind turbine setup for wind tunnel tests. *J. Wind. Eng. Ind. Aerodyn.* **2019**, *193*, 103949. [CrossRef]
36. Rainbird, J.M.; Bianchini, A.; Balduzzi, F.; Peiró, J.; Graham, M.; Ferrara, G.; Ferrari, L. On the influence of virtual camber effect on airfoil polars for use in simulations of Darrieus wind turbines. *Energy Convers. Manag.* **2015**, *106*, 373–384. [CrossRef]
37. Schlichting, H. *Boundary Layer Theory*; McGraw Hill: New York, NY, USA, 1979.
38. *Fluent Theory Guide*; Release 19; Ansys Inc.: Canonsburg, PA, USA, 2018.
39. Bianchini, A.; Balduzzi, F.; Ferrara, G.; Persico, G.; Dossena, V.; Ferrari, L. A Critical Analysis on Low-Order Simulation Models for Darrieus Vawts: How Much Do They Pertain to the Real Flow? *J. Eng. Gas Turbines Power* **2018**, *141*, 011018. [CrossRef]
40. Bianchini, A.; Balduzzi, F.; Ferrara, G.; Ferrari, L. Virtual incidence effect on rotating airfoils in Darrieus wind turbines. *Energy Convers. Manag.* **2016**, *111*, 329–338. [CrossRef]
41. Marten, D.; Bianchini, A.; Pechlivanoglou, G.; Balduzzi, F.; Nayeri, C.; Ferrara, G.; Paschereit, C.O.; Ferrari, L. Effects of Airfoil's Polar Data in the Stall Region on the Estimation of Darrieus Wind Turbine Performance. *J. Eng. Gas Turbines Power* **2016**, *139*, 022606. [CrossRef]

energies

Article

Design and 3D CFD Static Performance Study of a Two-Blade IceWind Turbine

Hamdy Mansour [1] **and Rola Afify** [2,*]

1 Mechanical Engineering Department, Faculty of Engineering, Alexandria University, Alexandria P.O. Box 21544, Egypt; hamdydabos@outlook.com
2 Mechanical Engineering Department, College of Engineering and Technology, Arab Academy for Science, Technology and Maritime Transport (AASTMT), Abu kir, Alexandria P.O. Box 1029-Miami, Egypt
* Correspondence: rola@aast.edu; Tel.: +20-1030576111

Received: 8 September 2020; Accepted: 11 October 2020; Published: 14 October 2020

Abstract: The IceWind turbine, a new type of Vertical Axis Wind Turbine, was proposed by an Iceland based startup. It is a product that has been featured in few published scientific research studies. This paper investigates the IceWind turbine's performance numerically. Three-dimensional numerical simulations are conducted for the full scale model using the SST K-ω model at a wind speed of 15.8 m/s. The following results are documented: static torque, velocity distributions and streamlines, and pressure distribution. Comparisons with previous data are established. Additionally, comparisons with the Savonius wind turbine in the same swept area are conducted to determine how efficient the new type of turbine is. The IceWind turbine shows a similar level of performance with slightly higher static torque values. Vortices behind the IceWind turbine are confirmed to be three-dimensional and are larger than those of Savonius turbine.

Keywords: aerodynamics; computational fluid dynamics (CFD); ansys fluent; savonius turbine; icewind turbine; static torque; three-dimensional simulation

1. Introduction

Many factors have led to increased interest in renewable energy including the reduction in conventional energy sources, the fact that conventional energy sources cause climate change, and the availability and hygiene of renewable energy sources. Lately, wind energy has become particularly important. The IceWind turbine is a new type of Vertical Axis Wind Turbine (VAWT) that converts wind energy into electricity. It is an attractive and cost effective energy source for electric generation in low velocity regions. The prefix "Ice" in "IceWind" comes from "Iceland", its home town [1]. Currently, there are two products: the CW IceWind turbine, shown in Figure 1, and the RW IceWind turbine, shown in Figure 2.

Figure 1. CW IceWind turbine.

Figure 2. RW IceWind turbine.

The IceWind turbine was first introduced by Aymane [2]. He mentioned that the IceWind turbine is not as simple to manufacture as the Savonius wind turbine, but its shape is better looking. Furthermore, the IceWind turbine produces less noise than the Savonius turbine. He confirmed that any product should not only show a good level of performance but should also have wide acceptance from the public. He invited participants to fill in a survey about the overall appearance, noise level, and efficiency of the turbine. Eighty-five percent of participants declared that the IceWind turbine produced less noise and had a better overall appearance than the Barrel Savonius. Moreover, Afify [3] investigated the turbine's performance experimentally to determine its optimum design. He concluded that a single stage, three-blade IceWind turbine with end plates, an aspect ratio of 0.38, and a blade arc angle of 112° performs better than the Savonius turbine.

Computational Fluid Dynamics (CFD) can predict fluid flow aerodynamic performance. Sarma et al. [4] mentioned that the intent of using CFD is to enable the study of velocity and torque distribution. Nasef et al. [5] numerically analyzed the aerodynamic performance of stationary and rotating Savonius rotors with several overlap ratios using four turbulence models. Their results indicated that the Shear Stress Transport (SST) k-ω turbulence model gives more accurate results than the other studied turbulence models. Kacprzak et al. [6] examined the performance of the Savonius wind turbine with fixed cross-sections using quasi 2D flow predictions through ANSYS CFX. Simulations were achieved in a way that allowed comparison with wind tunnel data documented in a related paper, where two designs were simulated: Classical and Bach-type Savonius rotors. The comparison detected the significance of applying a laminar-turbulent transition model. Dobrev and Massouh [7] aimed to consider the flow through a Savonius type turbine using a three-dimensional model by means of k-ω and DES (Detached Eddy Simulation). Due to the continuous variation of the flow angle with respect to the blades and turbine principles of operation, strong unsteady effects including separation and vortex shedding were observed. The flow analysis helped to validate their wind turbine design. McTavish et al. [8] developed a novel vertical axis wind turbine (VAWT) consisting of many asymmetric vertically stacked stages. The VAWT torque characteristics were computationally investigated using CFdesign 2010 software. Steady two-dimensional CFD simulations demonstrated that the new type had similar average static torque characteristics to present Savonius rotors. Additionally, rotating three-dimensional CFD simulations were performed.

In the present study, three-dimensional numerical simulations are used to calculate the static torque of the IceWind turbine and to show its air flow velocity distribution and streamlines and pressure distribution.

2. Physical Model

Figure 3 shows the steps used in SolidWorks to draw the IceWind blade. It consists of three circles and three lines arranged as shown in Figure 3a. Trimming was used and dimensions were applied, as shown in Figure 3b. The sketch was revolved 90° about its axis (right vertical line), as shown in Figure 3c, and then, the extrude function was used to cut the shape, as shown in Figure 3d. Figure 4 shows the final IceWind blade that was used in this study. Its dimensions are $d = 75$ mm and $H = 75$ mm with a 25 mm blade tip height and a swept area of $A_s = 4250.51$ mm^2. Figures 5 and 6 show the used IceWind turbine.

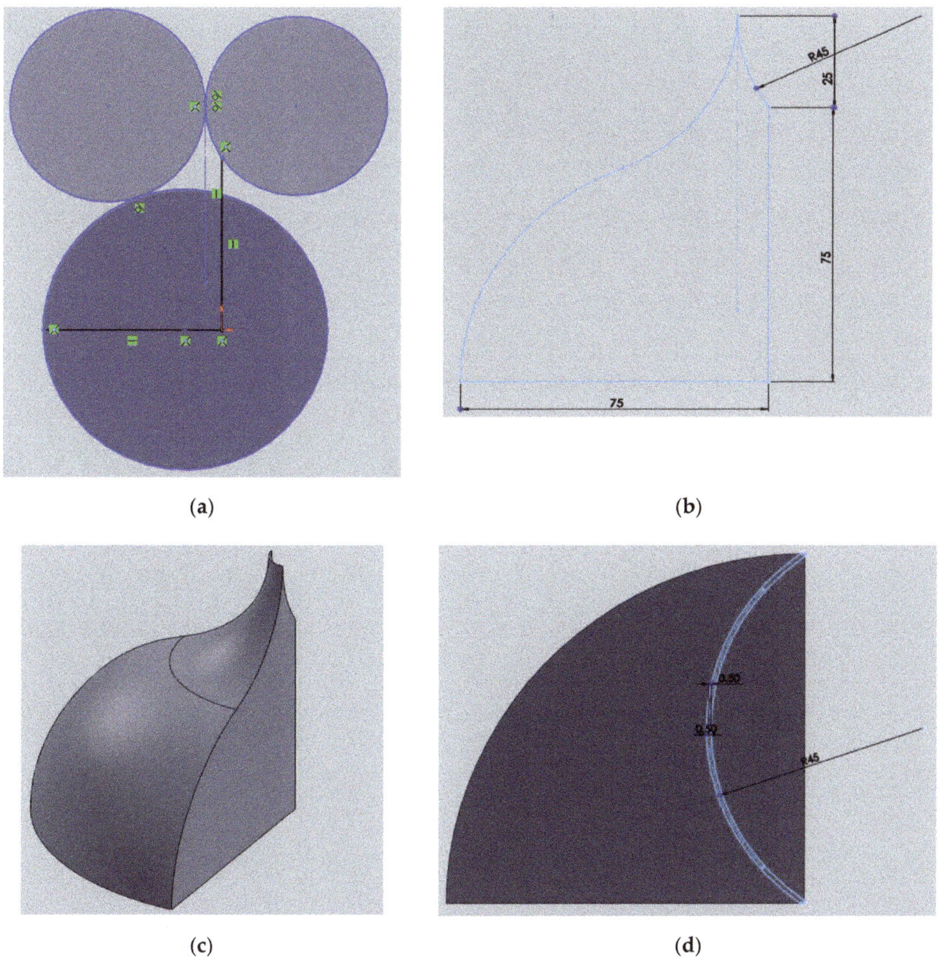

Figure 3. Steps used in SolidWorks to draw the IceWind blade (dimensions are in mm). (**a**) First shape; (**b**) trimmed shape; (**c**) revolved shape; (**d**) preparation of extruded cut shape.

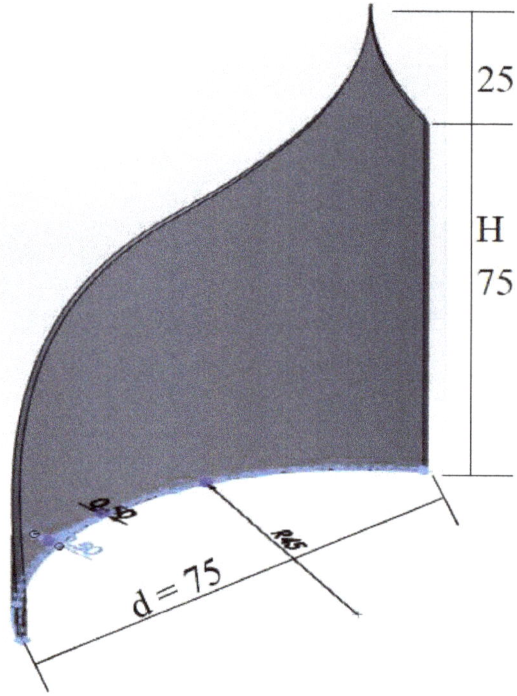

Figure 4. IceWind blade (dimensions are in mm).

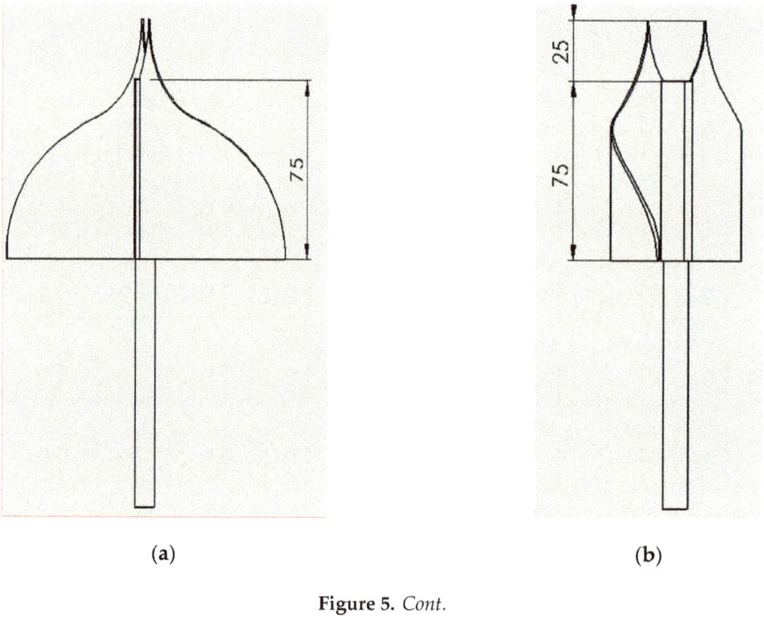

(**a**) (**b**)

Figure 5. *Cont.*

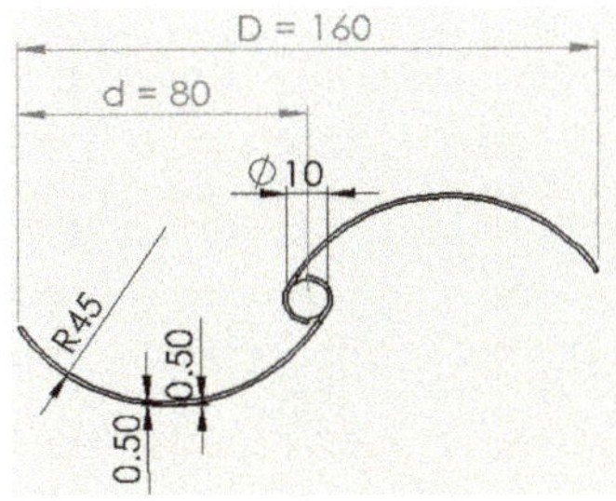

(c)

Figure 5. IceWind turbine's three views (dimensions are in mm): (**a**) elevation, (**b**) side view, and (**c**) plan.

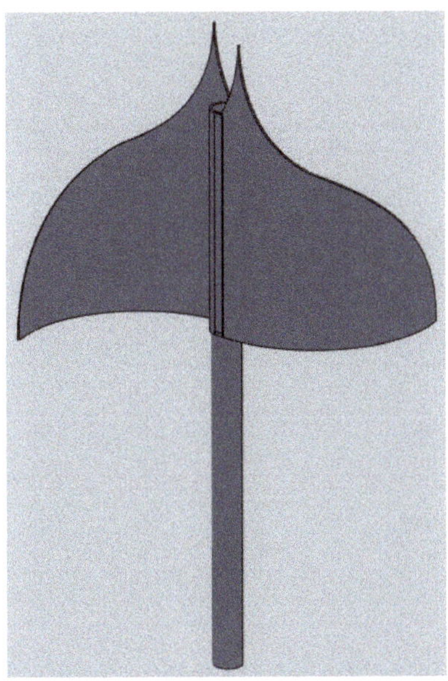

Figure 6. IceWind turbine.

3. Numerical Model

Simulations of air flow around the turbine were conducted using Ansys FLUENT after setting the test conditions to be similar to real conditions.

3.1. Domain Dimensions

The numerical domain included the wind tunnel space and the IceWind turbine, as shown in Figure 7, to simulate the flow around the turbine. The upstream, downstream, width, and height dimensions were 500, 500, 300, and 300 mm, respectively, as shown in Figure 7. The domain's overall dimensions were 300 × 300 × 1000 mm. These dimensions are the wind tunnel and turbine dimensions given in [3].

A position angle θ was defined as the angle where the blade tips were inclined to the direction of the wind tunnel's air flow, as shown in Figure 8.

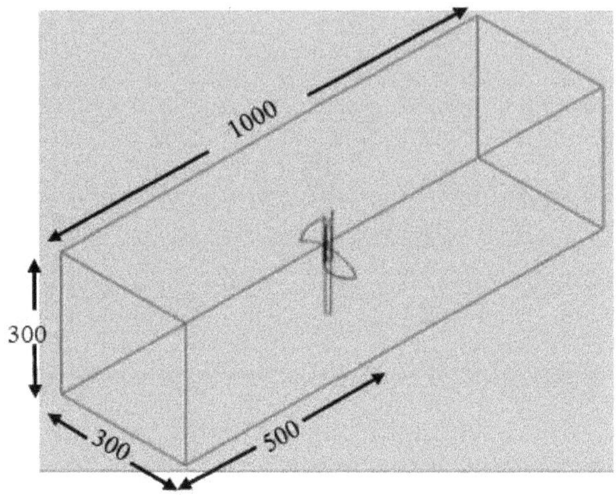

Figure 7. Wind tunnel space with the IceWind turbine and domain dimensions (mm).

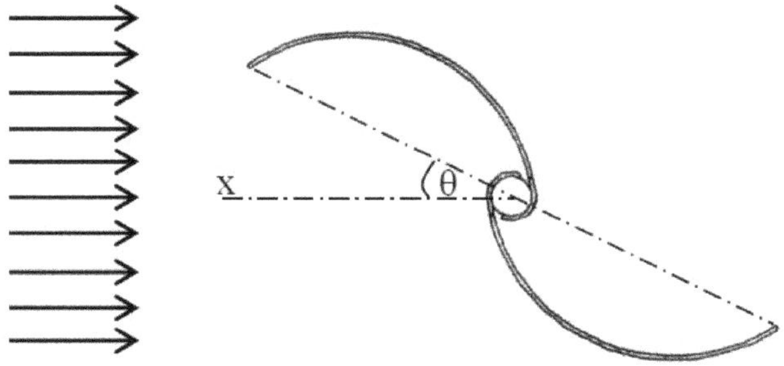

Figure 8. A position angle θ of the turbine's two blades.

3.2. Boundary Conditions

The boundary conditions are shown in Figure 9 and Table 1. They are exactly as used in [3]. The left boundary was defined as an inlet. The air flow was a wind velocity of 15.8 m/s. A pressure outlet boundary condition was assumed at the right boundary. Other boundaries and the turbine were considered to be walls.

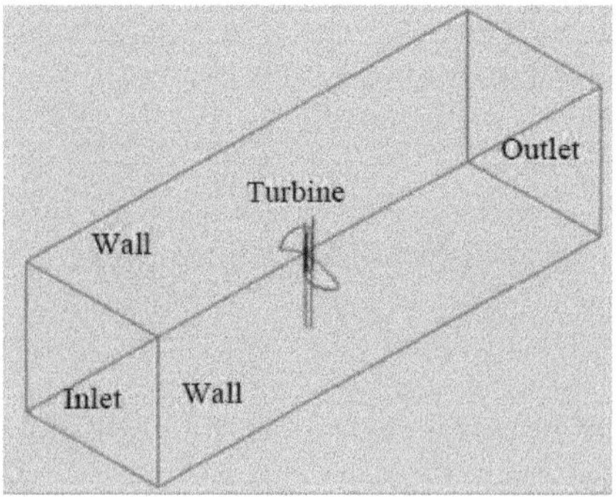

Figure 9. Boundary conditions.

Table 1. Boundary conditions.

Boundary Location	Boundary Condition
Inlet (Left)	Velocity inlet
Outlet (Right)	Pressure outlet
Top	Wall
Sides	Wall
bottom	Wall
Turbine's surfaces	Wall

3.3. Domain Meshing

The accuracy of the model results was sensitive to the size and distributions of the mesh. For this three-dimensional simulation study, two diverse zones—the rotating and stationary zones—were drawn. A vertical cylinder around the turbine was considered to be the rotating zone, and the whole wind tunnel test section excluding this cylinder was the stationary zone, as shown in Figure 10.

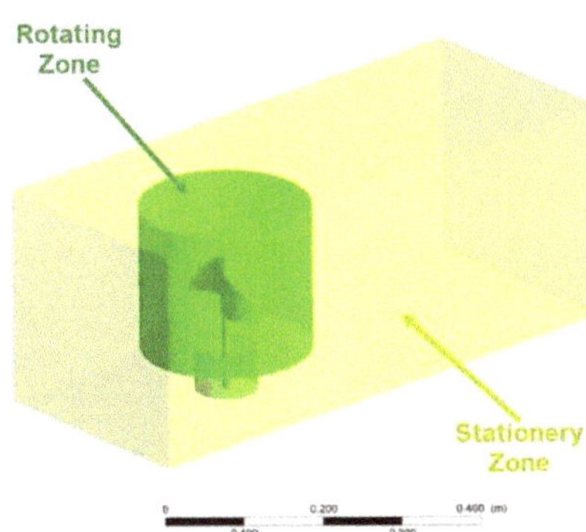

Figure 10. Two different zones of the IceWind turbine's domain: rotating and stationary zones.

A mesh independency study was carried out. Figure 11 shows the relation between static torque and the number of elements at an air velocity of 15.8 m/s for the IceWind turbine at $\theta = 90°$. For element numbers of 4.5×10^6 and 7.8×10^6, the value of static torque (N·m) was almost the same. To give a high level of accuracy, 7.8×10^6 elements were used.

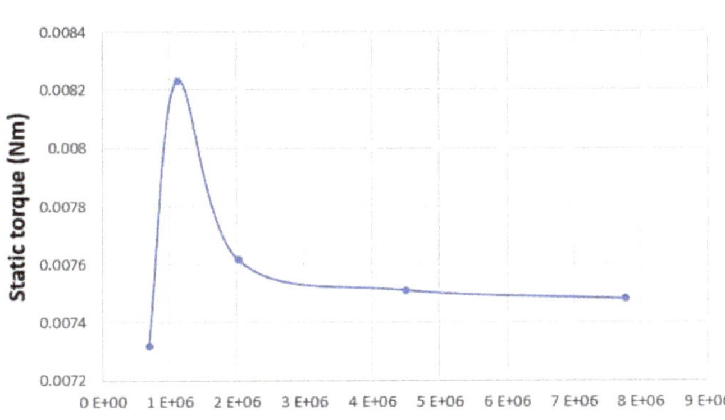

Figure 11. Relation between static torque and the number of elements at an air velocity of 15.8 m/s for the IceWind turbine at $\theta = 90°$.

Figures 12 and 13 show that computational mesh consists of tetrahedral cells. It is very fine around the blades and shaft (maximum y^+ below 2). A close-to-equilateral coarse mesh is generated in the stationary zone. The contact between these two zones is considered to be the interface boundary condition, and this guarantees that continuity in the flow field is acquired while minimizing numerical errors. Second order discretization was used for all solution variables [9].

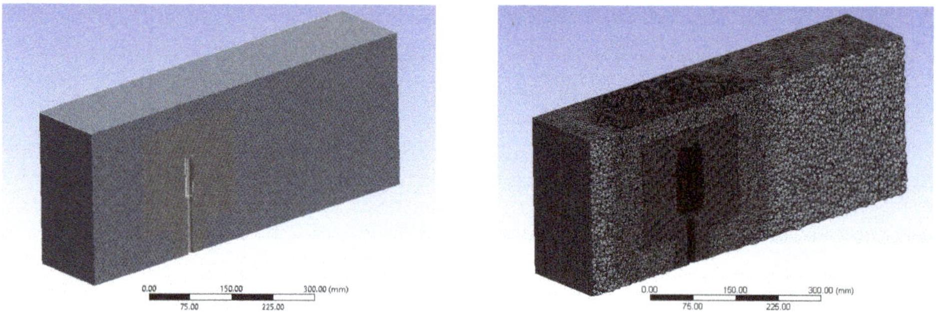

(a) (b)

Figure 12. Mesh of the IceWind turbine domain (section): (**a**) elevation and (**b**) side view.

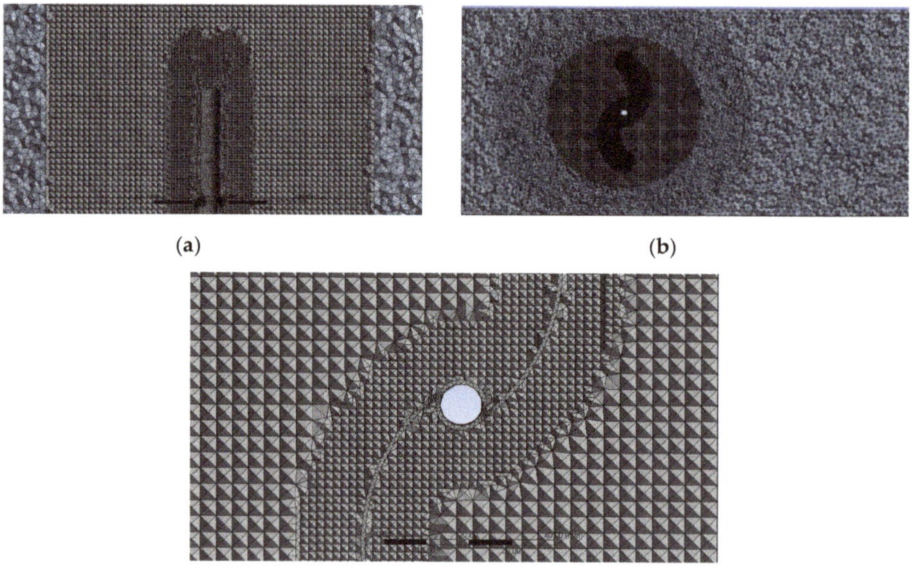

(a) (b)

(c)

Figure 13. Mesh of the IceWind turbine domain (zoomed in): (**a**) elevation, (**b**) plan, and (**c**) plan (zoomed in).

3.4. Turbulence Modeling Approach

The regime of the system was laminar. Previous studies showed that the k-ε and Spallart-Allmaras models cannot catch and predict the flow progress, especially in the laminar separation bubble [10,11]. Therefore, the SST k-ω model [12,13] can be utilized as a low Reynolds turbulence model with no additional damping functions. Shear Stress Transport (SST) formulation is created by combining the k-ω and k-ε models. This structure supports the use of the SST method to switch to the k-ε model to revoke the problems of k-ω in inlet free-stream turbulence properties and utilize the k-ω formulation in the internal parts of the boundary layer. The k-ω SST model is a commonly used turbulent model in VAWT simulations [14–20]. Furthermore, it is a good predictor of turbulence in adverse pressure gradients and separating flow.

Two mathematical formulas, k and ω equations, are proposed for use in SST methods below [9]:

$$\frac{\partial(\rho k)}{\partial t} + \frac{\partial(\rho\, k\, u_i)}{\partial x_i} = \frac{\partial}{\partial x_j}\left(\Gamma_k \frac{\partial k}{\partial x_j}\right) + G_k - Y_k + s_k \tag{1}$$

$$\frac{\partial(\rho \omega)}{\partial t} + \frac{\partial(\rho \omega\, u_i)}{\partial x_i} = \frac{\partial}{\partial x_j}\left(\Gamma_\omega \frac{\partial \omega}{\partial x_j}\right) + G_\omega - Y_\omega + s_\omega \tag{2}$$

where Γ_k and Γ_ω express the active diffusivity of k and ω. s_k and s_ω are user-defined source terms. G_k and G_ω show the turbulent kinetic energy generation due to the mean velocity gradients. Y_k and Y_ω mean the dissipation of k and ω due to turbulence.

The chosen fluid model for computation comprises air at 25 °C, pressure equal to one atmosphere, isothermal heat transfer, and a turbulent flow model.

For laminar steady flow, the simulations were run to reach steady state conditions, and the residuals reached a value of less than 6×10^{-5}, as shown in Figure 14.

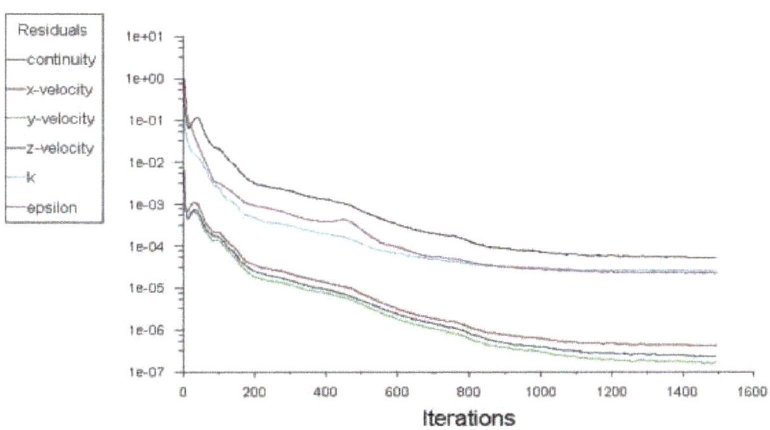

Figure 14. Relation between residuals and the number of iterations at an air velocity of 15.8 m/s for the IceWind turbine at $\theta = 90°$.

4. Savonius Turbine

Due to the lack of visualization results for the IceWind turbine, the traditional turbine type was compared with the performance of the new type (IceWind turbine). To ensure a good assessment, the authors used the Savonius turbine, as shown in Figures 15 and 16, with the same swept area: $A_s = 4250.51\ \text{mm}^2$. With the same d value of 75 mm, the Savonius blade height was equal to 56.67 mm. To ensure consistency, all IceWind turbine three-dimensional simulation conditions were applied for the Savonius turbine, including the domain dimensions, domain meshing, and turbulence modeling approach.

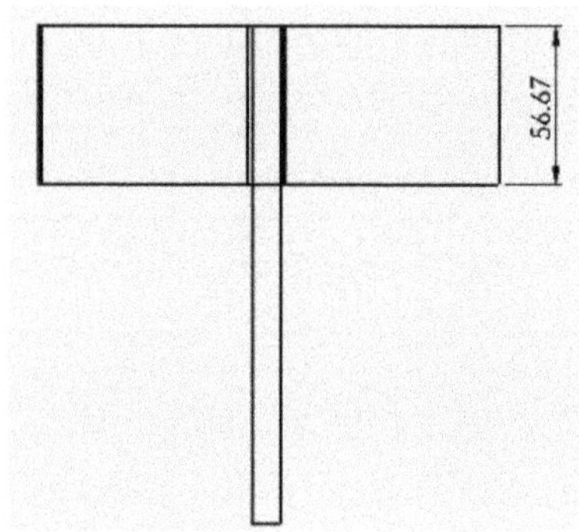

(a)

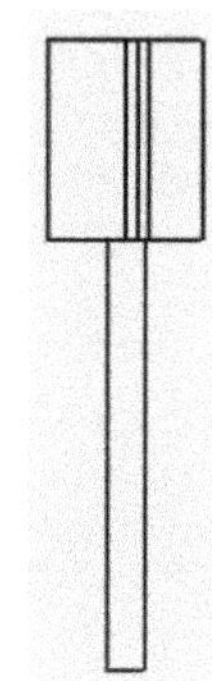

(b)

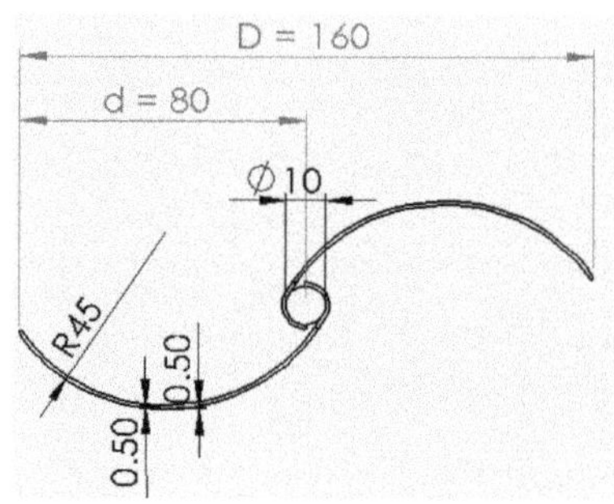

(c)

Figure 15. The Savonius turbine's three views (dimensions are in mm): (**a**) elevation, (**b**) side view, and (**c**) plan.

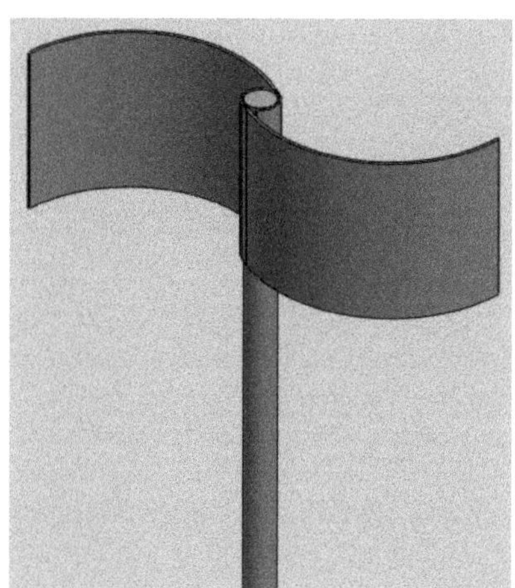

Figure 16. Savonius turbine.

5. Results

This considered case was a still rotor case. The tip speed ratio—tip speed divided by wind speed—was equal to zero because the rotor was fixed for each simulated angle.

Static torque—starting torque—is the torque required for starting turbine rotation. Figure 17 shows the relation between static torque and the rotation angle for the IceWind and Savonius turbines at an air velocity of 15.8 m/s. The static torque of the IceWind turbine was found to have two peaks at θ = 60° and 240°. The Savonius turbine also showed two peaks but not at the same angles. The IceWind and Savonius turbines reached maximum values of 0.055 and 0.052 N·m, respectively. These slight differences may be because the flow field is two-dimensional near the Savonius rotor, whereas near IceWind rotor, it is three-dimensional. This fact will be proved later in the present study. It was found that the torque performance is improved by the IceWind rotor shape.

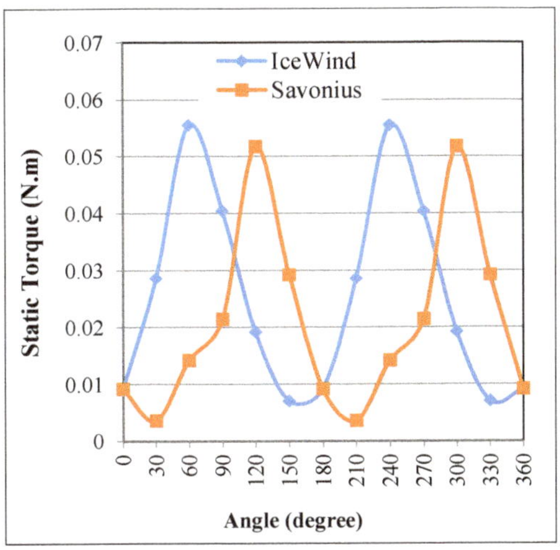

Figure 17. Relation between static torque and the rotation angle for the IceWind and Savonius turbines at an air velocity of 15.8 m/s.

Determination of the air flow velocity distribution and streamlines and pressure distribution around the turbine's surface enabled the air flow characteristics, disturbances, and locations of the highest pressure to be found. The air velocity distribution and streamline and pressure distribution results for both turbines at a wind velocity of 15.8 m/s are shown in Figures 18–20. The rotors of the two turbines showed similar flow patterns. The flow structure around the Savonius rotor was called "Coanda-like flow" by [21,22], and it controls flow separation on the convex side. A low pressure region forms on the side of the proceeding blade, contributing to the torque generation of the rotating rotor [23].

Figure 18a,b show the air flow velocity distribution around the still IceWind and Savonius turbines' rotors. These results were obtained at the plane that goes through the bottom of the turbines' blades as both have the same complete shape at this plane. The figures show similar velocity distributions. At θ = 0°, the high fluid flow velocity moves at a tangent to the convex sides, while two circulating low velocity zones in the two concave sides are established. The air flow velocity has a slightly different maximum velocity of 28 m/s for the IceWind turbine compared with 22 m/s for the Savonius turbine. At θ = 90°, the largest dead area is observed in the wake of the returning blade. The figures show similar velocity distributions. High fluid flow velocity touches the ends of both blades. Furthermore, a pair of asymmetric vortices develops behind both turbines. The smallest dead area is observed in the wake of the returning blade at θ = 0°. Moreover, a maximum velocity of 34 m/s is observed for the IceWind turbine at θ = 30°.

Figure 19a,b show air flow velocity streamlines around the still IceWind and Savonius turbines rotors at a velocity of 15.8 m/s. It is obvious from the figures that vortices behind the Savonius rotor are located between two imaginary planes that go through the top and bottom of the turbine blades. However, the rotor can be considered to be two-dimensional along the whole height. According to the top curvature of IceWind turbine, the plane that goes through the top of the turbine blades does not exist anymore. However, the plane that goes through the bottom of the turbine blades still exists. Vortices are located between the plane that goes through the bottom of the turbine blades and another inclined plane that follows the turbine top curvature. This provides IceWind turbine vortices with three-dimensionality. Vortices behind the IceWind turbine rotor appear to be larger than those behind the Savonius turbine.

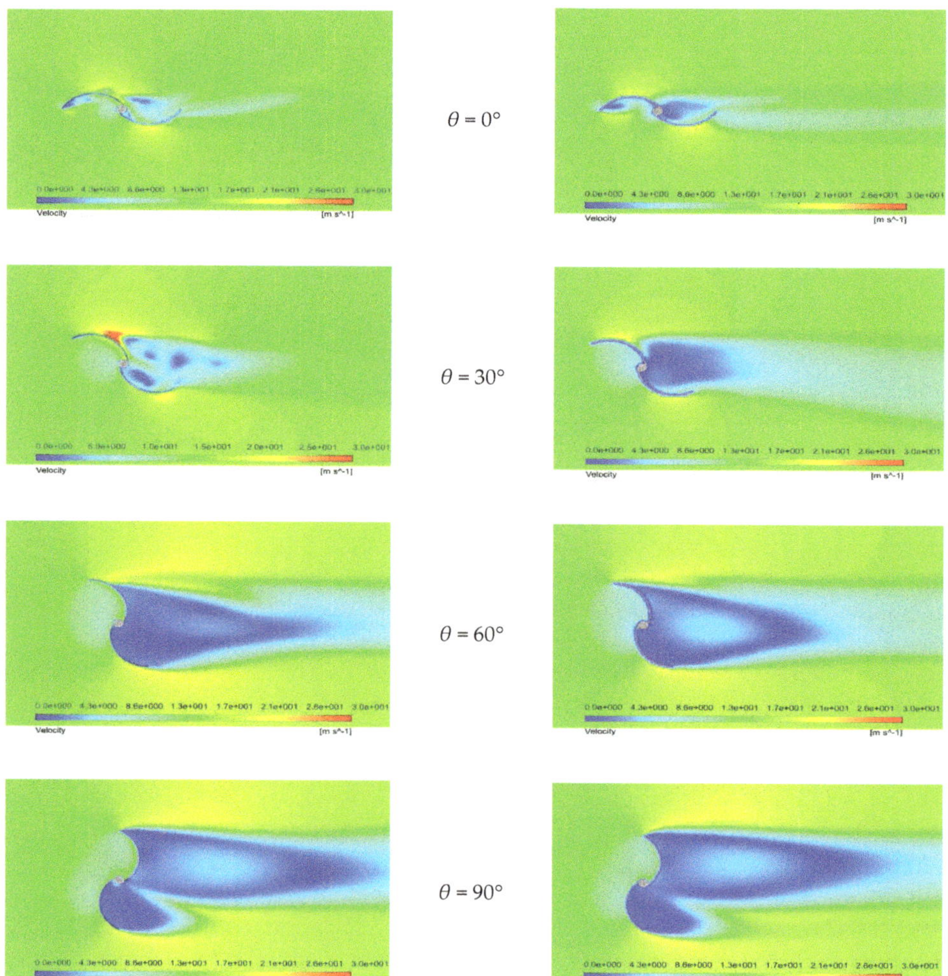

Figure 18. *Cont.*

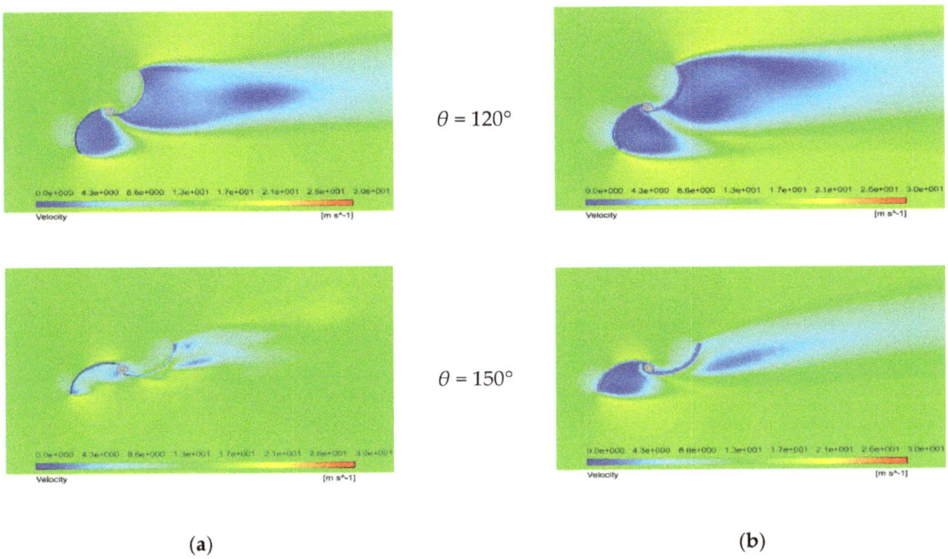

Figure 18. Air flow velocity distributions at an air velocity of 15.8 m/s around (**a**) the IceWind turbine and (**b**) the Savonius wind turbine.

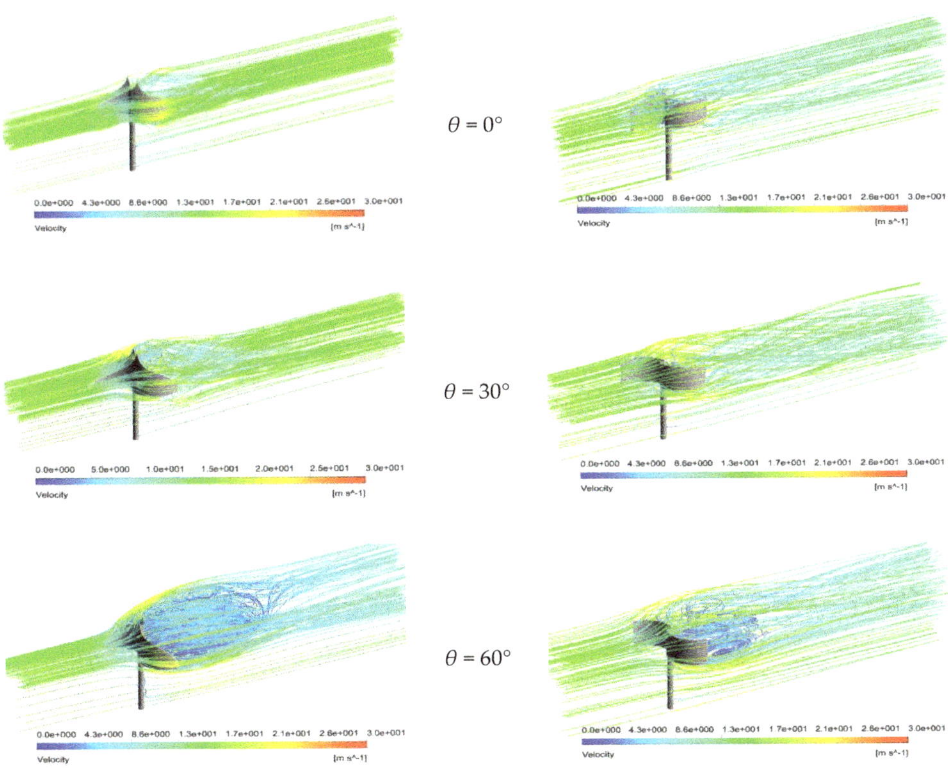

Figure 19. *Cont.*

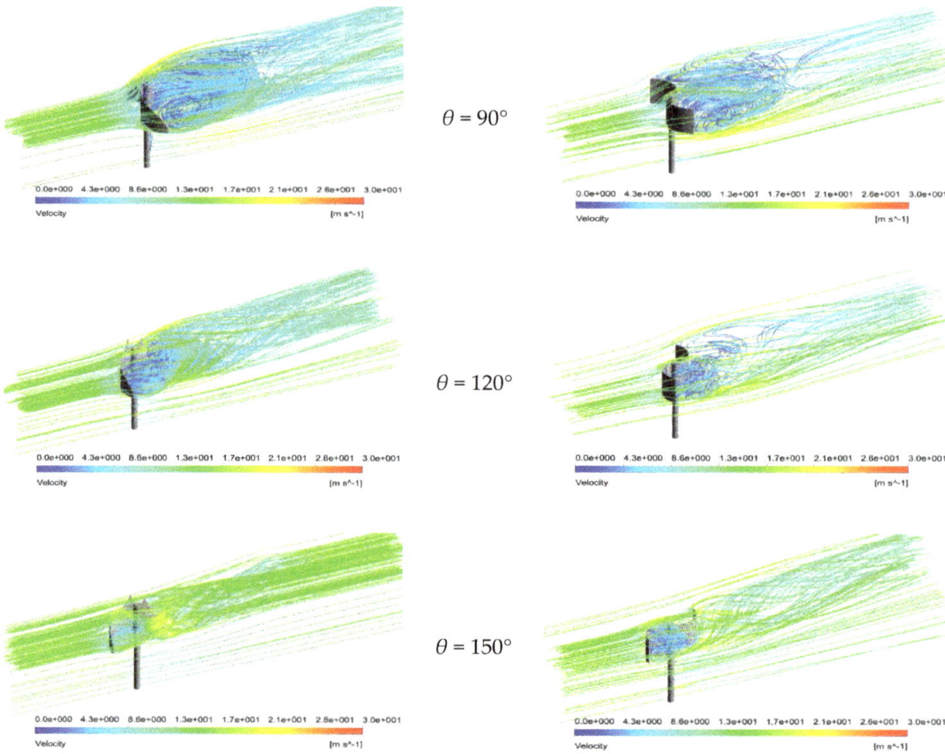

(a) (b)

Figure 19. Air flow velocity streamlines at an air velocity of 15.8 m/s around (**a**) the IceWind turbine and (**b**) the Savonius wind turbine.

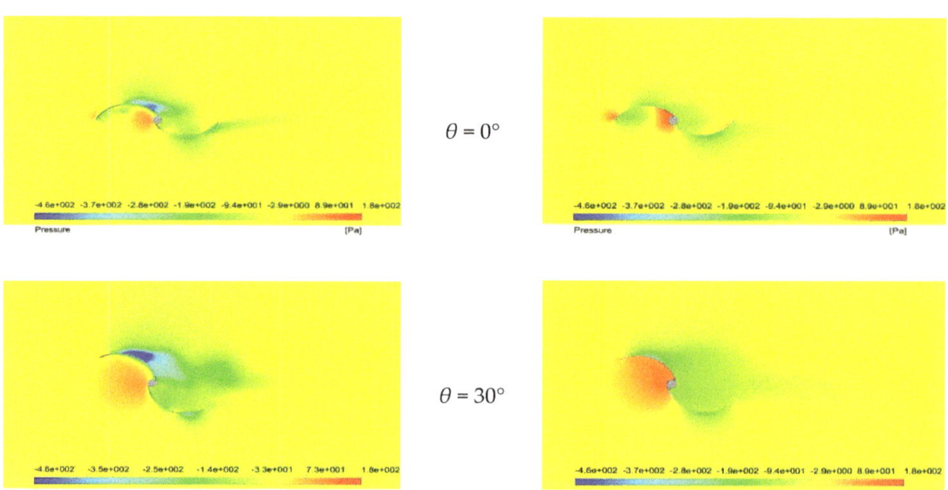

Figure 20. *Cont.*

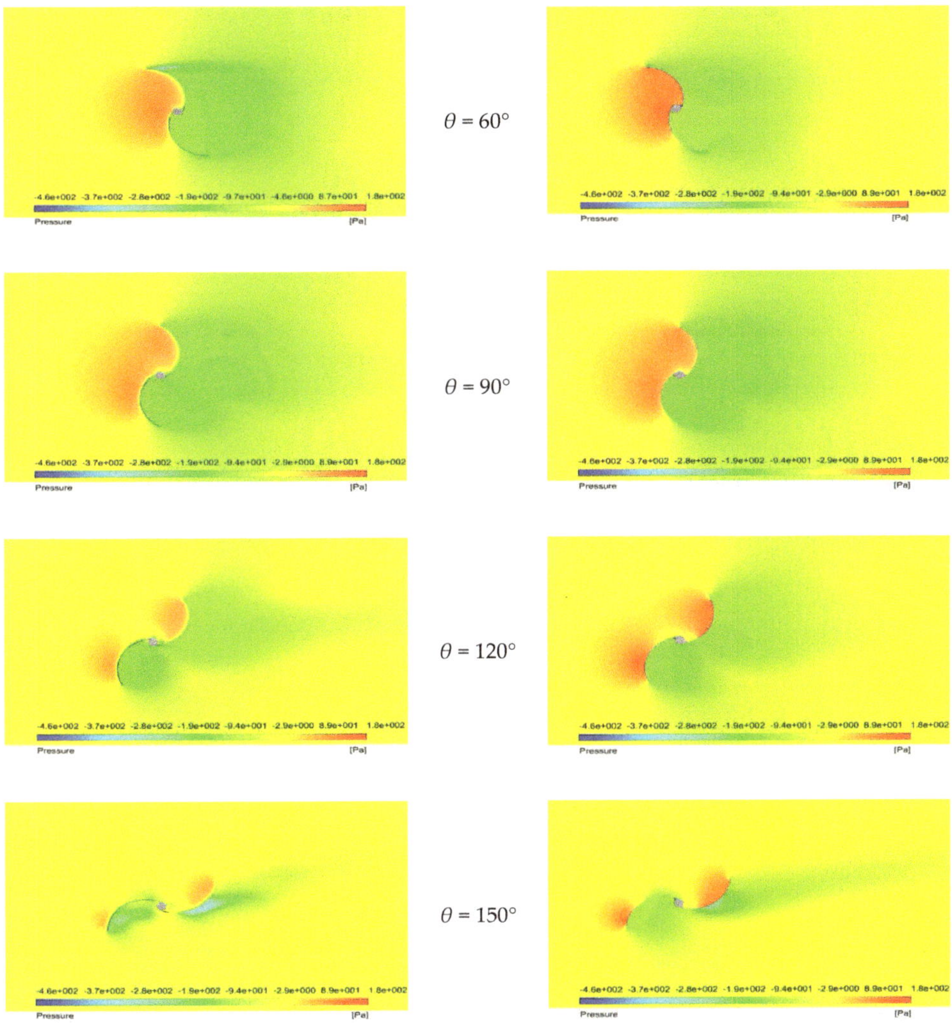

Figure 20. Pressure distributions at an air velocity of 15.8 m/s around (**a**) the IceWind turbine and (**b**) the Savonius wind turbine.

Figure 20a,b show air flow pressure distributions around the still IceWind and Savonius turbines' rotors. These results were obtained at the plane that goes through the bottom of the turbines' blades as both have the same complete shape at this plane. The concave side of the proceeding blade has a positive pressure, while the convex side of the same blade has a negative pressure. In contrast, the oncave side of the returning blade has a negative pressure, while the convex side of the same blade has a positive pressure. In other words, positive pressure appears on the turbine side facing the air, and the opposite sides have a negative pressure. At $\theta = 0°$, both figures show similar pressure distributions. The dead area is small, and the vortex producing a negative pressure almost disappears, whereas the largest dead area is observed in the wake of the returning blade at $\theta = 90°$. Separation occurs due to an adverse pressure gradient in the downstream direction. Both figures show similar pressure distributions. Moreover, the maximum negative pressure appears at $\theta = 30°$ for both turbines.

Three-dimensional numerical modeling was successfully used in the current case to visualize the flow around both turbines. This visualization comparison with the Savonius turbine showed noticeably similar performances for the two turbines.

6. Comparison with Previous Work

It would be better to use the coefficient of static torque than static torque for comparison, but in the present paper, the same outcome would have been achieved because the present study and [3] used the same dimensions and conditions.

Figure 21 shows the relation between static torque and the rotation angle at an air velocity of 15.8 m/s as determined in an experimental study [3] and in the present numerical study on the IceWind turbine. The same trend was observed. The static torque was found to have two peaks at $\theta = 60°$ and 240° in both studies. Moreover, the present numerical results gave slightly lower values than the experimental results. The IceWind turbine reached maximum values of 0.055 and 0.064 Nm for the numerical and experimental studies [3], respectively. This deviation may be due to the experimental conditions and numerical assumptions made.

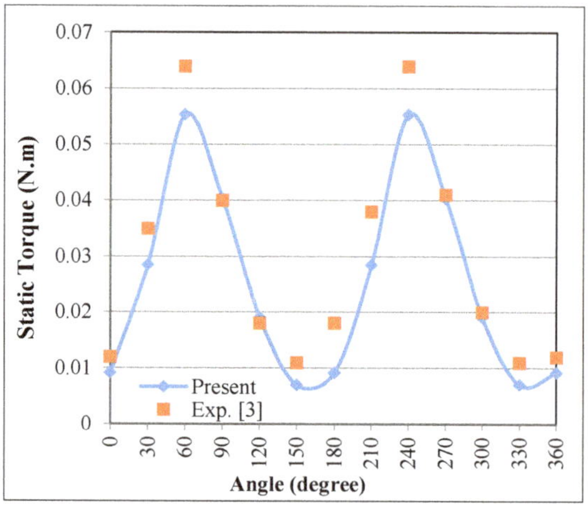

Figure 21. Relation between static torque and the rotation angle for the IceWind turbine at an air velocity of 15.8 m/s in the numerical and experimental [3] studies.

7. Conclusions

In the current work, three-dimensional simulations on a new Vertical Axis Wind Turbine type called the IceWind turbine were conducted. Ansys FLUENT was used to determine the static torque, velocity distributions and streamlines, and pressure distributions at a wind velocity of 15.8 m/s. From the numerical results, the following can be concluded:

- The comparison between the IceWind and Savonius turbines showed similar flow patterns. However, the IceWind turbine was found to be slightly better than the Savonius wind turbine with the same swept area. Although the IceWind turbine is not simple to manufacture, its shape has a better look and performance.
- The air flow velocity distribution of the IceWind turbine led to the maximal velocity and a larger wake area.

- The air flow velocity streamlines demonstrated that vortices behind the IceWind rotor are three-dimensional.
- The air flow pressure distributions showed that positive pressure appears on the turbine side facing the air. On the opposite side, the pressure is negative.
- The numerical method validated previous experimental works, and reasonable agreement was achieved.

Author Contributions: Conceptualization, R.A.; methodology, H.M. and R.A.; software, H.M. and R.A.; validation, H.M. and R.A.; formal analysis, H.M. and R.A.; investigation, H.M. and R.A.; resources, H.M. and R.A.; Writing—Original draft preparation, H.M. and R.A.; Writing—Review and editing, H.M. and R.A.; visualization, H.M. and R.A. All authors have read and agreed to the published version of the manuscript.

Funding: This research received no external funding.

Conflicts of Interest: The authors declare no conflict of interest.

References

1. IceWind Turbine Site. Available online: http://icewind.is/en/ (accessed on 27 November 2019).
2. Aymane, E. Savonius Vertical Wind Turbine: Design, Simulation, and Physical Testing. Master's Thesis, School of Science and Engineering, Alakhawayn University, Ifrane, Morocco, 2017.
3. Afify, R.S. Experimental Studies of an IceWind Turbine. *Int. J. Appl. Eng. Res.* **2019**, *14*, 3633–3645.
4. Sarma, N.K.; Biswas, A.; Misra, R.D. Experimental and computational evaluation of Savonius hydrokinetic turbine for low velocity condition with comparison to Savonius wind turbine at the same input power. *Energy Convers. Manag.* **2014**, *83*, 88–98. [CrossRef]
5. Nasef, M.H.; El-Askary, W.A.; AbdEL-hamid, A.A.; Gad, H.E. Evaluation of Savonius rotor performance: Static and dynamic studies. *J. Wind Eng. Ind. Aerodyn.* **2013**, *123*, 1–11. [CrossRef]
6. Kacprzak, K.; Liskiewicz, G.; Sobczak, K. Numerical investigation of conventional and modified Savonius wind turbines. *Renew. Energy* **2013**, *60*, 578–585. [CrossRef]
7. Dobrev, I.; Massouh, F. CFD and PIV investigation of unsteady flow through Savonius wind turbine. *Energy Procedia* **2011**, *6*, 711–720. [CrossRef]
8. McTavish, S.; Feszty, D.; Sankar, T. Steady and rotating computational fluid dynamics simulations of a novel vertical axis wind turbine for small-scale power generation. *Renew. Energy* **2012**, *41*, 171–179. [CrossRef]
9. Fluent, A. *12.0 Theory Guide*; Ansys Inc.: Canonsburg, PA, USA, 2009.
10. Marsh, P.; Ranmuthugala, D.; Penesis, I.; Thomas, G. Three-dimensional numerical simulations of straight-bladed vertical axis tidal turbines investigating power output, torque ripple and mounting forces. *Renew. Energy* **2015**, *83*, 67–77. [CrossRef]
11. McNaughton, J.; Billard, F.; Revell, A. Turbulence modelling of low Reynolds number flow effects around a vertical axis turbine at a range of tip-speed ratios. *J. Fluids Struct.* **2014**, *47*, 124–138. [CrossRef]
12. Wilcox, D.C. *Turbulence Modeling for CFD*, 2nd ed.; DCW Industries: La Canada Flintridge, CA, USA, 1998; ISBN 0-9636051-5-1.
13. Blazek, J. *Computational Fluid Dynamics: Principles and Applications*, 1st ed.; Elsevier Science Ltd.: Oxford, UK, 2001; ISBN 0080430090.
14. Menter, F.R. Two-equation eddy-viscosity turbulence models for engineering applications. *AIAA J.* **1994**, *32*, 1598–1605. [CrossRef]
15. Chong, W.; Fazlizan, A.; Poh, S.; Pan, K.; Hew, W.; Hsiao, F. The design, simulation and testing of an urban vertical axis wind turbine with the Omni-direction-guide-vane. *Appl. Energy* **2013**, *112*, 601–609. [CrossRef]
16. Nobile, R.; Vahdati, M.; Barlow, J.F.; Mewburn-Crook, A. Unsteady flow simulation of a vertical axis augmented wind turbine: A two-dimensional study. *J. Wind Eng. Ind. Aerodyn.* **2014**, *125*, 168–179. [CrossRef]
17. Lim, Y.; Chong, W.; Hsiao, F. Performance investigation and optimization of a vertical axis wind turbine with the Omni-direction-guide-vane. *Procedia Eng.* **2013**, *67*, 59–69. [CrossRef]
18. Danao, L.A.; Edwards, J.; Eboibi, O.; Howell, R. A numerical investigation into the influence of unsteady wind on the performance and aerodynamics of a vertical axis wind turbine. *Appl. Energy* **2014**, *116*, 111–124. [CrossRef]

19. Mohamed, M.H.; Ali, A.M.; Hafiz, A.A. CFD analysis for H-rotor Darrieus turbine as a low speed wind energy converter. *Eng. Sci. Technol.* **2015**, *18*, 1–13. [CrossRef]
20. Almohammadi, K.M.; Ingham, D.B.; Ma, L.; Pourkashan, M. Computational fluid dynamics (CFD) mesh independency techniques for a straight blade vertical axis wind turbine. *Energy* **2013**, *58*, 483–493. [CrossRef]
21. Fujisawa, N.; Gotoh, F. Visualization study of the flow in and around a Savonius rotor. *Exp. Fluids* **1992**, *12*, 407–412. [CrossRef]
22. Fujisawa, N. On the torque mechanism of Savonius rotors. *J. Wind Eng. Ind. Aerodyn.* **1992**, *40*, 277–292. [CrossRef]
23. Fujisawa, N.; Gotoh, F. Pressure measurements and flow visualization study of a Savonius rotor. *J. Wind Eng. Ind. Aerodyn.* **1992**, *39*, 51–60. [CrossRef]

Publisher's Note: MDPI stays neutral with regard to jurisdictional claims in published maps and institutional affiliations.

energies

MDPI

Article

Numerical Investigations of the Savonius Turbine with Deformable Blades

Krzysztof Sobczak *, Damian Obidowski, Piotr Reorowicz and Emil Marchewka

Institute of Turbomachinery, Lodz University of Technology, 90-924 Lodz, Poland;
damian.obidowski@p.lodz.pl (D.O.); piotr.reorowicz@p.lodz.pl (P.R.); emil.marchewka@dokt.p.lodz.pl (E.M.)
* Correspondence: krzysztof.sobczak@p.lodz.pl; Tel.: +48-42-631-2362

Received: 9 June 2020; Accepted: 16 July 2020; Published: 19 July 2020

Abstract: Savonius wind turbines are characterized by various advantages such as simple design, independence of wind direction, and low noise emission, but they suffer from low efficiency. Numerous investigations were carried out to face this problem. In the present paper, a new idea of the Savonius turbine with a variable geometry of blades is proposed. Its blades, made of elastic material, were continuously deformed during the rotor revolution to increase a positive torque of the advancing blade and to decrease a negative torque of the returning blade. In order to assess the turbine aerodynamic performance, a two-dimensional numerical model was developed. The fluid-structure interaction (FSI) method was applied where blade deformations were defined by computational solid mechanics (CSM) simulations, whereas computational fluid dynamics (CFD) simulations allowed for transient flow prediction. The influence of the deformation magnitude and the position of maximally deformed blades with respect to the incoming wind direction were studied. The aerodynamic performance increased with an increase in the deformation magnitude. The power coefficient exceeded $Cp = 0.30$ for the eccentricity magnitude of 10% and reached 0.39 for the highest magnitude under study. It corresponded to 90% improvement in comparison to $Cp = 0.21$ in the case of the fixed-shape Savonius turbine.

Keywords: vertical axis wind turbine (VAWT); Savonius turbine; deformable blades; power coefficient; blade load; computational fluid dynamics (CFD); fluid-structure interaction (FSI)

1. Introduction

Vertical axis wind turbines (VAWTs) are typically characterized by lower wind energy-conversion efficiency than commonly used horizontal axis wind turbines (HAWTs). However, they are often favored in micro power generation due to their simple design, a possibility to locate a generator near the ground, and to accept the wind blowing from different directions [1,2]. They are quieter and safer than small-scale HAWTs and, thus, suitable for applications in urbanized areas [3–5].

A vertical axis wind turbine, referred to as the Savonius turbine, was invented by S.J. Savonius [6]. In the top view, it resembles the letter "S" with two typically semi-cylindrical blades, often slightly overlapping. Its primary advantage lies in simple and, thus, cheap and robust design [7,8]. Similar to other VAWTs, Savonius turbines are independent of the wind direction. However, contrary to Darrieus wind turbines, they are characterized by a high starting torque for selected rotor positions. They are classified as drag-driven turbines and operate at low rotational speeds, with tip speed ratios not exceeding 2, which makes them safer than HAWTs at strong winds [9]. They perform well at low wind speeds most often encountered close to the ground and they are characterized by a low level of noise emission. Thus, Savonius turbines are suitable for application in urbanized areas [3].

Unfortunately, the primary disadvantage of Savonius turbines is their low efficiency. Typically reported values of the power coefficient for designs with semi-cylindrical blades fall within the range $Cp = 0.15$–0.20 [7,10,11]. Therefore, this kind of turbine in its basic configuration is not usually a reasonable alternative when compared to other types of wind turbine. Nonetheless, owing to their advantages, Savonius turbines were subject to numerous investigations aimed at increasing their efficiency. Many works were focused on the search for optimal dimensions of geometrical parameters of rotors. The influence of the number of blades, overlap ratio, aspect ratio, end plates and other factors was studied both with experimental and numerical methods. Results of those investigations were presented in numerous papers, which further were summarized in thorough reviews [7,12].

Many studies were focused on modifications of the blade shape or an application of additional elements to direct the air flow towards blades. The replacement of conventional semi-cylindrical blades by more sophisticated shapes allowed one to increase the Savonius turbine performance significantly, with maximal values of the power coefficient up to $Cp = 0.25$–0.30. A substantial part of the research concentrates on two-dimensional (2D) thin blade configurations, i.e.,: Bach, Benesh, elliptical and spline [11,13,14]. Airfoil shape blades were studied in [15,16]. Optimization methods were also applied in order to search for an optimal blade shape in [16–19], taking advantage of 2D simplifications in numerical simulations. Three-dimensional (3D) blade arrangements with twisted or helical blades were tested as well. In this case, it was possible to reduce static and dynamic torque variations for different angular positions of the rotor with respect to the incoming wind, however, no significant improvement was reported as far as the turbine performance is considered [12,14].

Different augmentation systems can be used to change the wind flow path around and in the Savonius rotor. Its power output was increased by 20% up to 50% if the turbine rotor was equipped with flat plate deflectors [18,20,21], v-shaped deflectors [22] or a combination of flat and circular deflectors [23], shielding the returning blade and reducing its negative moment. A similar effect was achieved if the wind was directed towards the advancing blade with a curtain-deflector system [24], self-adjusting conveyor-deflector curtains [15], a system of adjustable shielding plates for twin rotors [9] or even a rectangular guide-box tunnel surrounding the rotor [25]. A comprehensive summary of different augmentation systems with the power coefficient exceeding considerably $Cp = 0.3$ can be found in [11,26]. However, a disadvantage of such approaches consists in larger dimensions of the turbine, an increase in the complexity of its geometry and dependence on the wind direction.

An idea of the Savonius turbine with a variable geometry of blades is proposed in order to enlarge the projected area of the advancing blade (increase the positive moment) and, at the same time, to diminish the area of the returning blade (decrease the negative moment). Elaborate two-dimensional (2D) computational fluid dynamics (CFD) simulations were performed to assess the output power gain for different arrangements of blade deformations.

The deformations of blades in this case were determined with a structural solver and then the geometry was transferred to a fluid solver for the aerodynamic analysis. This approach is referred to as the fluid-structure interaction (FSI). The FSI is a very wide concept of solvers coupling in order to obtain high-fidelity numerical solutions. Solvers can be one-way coupled once the data, i.e., loads from the fluid acting on the wall, are transferred to the structural solver where stress and strains are determined [27,28]. Another example of one-way coupling takes place where the deformation of the structure influences the flow structure and loads determined in the fluid solver. The most advanced method, called the two-way FSI, requires co-simulation between computational fluid dynamics and structural mechanics. Both CFD and structural solvers are coupled and synchronized to attain converged solutions. The two-way FSI is applied to highly dynamical systems as in the case of the aeroelastic response analysis [29] or whenever the structure is flabby [30]. The FSI strategy is typically used in horizontal axis wind turbines (HAWTs). It is also used in the analysis of vertical axis wind turbines (VAWTs), e.g., in the case of an H-rotor [31,32], but no reference reporting an analysis of the Savonius-based turbine is known to the authors. It is due to much lower blade loads than in the case of lift-driven HAWTs or VAWTs.

Idea of the Savonius Rotor with Deformable Blades

An idea of the novel turbine with a variable geometry of blades is shown in Figure 1 for subsequent phases of the rotor revolution [33]. Shapes of the blades made of a flexible material change constantly during its rotation. This is achieved by guiding the outer edges of the blades (tips marked as dots in the color of the blades) along the guide ring (red) placed eccentrically with respect to the rotor shaft. Rods attached to the outer edges of the blades can move linearly with respect to sliders, which are fixed to the shaft. The rode–slider mechanisms, marked by a red dashed line, are applied at the top and bottom of the turbine and they transfer the torque generated by the blades to the shaft. The inner edges of the blades are also attached to the rotor shaft.

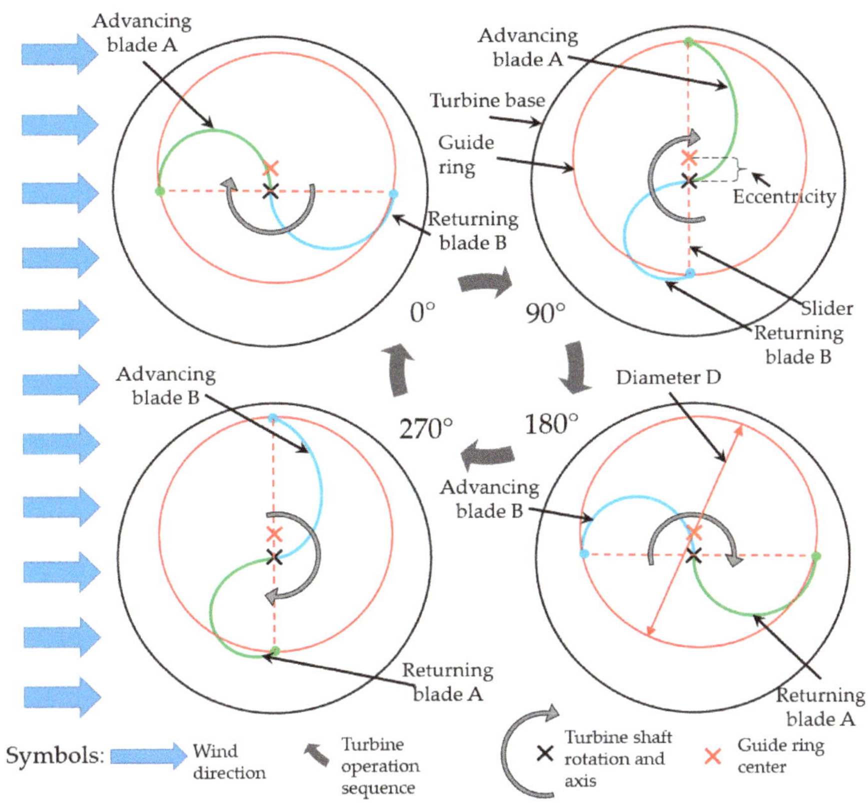

Figure 1. Principle of the turbine operation.

The phases 90° and 270° presented in Figure 1 illustrate the maximal deformation of the blades. If the extended, advancing blade is located in such a way that its concave side is exposed to the wind and, simultaneously, the wind blows at the convex side of the contracted, returning blade, one can expect that the turbine will be driven with the wind energy more efficiently than in the case of both blades having the same, fixed shape. This implies that the turbine needs to be properly located with respect to the incoming wind. Therefore, the guide ring needs to change its position by rotation around the axis of the base of the turbine, which is coaxial with the rotor shaft axis, marked as a black X in Figure 1. The guide ring mechanism has to be equipped with an aerodynamic or mechanical system in order to adjust its position with respect to the wind direction. The turbine generator is fixed to the frame and a gear or a transmission has to be applied to transfer the mechanical energy from the turbine blades.

The additional mechanisms make the design of the proposed turbine more complex than the original Savonius. However, they consume a part of the energy generated by the turbine. Thus, all elements need to be carefully designed to be resistant and efficient. The present study focuses on the aerodynamic performance of a novel, efficient design but neither mechanical losses due to friction nor energy losses due to blade deformations are considered. Thus, one must bear in mind that the overall performance of the turbine will depend on the mechanical design and that a portion of energy will be consumed by the system itself.

2. Materials and Methods

The numerical simulation of the Savonius turbine with deformable blades was performed within the FSI method available in Ansys Workbench v19.2. The geometry and position of the rotor blades change continuously and, thus, transient simulations were performed both for a structural analysis with the ANSYS solver and a fluid flow analysis with the ANSYS fluent solver.

Full 3-dimensional (3D) simulations of the turbine were not possible due to enormous requirements of the coupled fluid flow and structural solvers. Therefore, it was decided to solve the problem as the 2D one, actually quasi-2D, as the structural solver demanded a three-dimensional model to be applied. Despite its limitations as 3D effects at the blade ends are disregarded, a 2D approach is a frequently applied simplification, which allows one to learn about performance of the turbine configuration. It is especially useful to observe changes in the performance if different configurations of the blades are compared. However, one can keep in mind that significant differences can be obtained comparing a 2D prediction with the full three-dimensional one, especially when the aspect ratio of Savonius turbines is low [34,35].

The geometry of the blades, the guidance system and the fluid domain were prepared in SolidWorks. In general, the guide ring can be of any arbitrary smooth shape, preferably elliptical. However, in this first study the guide ring of the constant diameter $D = 1$ m was selected. It was also the value of the diameter of the reference turbine rotor with fixed-shape blades, where the eccentricity was equal to zero. The blades of that reference rotor had a semi-circular shape. In the case of the rotor with eccentricity, the arc length of the blades was the same as in the reference rotor, but the distance between outer edges of the deformable blades was variable during the rotor revolution. In order to simplify the numerical model, it was decided to disregard the rotor shaft. Its impact on the flow around the rotor blades is rather limited and it can be assumed to be similar in all configurations. The blade overlap can have a positive effect on the turbine performance. However, in those investigations, it was decided not to overlap the blades.

As one can see in Figure 2, the fluid domain was divided into two regions. The internal domain of the diameter $1.5D$ including turbine blades was surrounded by the external one. The total length of the domain was $60D$ in the flow direction, with the turbine axis located $20D$ from its inlet and in the middle of the domain height, which was set to $40D$. The domain blockage was similar to our previous studies [34] or in [18] and it did not affect comparisons between different turbine configurations.

Due to the complexity of blade deformations, combined with their rotation, a tool outside the fluid flow solver was needed to define the instantaneous rotor geometry. It was decided to use the structural ANSYS solver in order to take advantage of the FSI method implemented in the Ansys Workbench. A one-way system coupling was defined between simulation components, where the deformation of blades obtained in the structural analysis was transferred to the fluid flow solver. Because the pressure variation around the blade for the considered wind speed ($v = 4$ m/s) was less than 100 Pa, the two-way coupling, where the pressure load on blades would be transferred from the flow to the structural analysis, was disregarded. The one-way system coupling method was successfully used and presented in [27].

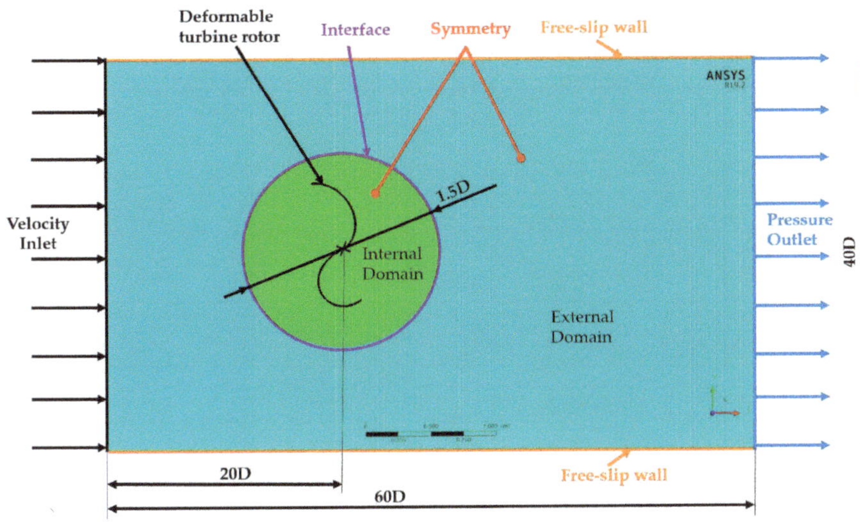

Figure 2. Computational domain scheme (external domain dimensions do not correspond to the scale).

In order to satisfy a good coupling of the structural and flow parts of the problem, the same timestep of transient simulations was applied to both of them. The value of the timestep was selected on the basis of solution stability tests performed for the rotors with the highest magnitude of blade deformation. The typically accepted timestep corresponding to the revolution of the rotor by 1° was selected initially in the Savonius rotor simulations. However, due to solution instabilities during the remeshing procedure, it was successively reduced. In the case of the timestep equal to 0.001 s, the numerical errors resulting from the mesh deformation and the remeshing algorithm were very limited and the computations were successful. The simulations were carried out for the tip speed ratio $TSR = 0.8$, for which Savonius rotors of typical aspect ratios ($AR = H/D = 0.8$–1.5) reach the maximal value of the power coefficient Cp [34,36]. For this TSR, the angular velocity of the turbine was 6.4 rad/s. Thus, the selected timestep of 0.001 s corresponded to a revolution of the rotor by approximately 0.367°, resulting in 982 steps per one revolution, which is sufficiently low as far as the time discretization is concerned [16]. The total time of simulations was 10 s, which corresponds to more than 10 full revolutions, thus eliminating an influence of the initial conditions onto the simulation results.

The tip speed ratio TSR Equation (1) and the power coefficient Cp Equation (2), i.e., the energy extracted by the turbine to the available wind energy, were defined as follows:

$$TSR = \frac{\omega R}{v}, \tag{1}$$

$$Cp = \frac{T\omega}{0.5\rho v^3 A}, \tag{2}$$

where: ω—angular velocity [rad/s], R—turbine radius, v—wind speed [m/s], T—output torque [Nm], ρ—air density [kg/m^3], A—projected area of the rotor (DH) [m^2], D—rotor diameter ($2R$) [m], H—rotor height [m].

2.1. Definition of Structural Simulations

As mentioned above, the structural solver was used to determine the deformation of the blades for each timestep. One structural model, whose scheme is presented in Figure 3, was designed to perform a series of simulations. Different eccentricity magnitudes and various angular positions of the eccentricity line with respect to the incoming wind were attained by changing the position of the guide ring center (red X) with respect to the rotor shaft (black X) as indicated by white arrows. That position was fixed in particular simulations.

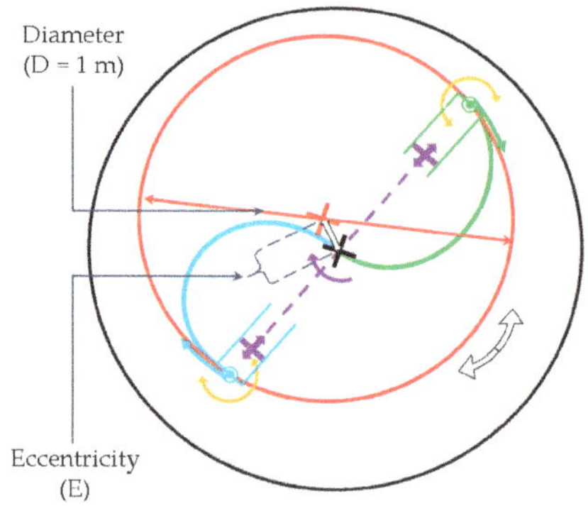

Figure 3. Turbine motion constrains in the structural simulation.

The kinematics of the blades (blue and green arcs) was restricted by several constraints indicated schematically in Figure 3. The turbine shaft rotated around the axis indicated by a black X, where the blades were tangent. Their outer tips marked with green and blue points were constrained by frictionless movement along the guide ring (red circle) as indicated by green and blue arrows. The torque generated by the blades was transferred to the shaft by rode–slider mechanisms. The rode marked with a purple dashed line was fixed to the turbine shaft. The sliders, marked as blue and green cylinders, were connected to the blade tips. The rode–slider mechanisms allowed for frictionless linear movement of the blade outer tips with respect to the axis of rotation, but the mechanisms kept them in the same relative angular position during the turbine rotation. Additionally, the blade outer tips could rotate without friction with respect to the rods as indicated by yellow arrows.

The blade was made of structural steel from the standard ANSYS material library, with the Young modulus and Poisson's ratio equal to 210 GPa and 0.3, respectively. The material was selected to limit deformation to the elastic range to ensure stability of the structural solution for selected eccentricity magnitudes.

2.2. Definition of Flow Simulations

In comparison to typical simulations of Savonius turbines with blades of fixed shapes (where the internal domain rotates), deformations of the blades enforced an application of an advanced meshing approach in that fluid flow analysis. In the case under study, both flow domains are in the stationary frame of reference.

The computational mesh was generated in ANSYS Meshing. Following the mesh dependence test, described in the next subsection, a mesh composed of 1.16 million control volumes was used in the simulations. The external domain did not change during the computational campaign, thus, it was meshed with a single layer of hexahedral elements through the slice. Refinements were applied around the interface with the internal domain and in the wake downstream of the turbine (Figure 4). The internal domain consisted of two regions marked in green and in red. Due to the requirements of the ANSYS Fluent concerning the dynamic mesh options of deformation and remeshing, tetrahedral elements were used in those regions. In order to ensure a high-quality mesh in the region around the blades (marked in red), a highly refined mesh with 12 layers of prismatic elements at the wall was generated. That number of layers was sufficient to satisfy the condition $y+ < 1$ of the first mesh element at the wall for almost all the simulations. The value was exceeded only for very limited regions at the blade tips. Thus, the mesh was sufficient to solve fully the flow in the boundary layer, which is very important as the flow is characterized by numerous boundary layer separations, especially at convex sides of the blades. Refinement of the mesh around the blades (especially at their tips) caused the number of elements through its width to be increased to two and even four elements at the blade tips in this region of the domain. In order to avoid numerical errors, the sizes of elements at the interface between the outer and inner domains did not differ significantly.

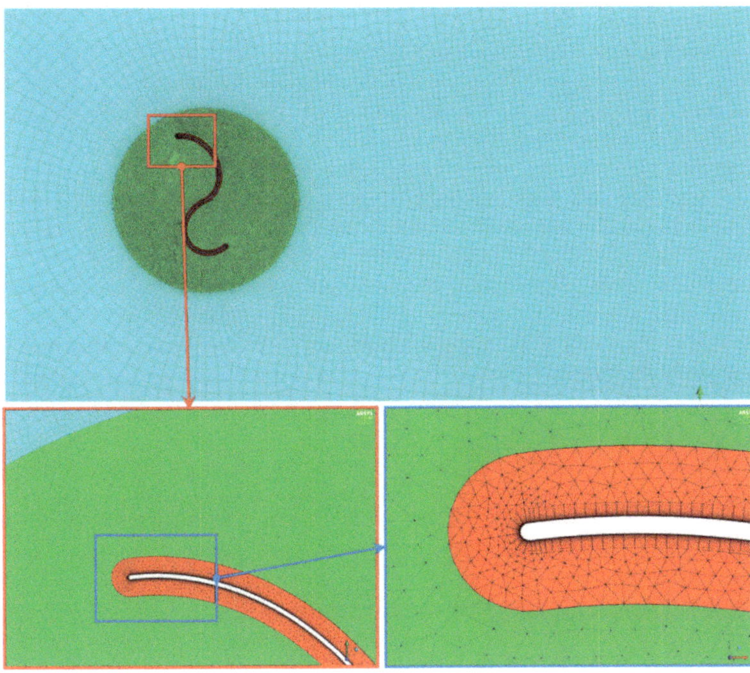

Figure 4. Computational mesh in the external and internal domains with details of its refinement in the blade vicinity presented in the successive magnifications.

The problem of rotation of turbine blades with a constant angular velocity as well as their deformation due to eccentricity was addressed by deforming, remeshing and smoothing algorithms of the dynamic mesh offered by ANSYS Fluent [37]. The instantaneous blade geometry was transferred from the structural analysis every timestep. In the region marked in red, the mesh was following the blade and the elements were deforming as the blade was changing its shape. In the region marked in green, the continuous motion of the blade caused deformation of mesh elements and remeshing was launched wherever they degenerated and the quality measures were not satisfied, namely, the minimal and maximal length scale (1 and 5 mm) with the maximum face skewness of 0.6 and maximum cell skewness of 0.8 were set. The highest skewness of the grid cells was in the region of the deformed and remeshed grid, with the maximal value of 0.84. In the remaining regions, the skewness was below 0.6. The maximal cell aspect ratio at the blades was equal to 68 and was almost constant during the solution as the deformation of the prismatic elements was limited.

The flow simulations were performed in a transient mode with the pressure-based ANSYS Fluent solver. Reynolds-averaged Navier–Stokes (RANS) equations were calculated with the k-ω SST (Shear Stress Transport) turbulence model of Menter [38], which is one of the most frequently used in Savonius turbine simulations [16,39,40], due to its good performance in the adverse pressure gradient and separated flows. Despite the fact that changes in air density are negligible, the fluid was defined as compressible according to the ideal gas law, in order to compensate for numerical instabilities resulting from mesh deformation and remeshing.

A set of the boundary conditions applied in the simulations is indicated in Figure 2. At the inlet, the velocity of 4 m/s was specified with 1% of the turbulence intensity and the turbulent viscosity ratio equal to 1. This velocity corresponds to the Reynolds number based on the turbine diameter equal to 2.5×10^5. The pressure outlet condition (absolute pressure of 101,325 Pa) was defined at the opposite end of the domain. At the upper and lower surfaces of the domain in Figure 2, a free-slip wall condition was imposed. As the task was solved in a quasi-2D way, the symmetry condition was defined on the side surfaces, perpendicular to the axis of turbine rotation. The motion of the blades was governed in the structural simulation and was transferred to the Fluent solver by system coupling. The blades of the turbine were defined as non-slip and smooth walls. The interface between both domains was applied to exchange data. The initial conditions were determined on the basis of the boundary conditions.

The SIMPLE (semi-implicit method for pressure-linked equations) pressure-velocity coupling method was used. The second-order spatial discretization schemes were applied for mass, momentum and energy equations. A first-order implicit transient scheme was also used. Maximally 20 iteration loops were solved in every timestep of transient simulations. Typically, it was enough to reach the residual target of 1e-3, selected as the convergence criterion for all equations solved in Fluent. This criterion was not reached only for a few timesteps, due to numerical problems resulting from mesh deformation or remeshing.

2.3. Validation of the Numerical Procedure

In order to assess the precision of the numerical method with mesh deformation and remeshing used in this investigation, it was decided to compare its results with the results of a typical simulation of the Savonius turbine, applied, e.g., in [13,16,17]. It was possible to do it for the eccentricity magnitude equal to zero, i.e., when rotor blades were not deformed. In this case, the flow was additionally simulated with a fixed mesh, where continuous changes of the rotor position were obtained by rotation of the circular inner domain and exchange of data at the sliding mesh interface between the stationary and rotational domains. The rest of the simulation conditions was preserved.

Taking advantage of much faster simulations for such a task arrangement, the mesh size dependence was also verified by means of Richardson's extrapolation, similarly as in [41]. The same task was solved on three different grids with a mesh refinement ratio of 2. The numbers of control volumes of the grids and the average power coefficient values obtained for them are presented in Table 1. According to the Richardson's extrapolation procedure described in [35], the extrapolated value (*RE*), the apparent order p, the ratio of error R and the fine-grid convergence index (*FGCI*) were determined. The negative value of R indicates an oscillatory convergence. Only a slight variation of the average power coefficient was observed (1.2%), however, the $y+ < 1$ condition was fulfilled for the mesh composed of 1.16 million control volumes shown in Figure 4. The uncertainty due to discretization determined for the average Cp on the basis of the Richardson's extrapolation procedure described in [42] for this mesh was low, i.e., 1.7%. Thus, it was decided to use it in the further numerical investigations.

Table 1. Mesh size dependence study.

Mesh.			Richardson's Extrapolation			
Coarse	Medium	Fine	*RE*	*p*	*R*	*FGCI*
261,000	542,000	1,161,000				
0.1956	0.1944	0.1967	0.1994	2.44	−0.527	1.70%

In the cases of typical 2D simulations of fixed-shape Savonius rotors within the RANS method [16,17], after simulations of a few or more of rotor revolutions, one can reach the limit cycle of the blade load. In this case, a change in the period-averaged value of the power coefficient was lower than 1% of its value after 7 revolutions. The same criterion for this coefficient for particular instants of the rotor revolution (limit cycle) was reached after 10 revolutions. This limit cycle is marked in Figure 5 with a dashed line.

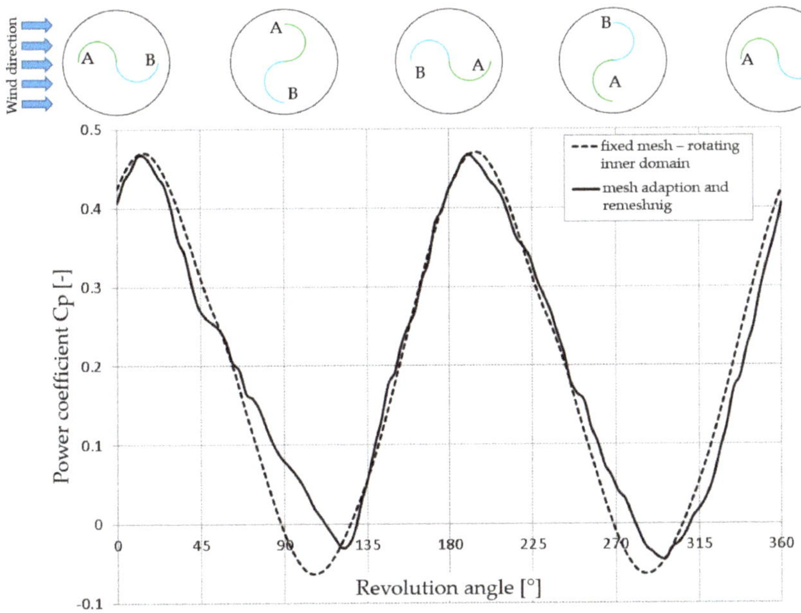

Figure 5. Comparison of the power coefficient *Cp* for one revolution of the non-deformable rotor from simulations with the rotating inner domain and the deformed/remeshed one. Subsequent positions of the turbine illustrated in the schemes above the graph correspond to the turbine revolution angle.

In the case of the simulations with mesh deformation and remeshing applied in the further investigations, such a limit cycle was not obtained, mostly due to numerical instabilities. They resulted from the continuous deformation of mesh control volumes, which successively deteriorated in quality up to the moment when the specific regions were remeshed. Remeshing introduced some further numerical errors due to the data interpolation onto a new mesh. Thus, slight random fluctuations of the blade load were present. Therefore, the data presented in Figure 5 with a solid line were averaged for the last 5 out of 15 rotor revolutions. This number of periods was sufficient to reach no significant changes of the data averaged for the same angular positions during the rotor revolution. The period-averaged values of the power coefficient changes were lower than 1% after 10 revolutions.

As one can see, no significant differences can be observed for two parts of the period for the fixed-mesh simulations (dashed line in Figure 5). In the case of simulations with mesh deformation and remeshing (solid line), differences for two parts of the period are significant. They are observed especially for the revolution angles where the rotor performance is the lowest (75–120° and 255–300°). In these particular ranges, random fluctuations of the blade loads were the highest. The most probable explanation is a very complex flow pattern with numerous vortex structures at the convex side of the advancing blade. The vortex in particular at its tip is very strong with high velocity gradients and, thus, it can be vulnerable to some fluctuations due to the mesh quality deterioration or the remeshing procedure. The poorer flow prediction in those ranges caused that the period-averaged value of the power coefficient was 4.6% higher with respect to the fixed mesh ($Cp = 0.206$ for remeshed, $Cp = 0.197$ for fixed-mesh simulations). This difference is definitely not negligible, but it is an order of magnitude lower than the pressure coefficient increase predicted for rotors with deformable blades. Thus, despite its drawbacks, the method with mesh deformation and remeshing can be considered as sufficiently reliable for the needs of these investigations.

3. Results and Discussion

The influence of the deformation magnitude and the position of maximally deformed blades with respect to the incoming wind direction was studied. A detailed analysis was performed for the turbine with deformable blades in comparison to the non-deformable (fixed-shape) Savonius rotor.

The presentation and discussion of the results was divided into three sections. In the first section, the turbine performance was analyzed for different values of the deformation magnitude in a full range of the position of maximal deformation with respect to the incoming wind. The favorable range of the position was identified. In the second section, changes in the power coefficient during the rotor revolution were presented for the rotor with deformable blades in the optimal position and compared to the non-deformable Savonius rotor. The parts of the revolution cycle where the power output of the deformable rotor was significantly increased were identified and contributions of the individual blades were revealed. In the third section, a detailed analysis of blade loads was performed for three instants of the rotor revolution to display the main differences between the rotors with deformable and non-deformable blades.

3.1. Influence of Eccentricity of Deformed Blades on the Turbine Performance

Numerical simulations of the air flow in the Savonius rotor with deformable blades were carried out for different magnitudes of eccentricity and different angular positions of the eccentricity line with respect to the direction of the incoming wind. The eccentricity magnitude E was defined as a distance between the axis of turbine rotation (marked with a black X in Figure 6) and the center of the blade tip trajectory (marked with a red X), whereas the eccentricity line connects these centers. Selected angular positions of the eccentricity line marked by purple arrows are shown in Figure 6. The rotor blades are presented for their highest deformation. Three magnitudes of eccentricity $E = 50$ mm (the ratio of eccentricity to the guide ring diameter $E/D = 5\%$), 100 mm (10%) and 150 mm (15%) were investigated for the whole 360° range of angular positions of the eccentricity line with respect to the direction of the incoming wind.

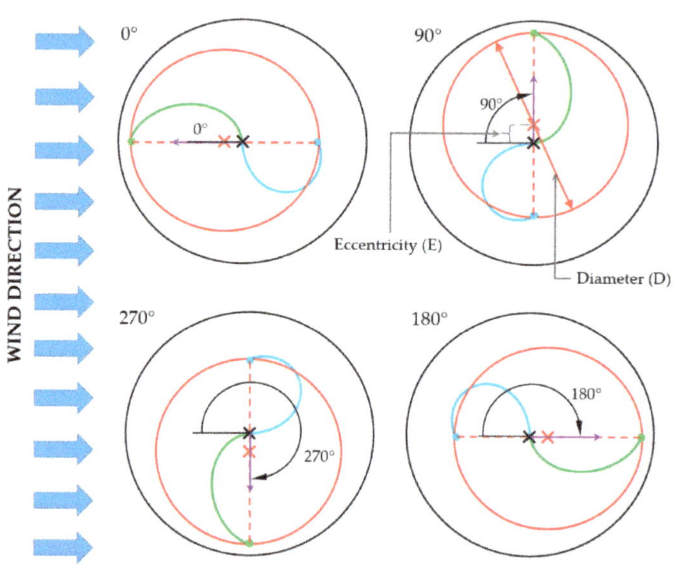

Figure 6. Angular positions of the eccentricity line.

The polar plot of the power coefficient distribution is shown in Figure 7, where data markers indicate all the simulations performed. The black line in the figure indicates the case with a non-deformable rotor. The aim of the investigations was to find the optimal eccentricity position. Therefore, the initial screening calculations were performed with eccentricity changes every 45°. Then, the additional calculations were conducted for the range of 75–180°. In every case 10 revolutions were simulated and average values of the power coefficient out of the last three revolutions were calculated and shown in the figure. Finally, for the position of 105°, an additional 5 revolutions were simulated to determine more precisely the average values for further analysis. However, no significant difference (less than 2%) of the averaged pressure coefficient was obtained with respect to the previously obtained value.

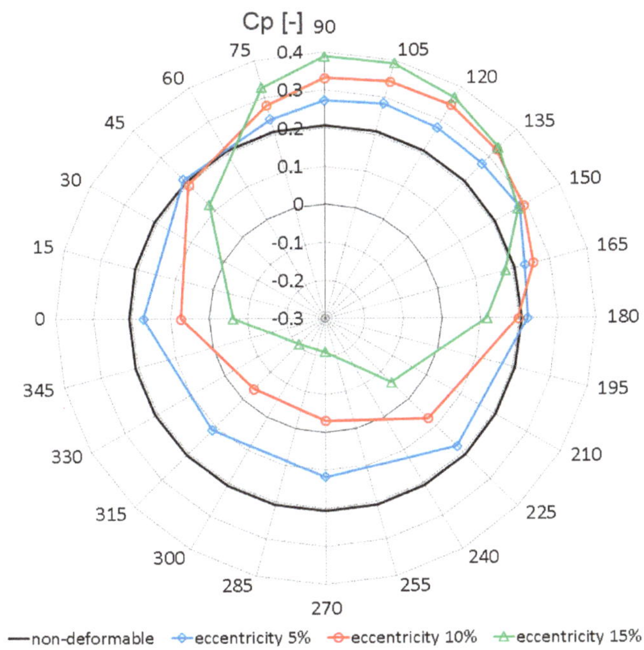

Figure 7. Polar plot of the power coefficient *Cp* for deformable blade rotors with different magnitudes and positions of eccentricity.

As one can see in Figure 7, some gain of the Savonius rotor aerodynamic performance due to blade deformation is obtained for the eccentricity angular position in the range 45–180° (with respect to the direction of the incoming wind—0°). This range diminishes with a growth in the eccentricity magnitude. The optimal position of eccentricity is around 105° for all eccentricity values, however, differences in performance in the range 90–120° are low. In these cases, the advancing blade is expanded and the returning blade is contracted. For the eccentricity position of 105°, the power coefficient values are 0.284, 0.344 and 0.393 for 5%, 10% and 15% of the eccentricity magnitude, respectively. It provides a 37%, 66% and 90% increase with respect to the power coefficient of the non-deformable rotor ($Cp = 0.207$).

A proper angular position of the eccentricity is crucial. It can be easily noticed that for misplaced eccentricity, negative effects are very strong. For eccentricity magnitudes higher than 5%, the power coefficient is negative in a wide range of angular positions, which means that the turbine cannot convert the wind energy. Therefore, the mechanism to position the rotor with respect to the incoming wind is of a key importance in the case of this design.

The results presented in the next figures and their detailed analysis refer to the rotor with deformable blades with the eccentricity of 100 mm (10% of the rotor diameter) in the position of 105°. In this angular position (or in its close vicinity), the investigated turbine reached the maximum of its performance at any eccentricity. The eccentricity value selected for this analysis is high enough to reach significant improvement in the aerodynamic performance of the rotor and to see clearly deformation of the blades in the pressure field plots. On the other hand, deformation is not excessively high as far as the fatigue of the blade and the form of the blade deformation due to mechanical constraints are concerned. The results for this rotor configuration are compared with the results for the rotor with non-deformable blades (eccentricity 0).

3.2. Changes in the Power Coefficient during Rotor Revolution

Changes in the power coefficient during one revolution of the rotors for deformable blades and non-deformable blades are presented in Figure 8. They are sums of contributions of the individual blades shown in Figure 9. The hatched areas between deformable and non-deformable blade lines in both figures indicate positive (+) or negative (−) effects of deformable blade geometries onto the power coefficient readings. Both blades were analyzed separately for better understanding of the occurring phenomena. Blade A is marked blue and Blade B is indicated by red lines in Figure 9. Small changes can be distinguished between characteristics for each blade of the particular rotor type. They resulted from not fully repeatable loads of each blade due to random structures influenced by numerical errors (mesh deformation and remeshing). Still, 5 rotor revolutions selected for data averaging seem to be sufficient to reveal differences between the deformable and non-deformable rotors.

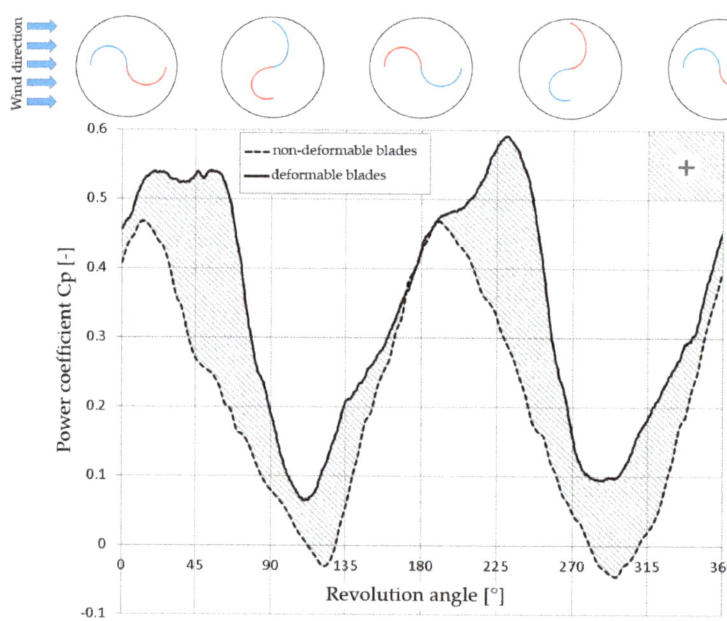

Figure 8. Comparison of the averaged values of the power coefficient Cp for one revolution of the rotors with non-deformable and deformable blades. Subsequent positions of the turbine illustrated in the schemes above the graph correspond to the turbine revolution angle.

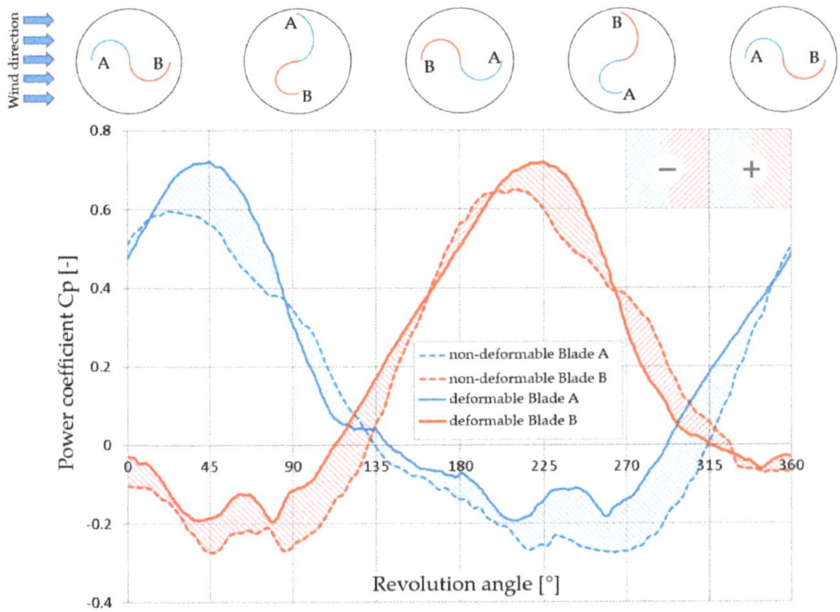

Figure 9. Comparison of the averaged values of the power coefficient *Cp* for each rotor blade for one revolution of the rotors with non-deformable and deformable blades. Subsequent positions of the turbine illustrated in the schemes above the graph correspond to the turbine revolution angle.

An increase in the power coefficient of the deformable blade rotor with respect to the one with fixed shape can be observed in Figure 8 during all their revolutions. The most significant increase is observed for 30–70° and 210–250° parts of the cycle. However, one can see that for the part of the period when *Cp* reaches its minimum (90–135° and 270–330°), a significant increase is also observed. This positive effect results from an increase in the performance of both blades for almost all the revolutions, as one can see in Figure 9. A slight negative effect can be spotted for the advancing blade in the ranges 85–130° and 265–325° and a very minor one for the returning blade at 160–220° and 340–15°.

3.3. Blade Loading and Torque Generation

In order to better understand the contribution of the particular blade loading to the torque generation, torque distributions along the blades (from the axis towards the blade tips) are shown in Figure 10 and pressure fields in Figure 11, correspondingly. The dT/dL derivative per 1 m of the blade span (actually it is $\Delta T/\Delta L$ as it is based on the finite volume simulations) is shown in Figure 10, thus, in order to obtain the torque value, this parameter needs to be integrated along the blade. Three particular instants of the rotor revolution were selected (15°, 55° and 105°), which is sufficient to display the main differences between the rotors with deformable and non-deformable blades. The data were presented for a half of the cycle, because no significant differences can be distinguished for its second part. The data in Figures 10 and 11 are attained for the 10th rotor revolution, thus there is no perfect agreement with the results in Figure 9 obtained from averaging over 5 revolutions. Animations of changes in parameters for one rotor revolution are presented in Supplementary Materials (Video S1 presents the non-deformable rotor, whereas Video S2 shows the deformable one).

In the 15° (also 195°) instant of the revolution cycle, the non-deformable rotor reaches the maximum of its power output (Figure 8). In this case, one can notice in Figure 11 that a high pressure at the concave side of the advancing blade (A) and a low pressure at its convex side (due to flow acceleration) result in the highest positive torque generated by this blade. It is diminished by the retarding contribution of the returning blade (B), mostly due to a pressure difference for its part near the axis of the turbine. In this instant of the revolution cycle, the blades of the deformable rotor have a semi-circular shape like the blades of the non-deformable rotor. No significant increase in the power coefficient is observed at this moment for the rotor in reference to the non-deformable one (Figure 8). The data presented in Figure 9 reveal slight positive contributions both of the advancing (A) and returning (B) blades. The positive contribution of the advancing blade (blue lines) can be distinguished also in Figure 10. An increase of the torque can be seen for a substantial portion of the blade starting from the rotor axis (internal part), whereas the tip (external) part has slightly lower loading. In the case of the returning blade (red lines), positive and negative contributions of the blade deformation can be observed locally for the internal part of the blade, whereas a positive effect is clearly visible for the external part.

The highest positive effect of the blade deformation with respect to the non-deformable turbine is observed for the 55° (also 235°) instant of the revolution cycle (Figure 8). In this position the deformable rotor reaches the highest power output, whereas it is already significantly diminished for the non-deformable one. In Figure 9 it can be seen that both the blades contribute to the power gain. It is due to the blade deformation, i.e., expansion of the advancing blade (A) and contraction of the returning one (B), as shown in Figure 11. The torque of the advancing blade is significantly higher at its external part (Figure 10). It results from higher velocity at the convex side in a very limited way. The main reason of the higher torque is just a higher arm (radius) due to the blade expansion. Despite the fact that the pressure build-up (stagnation zone) at the convex side of the returning blade is higher than for the non-deformable blade, its contraction and higher pressure at the concave side do not reduce further the torque at its central part. On the other hand, much higher fluid acceleration at the convex side of the deformed blade for its external part results in lower pressure and locally changes its torque to the positive one. Thus, the retarding effect of the contracted returning blade is reduced significantly.

A significantly positive effect of the blade deformation can be observed also for the 105° (and 285° as well) instant of the revolution cycle (Figure 8). In this position both the deformable and non-deformable rotors reach the lowest power output. The blade deformation in this case is the highest as presented in Figure 11. In Figure 9 one can see that the deformation of the advancing blade (A) decreases the pressure coefficient, nevertheless, it remains positive. A beneficial effect of the higher arm in the torque definition due to blade expansion can be noticed for the external part of this blade (Figure 10). It is lowered by higher pressure at its convex side due to lower intensity of the vortex structure at the blade tip in comparison to the non-deformable rotor (Figure 11). This positive effect does not compensate for a decrease in the torque in the internal part of the blade. This decrease is mainly due to much lower pressure at the concave side of the blade, which is an effect of the flow acceleration in this region. Deformation of the returning blade (B) improves considerably its performance almost along its whole length, excluding the tip only (Figure 10). It is partially due to a much lower torque arm of the contracted blade. Additionally, the higher curvature of the deformed blade yields higher flow acceleration and a significant reduction in the high pressure stagnation zone at its convex side in comparison to the non-deformable rotor (Figure 11). The blade contraction is also followed by a pressure increase at the convex side of the returning blade. Thus, the pressure difference between the convex and concave sides of the blade is significantly diminished. All these aspects contribute to a reduction in the retarding torque of the deformable blade almost to zero (Figure 9).

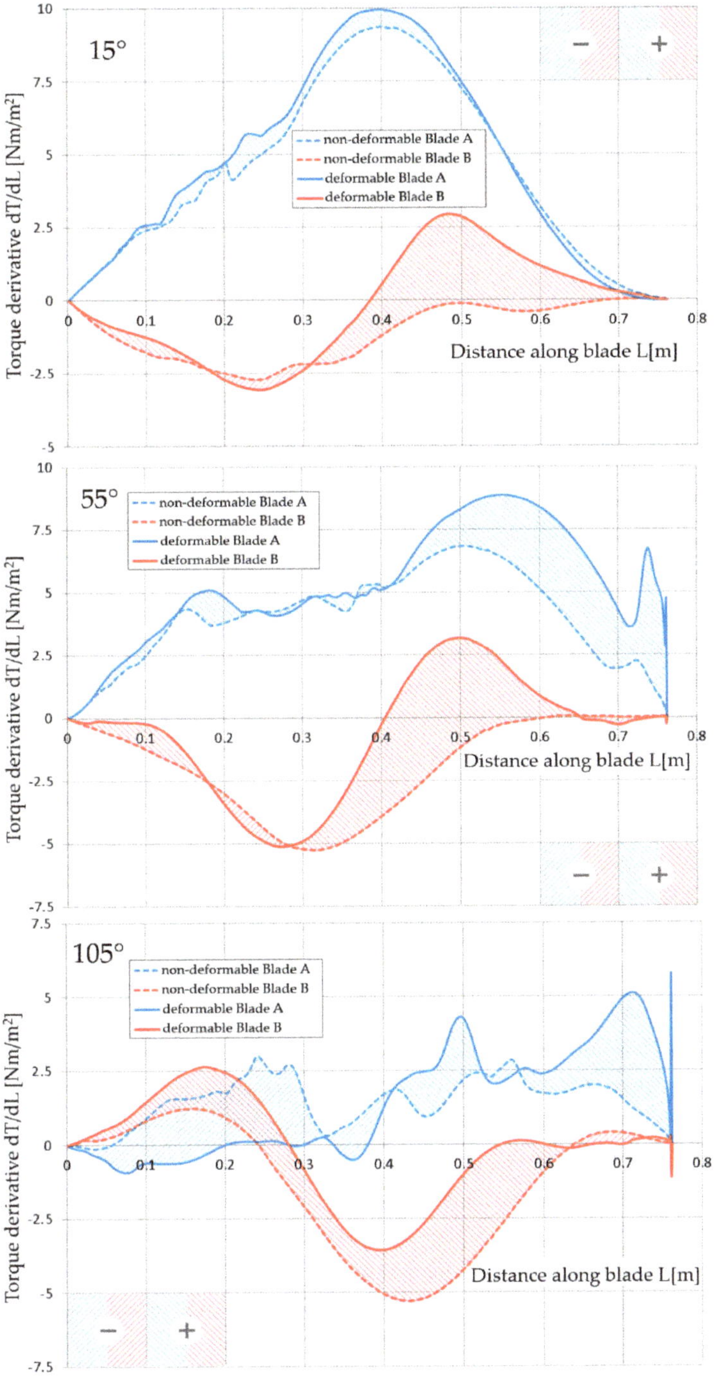

Figure 10. Torque derivative (*dT/dL*) variation along the blade for selected instants (15°, 55° and 105°) of the rotor revolution with non-deformable and deformable blades.

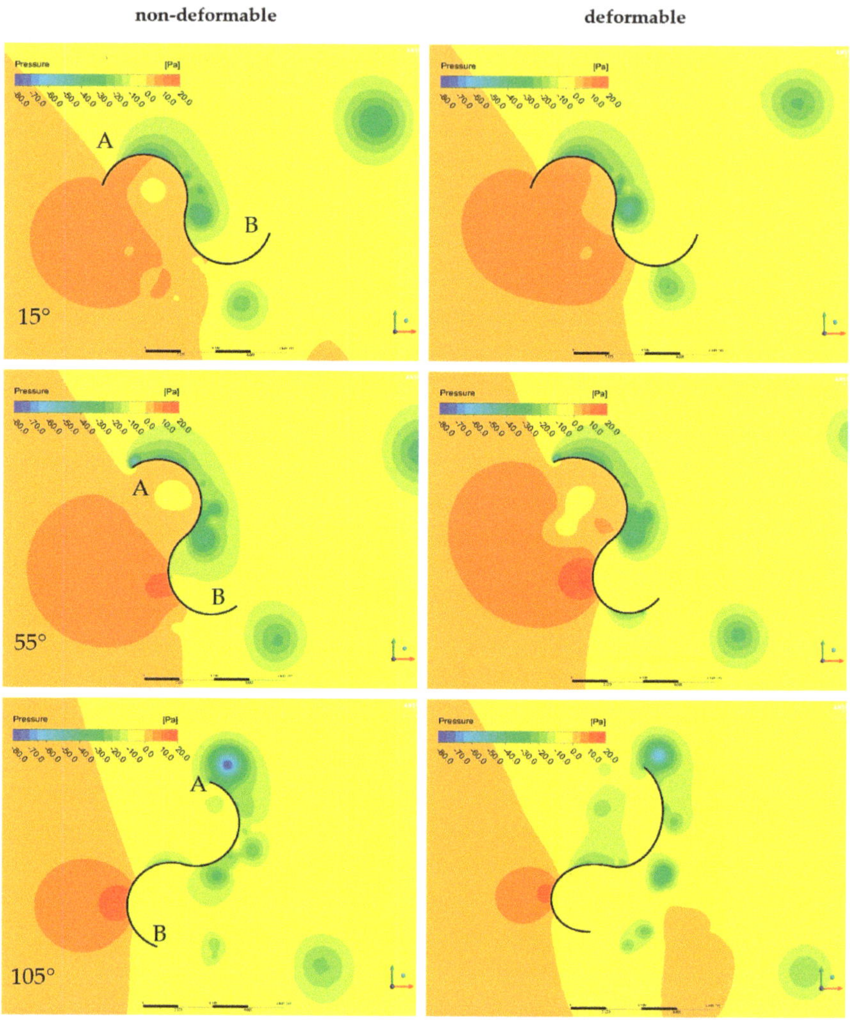

Figure 11. Pressure fields for selected instants (15°, 55° and 105°) of one revolution of the rotors with non-deformable and deformable blades.

4. Summary and Conclusions

An idea of the Savonius turbine with a variable geometry of blades was proposed. Blades made of an elastic material were continuously deformed during the rotor revolution in order to increase a positive torque of the advancing blade, and, at the same time, to decrease a negative moment of the returning blade. The main outcomes of the performed investigations are outlined below:

- An elaborate two-dimensional numerical model was developed to simulate a transient flow in the variable-geometry rotor in order to assess its aerodynamic performance. The shape and position of the rotor blades were subject to continuous changes according to the constraints defined in the structural analysis. The rotational motion and deformations of the blades were transferred to the fluid flow (CFD) analysis, where deformations of grid elements and remeshing options were applied. This method yielded a satisfactory agreement with a typical method of simulations, which consists in rotation of the internal domain surrounding the fixed-shape Savonius rotor.

- An improvement of the aerodynamic turbine performance in comparison to the non-deformable Savonius turbine was obtained. It was achieved when the blades were maximally deformed in the range of angular position 45–180° with respect to the direction of the incoming wind. The maximum of *Cp* was attained at the angle of 105°, i.e., when the blade chords were almost perpendicular to the wind direction.
- An increase in the blade deformation increased the rotor performance. For the eccentricity position of 105°, the power coefficient increased by 37%, 66% and 90% for 5%, 10% and 15% of the eccentricity magnitude, respectively. However, the range of angular positions for which a gain with respect to the non-deformable Savonius turbine occurred decreased.
- A detailed flow analysis for the rotor with eccentricity in the 105° position showed an increase in the power coefficient of the deformable blade rotor with respect to the one having the fixed shape during all their revolutions. The radial expansion of the advancing blade increased the positive torque and the contraction of the returning blade decreased the negative one, with slight changes in the flow patterns.

A significant increase in the aerodynamic performance of the Savonius turbine with continuously deformed blades was confirmed. The power coefficient exceeded $Cp = 0.3$ and it reached almost $Cp = 0.4$ for the highest eccentricity magnitude. Thus, this design is at least comparable to Savonius turbines equipped with augmentation systems presented in the literature [11,26]. However, additional mechanisms applied to deform blades and locate the rotor in the proper position with respect to the incoming wind are required. They will make the design of the turbine more complex than the original Savonius and they will consume a part of the energy generated by the turbine. Thus, all turbine elements need to be carefully designed to be resistant and efficient. Additionally, as the turbine performance depends considerably on the blade deformation, it is necessary to select an easily deformable material with high fatigue resistance. The numerical model of the deformable Savonius rotor needs to be further developed to limit the numerical instabilities during the solution. The mechanical losses due to friction and blade deformations are to be included. Also the problem of time consuming simulations has to be addressed.

Supplementary Materials: The following are available online at http://www.mdpi.com/1996-1073/13/14/3717/s1: Video S1: NonDeformable_Savonius.wmv; Video S2: Deformable_Savonius.wmv.

Author Contributions: Conceptualization, K.S. and D.O.; methodology, K.S., D.O., P.R.; investigation, K.S., D.O., P.R., E.M.; writing—original draft preparation, K.S. and D.O.; All authors have read and agreed to the published version of the manuscript.

Funding: The investigations have been financially supported by the research project POWR.03.02.00-00-I042/16-00 of the National Centre for Research and Development and Innovation Incubator 2.0 project MNISW/2019/157/DIR of the Ministry of Science and Higher Education.

Acknowledgments: We would like to thank Malgorzata Jozwik for her significant linguistic help during the preparation of the manuscript.

Conflicts of Interest: The authors declare no conflict of interest.

References

1. Tummala, A.; Velamati, R.K.; Sinha, D.K.; Indraja, V.; Krishna, V.H. A review on small scale wind turbines. *Renew. Sustain. Energy Rev.* **2016**, *56*, 1351–1371. [CrossRef]
2. Aslam Bhutta, M.M.; Hayat, N.; Farooq, A.U.; Ali, Z.; Jamil, S.R.; Hussain, Z. Vertical axis wind turbine—A review of various configurations and design techniques. *Renew. Sustain. Energy Rev.* **2012**, *16*, 1926–1939. [CrossRef]
3. Kumar, R.; Raahemifar, K.; Fung, A.S. A critical review of vertical axis wind turbines for urban applications. *Renew. Sustain. Energy Rev.* **2018**, *89*, 281–291. [CrossRef]

4. Toja-Silva, F.; Colmenar-Santos, A.; Castro-Gil, M. Urban wind energy exploitation systems: Behaviour under multidirectional flow conditions—Opportunities and challenges. *Renew. Sustain. Energy Rev.* **2013**, *24*, 364–378. [CrossRef]
5. KC, A.; Whale, J.; Urmee, T. Urban wind conditions and small wind turbines in the built environment: A review. *Renew. Energy* **2019**, *131*, 268–283. [CrossRef]
6. Savonius, S.J. The S-rotor and its applications. *Mech. Eng.* **1931**, *53*, 333–338.
7. Akwa, J.V.; Vielmo, H.A.; Petry, A.P. A review on the performance of Savonius wind turbines. *Renew. Sustain. Energy Rev.* **2012**, *16*, 3054–3064. [CrossRef]
8. Roy, S.; Saha, U.K. Review on the numerical investigations into the design and development of Savonius wind rotors. *Renew. Sustain. Energy Rev.* **2013**, *24*, 73–83. [CrossRef]
9. Doerffer, P.; Doerffer, K.; Ochrymiuk, T.; Telega, J. Variable Size Twin-Rotor Wind Turbine. *Energies* **2019**, *12*, 2543. [CrossRef]
10. Abraham, J.P.; Plourde, B.D.; Mowry, G.S.; Minkowycz, W.J.; Sparrow, E.M. Summary of Savonius wind turbine development and future applications for small-scale power generation. *J. Renew. Sustain. Energy* **2012**, *4*, 042703. [CrossRef]
11. Alom, N.; Saha, U.K. Four Decades of Research into the Augmentation Techniques of Savonius Wind Turbine Rotor. *J. Energy Resour. Technol. Trans. ASME* **2018**, *140*, 1–14. [CrossRef]
12. Kumar, A.; Saini, R.P. Performance parameters of Savonius type hydrokinetic turbine—A Review. *Renew. Sustain. Energy Rev.* **2016**, *64*, 289–310. [CrossRef]
13. Kacprzak, K.; Liskiewicz, G.; Sobczak, K. Numerical investigation of conventional and modified Savonius wind turbines. *Renew. Energy* **2013**, *60*, 578–585. [CrossRef]
14. Chen, L.; Chen, J.; Zhang, Z. Review of the Savonius rotor's blade profile and its performance. *J. Renew. Sustain. Energy* **2018**, *10*, 013306. [CrossRef]
15. Tartuferi, M.; D'Alessandro, V.; Montelpare, S.; Ricci, R. Enhancement of savonius wind rotor aerodynamic performance: A computational study of new blade shapes and curtain systems. *Energy* **2015**, *79*, 371–384. [CrossRef]
16. Kerikous, E.; Thévenin, D. Optimal shape of thick blades for a hydraulic Savonius turbine. *Renew. Energy* **2019**, *134*, 629–638. [CrossRef]
17. Tian, W.; Mao, Z.; Zhang, B.; Li, Y. Shape optimization of a Savonius wind rotor with different convex and concave sides. *Renew. Energy* **2018**, *117*, 287–299. [CrossRef]
18. Mohamed, M.H.; Janiga, G.; Pap, E.; Thévenin, D. Optimal blade shape of a modified Savonius turbine using an obstacle shielding the returning blade. *Energy Convers. Manag.* **2011**, *52*, 236–242. [CrossRef]
19. Zhang, B.; Song, B.; Mao, Z.; Tian, W.; Li, B.; Li, B. A Novel Parametric Modeling Method and Optimal Design for Savonius Wind Turbines. *Energies* **2017**, *10*, 301. [CrossRef]
20. Ogawa, T.; Yoshida, H. The effects of a deflecting plate and rotor end plates on performances of Savonius-type wind turbine. *Bull. JSME* **1986**, *29*, 2115–2121. [CrossRef]
21. Golecha, K.; Eldho, T.I.; Prabhu, S.V. Influence of the deflector plate on the performance of modified Savonius water turbine. *Appl. Energy* **2011**, *88*, 3207–3217. [CrossRef]
22. Shaughnessy, B.M.; Probert, S.D. Partially-blocked savonius rotor. *Appl. Energy* **1992**, *43*, 239–249. [CrossRef]
23. Alexander, A.J.; Holownia, B.P. Wind tunnel tests on a savonius rotor. *J. Wind Eng. Ind. Aerodyn.* **1978**, *3*, 343–351. [CrossRef]
24. Altan, B.D.; Atılgan, M. The use of a curtain design to increase the performance level of a Savonius wind rotors. *Renew. Energy* **2010**, *35*, 821–829. [CrossRef]
25. Irabu, K.; Roy, J.N. Characteristics of wind power on Savonius rotor using a guide-box tunnel. *Exp. Therm. Fluid Sci.* **2007**, *32*, 580–586. [CrossRef]
26. Wong, K.H.; Chong, W.T.; Sukiman, N.L.; Poh, S.C.; Shiah, Y.C.; Wang, C.T. Performance enhancements on vertical axis wind turbines using flow augmentation systems: A review. *Renew. Sustain. Energy Rev.* **2017**, *73*, 904–921. [CrossRef]
27. Lipian, M.; Czapski, P.; Obidowski, D. Fluid–Structure Interaction Numerical Analysis of a Small, Urban Wind Turbine Blade. *Energies* **2020**, *13*, 1832. [CrossRef]
28. Karczewski, M.; Sobczak, K.; Lipian, M.; Jozwik, K. Numerical and experimental tools for small wind turbine load analysis. In *Structural Control and Fault Detection of Wind Turbine Systems*; Institution of Engineering and Technology: London, UK, 2018; pp. 45–79, ISBN 9781785613944.

29. Shkara, Y.; Cardaun, M.; Schelenz, R.; Jacobs, G. Aeroelastic response of a multi-megawatt upwind horizontal axis wind turbine (HAWT) based on fluid–structure interaction simulation. *Wind Energy Sci.* **2020**, *5*, 141–154. [CrossRef]
30. Löhner, R.; Haug, E.; Michalski, A.; Muhammad, B.; Drego, A.; Nanjundaiah, R.; Zarfam, R. Recent advances in computational wind engineering and fluid-structure interaction. *J. Wind Eng. Ind. Aerodyn.* **2015**. [CrossRef]
31. Bazilevs, Y.; Korobenko, A.; Deng, X.; Yan, J.; Kinzel, M.; Dabiri, J.O. Fluid–Structure Interaction Modeling of Vertical-Axis Wind Turbines. *J. Appl. Mech.* **2014**, *81*. [CrossRef]
32. MacPhee, D.W.; Beyene, A. Fluid-structure interaction analysis of a morphing vertical axis wind turbine. *J. Fluids Struct.* **2016**. [CrossRef]
33. Obidowski, D.; Sobczak, K.; Jozwik, K.; Reorowicz, P. Vertical Axis Wind Turbine with a Variable Geometry of Blades. European Patent Application 19199085.2, 24 September 2019.
34. Sobczak, K. Numerical investigations of an influence of the aspect ratio on the Savonius rotor performance. *J. Phys. Conf. Ser.* **2018**, *1101*, 012034. [CrossRef]
35. Kacprzak, K.; Sobczak, K. Numerical analysis of the flow around the Bach-type Savonius wind turbine. *J. Phys. Conf. Ser.* **2014**, *530*, 012063. [CrossRef]
36. Kamoji, M.A.; Kedare, S.B.; Prabhu, S.V. Experimental investigations on single stage modified Savonius rotor. *Appl. Energy* **2009**. [CrossRef]
37. ANSYS. *ANSYS Fluent 19.2 Theory Guide*; Ansys Inc.: Canonsburg, PA, USA, 2018.
38. Menter, F.R. Two-equation eddy-viscosity turbulence models for engineering applications. *AIAA J.* **1994**, *32*, 1598–1605. [CrossRef]
39. Ferrari, G.; Federici, D.; Schito, P.; Inzoli, F.; Mereu, R. CFD study of Savonius wind turbine: 3D model validation and parametric analysis. *Renew. Energy* **2017**, *105*, 722–734. [CrossRef]
40. Kacprzak, K.; Sobczak, K. Computational assessment of the influence of the overlap ratio on the power characteristics of a Classical Savonius wind turbine. *Open Eng.* **2015**, *5*, 314–322. [CrossRef]
41. Aramendia, I.; Fernandez-Gamiz, U.; Zulueta, E.; Saenz-Aguirre, A.; Teso-Fz-Betoño, D. Parametric Study of a Gurney Flap Implementation in a DU91W(2)250 Airfoil. *Energies* **2019**, *12*, 294. [CrossRef]
42. Celik, I.B.; Ghia, U.; Roache, P.J.; Freitas, C.J.; Coleman, H.; Raad, P.E. Procedure for Estimation and Reporting of Uncertainty Due to Discretization in CFD Applications. *J. Fluids Eng.* **2008**, *130*, 078001. [CrossRef]

energies

MDPI

Article

Uncertainty Quantification of the Effects of Blade Damage on the Actual Energy Production of Modern Wind Turbines

Francesco Papi [1], Lorenzo Cappugi [2], Simone Salvadori [3], Mauro Carnevale [4] and Alessandro Bianchini [1,*]

1 Department of Industrial Engineering, Università degli Studi di Firenze, 50139 Florence, Italy; fr.papi@unifi.it
2 Department of Engineering, Lancaster University, Lancaster LA1 4YW, UK; l.cappugi@lancaster.ac.uk
3 Department of Energy, Politecnico di Torino, 10129 Torino, Italy; simone.salvadori@polito.it
4 Department of Mechanical Engineering, University of Bath, Bath BA2 7AY, UK; mc2497@bath.ac.uk
* Correspondence: alessandro.bianchini@unifi.it

Received: 30 June 2020; Accepted: 22 July 2020; Published: 23 July 2020

Abstract: Wind turbine blade deterioration issues have come to the attention of researchers and manufacturers due to the relevant impact they can have on the actual annual energy production (AEP). Research has shown how after prolonged exposure to hail, rain, insects or other abrasive particles, the outer surface of wind turbine blades deteriorates. This leads to increased surface roughness and material loss. The trailing edge (TE) of the blade is also often damaged during assembly and transportation according to industry veterans. This study aims at investigating the loss of AEP and efficiency of modern multi-MW wind turbines due to such issues using uncertainty quantification. Such an approach is justified by the stochastic and widely different environmental conditions in which wind turbines are installed. These cause uncertainties regarding the blade's conditions. To this end, the test case selected for the study is the DTU 10 MW reference wind turbine (RWT), a modern reference turbine with a rated power of 10 MW. Blade damage is modelled through shape modification of the turbine's airfoils. This is done with a purposely developed numerical tool. Lift and drag coefficients for the damaged airfoils are calculated using computational fluid dynamics. The resulting lift and drag coefficients are used in an aero-servo-elastic model of the wind turbine using NREL's code OpenFAST. An arbitrary polynomial chaos expansion method is used to estimate the probability distributions of AEP and power output of the model when blade damage is present. Average AEP losses of around 1% are predicted mainly due to leading-edge blade damage. Results show that the proposed method is able to account for the uncertainties and to give more meaningful information with respect to the simulation of a single test case.

Keywords: uncertainty quantification; wind energy; wind turbine; blade damage; AEP

1. Introduction

Wind turbine damage has in recent years gained interest from industry and academia in an effort to keep aging wind parks around the globe productive. According to Rempel [1], in the early days of the wind energy industry there was the general misconception that once the blade is in operation, no further maintenance is required. This has changed, partly due to a considerable number of field reports that have started to surface in recent years highlighting extreme and worrying examples of early blade deterioration. For instance, Rempel states that blades as young as three years of age can show signs of wear and that blades of 87 out of 111 wind turbines in a wind farm off the shores of Denmark had to be dismantled and brought to shore after less than five years in operation due to severe leading-edge (LE) damage, as shown in Røndgaard [2].

In addition to the more common LE issues, blade's trailing edge often suffers from damage. In particular, Decoret [3] states that debonding is commonly observed at the trailing edge (TE). This phenomenon occurs when the composite layers of the blade shell separate. If this happens at the TE of the blade, its dimension is expected to greatly increase in thickness, thus decreasing the aerodynamic performance. According to Wood [4], another common source of damage at the TE happens during blade transportation and turbine assembly. Crushing of the laminate may occur as well as chipping of the TE itself, especially in the tip region where the rear of the blades is typically very thin. The impact of LE damage on AEP has been studied by various authors. Amongst the most influential research in the field, Sareen et al. [5] test in a wind-tunnel a series of LE-damaged wind turbine airfoil configurations that mimic pictures of blades that were brought in for repair. They predict massive maximum losses in AEP of up to 25%. Han et al. [6] develop a computational fluid dynamics (CFD) model of an eroded airfoil based on their inspection of a 14-year-old Vestas V47 blade using the commercial software Star CCM+. They then simulate the NREL 5 MW [7] rotor with erosion applied from 70.7% of the rotor span outwards and find AEP reductions of 3.7%. Castorrini et al. [8] develop a numerical tool to predict airfoil performance degradation due to LE erosion. The tool is tuned based on photographic evidence of damaged blades and tested on the NREL 5 MW rotor, predicting power decreases of around 8%. As also noted by Herring et al. [9], these values, and others that can be found in published literature, quantitively greatly differ between each other. This could be due to the fact that erosion has a variable impact on different airfoil shapes, turbine sizes and operating conditions, thus leading to different results. Moreover, as far as the authors are aware, no study assesses the impact of TE damage on AEP at the present time, while some of the authors recently analyzed its effects on aerodynamic performance and loads under realistic inflow conditions [10]. Some light can be shed on the discrepancies highlighted between the work of many authors by approaching the problem in a probabilistic manner rather than in a deterministic way, as done until now.

This is done in the present study by introducing two aleatory variables that model leading-edge and trailing-edge damage, respectively. That blade damage is propagated through an aero-servo-elastic model of the DTU 10 MW RWT [11] as this is a modern reference rotor design. The model response in terms of AEP and power is approximated using an arbitrary polynomial chaos (aPC) expansion. The numerical procedure that is followed will be detailed in the following sections; however, a brief rundown can be provided as follows. The damage is applied to a give airfoil through geometry modification. The lift and drag coefficients are then obtained using computational fluid dynamics (CFD). The obtained coefficients are applied to the DTU10MW blade. The turbine is then simulated using NREL's open-source code OpenFAST [12]. Finally, the model regression can be performed and response surfaces of the outputs of interest, as well as associated probability density functions (PDFs), can be estimated. An overview of the entire modelling process is provided in Figure 1.

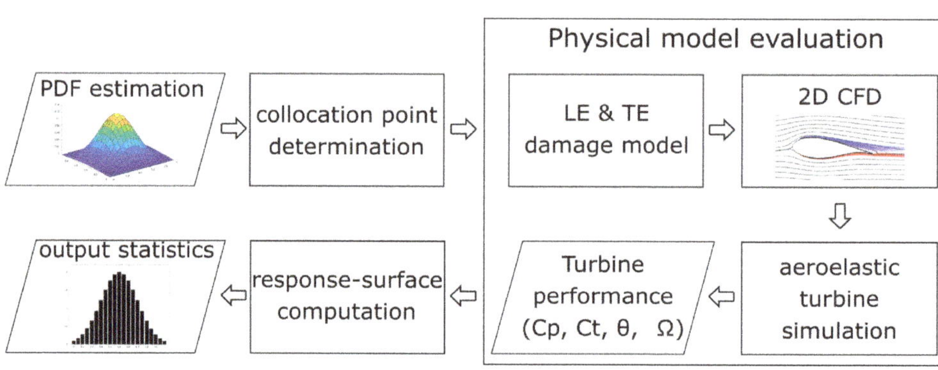

Figure 1. Flowchart of the uncertainty-quantification procedure.

It should be noted that blade damage is undoubtedly not the only source of uncertainty that affects the power production of a wind farm. Other common sources of uncertainty are related to environmental conditions, with uncertainties in wind speed and turbulence intensity being the main ones. As these are not the topic of the present study, which focuses specifically on the effects of blade damage, they are not included in the uncertainty quantification; however, in order to ensure that the study is up to the present simulation standards, they are accounted for using the standard procedures of the International Electrotechnical Commission (IEC).

2. Materials and Methods

In this section, the main details regarding the numerical modelling tools and choices that were made are given. Firstly, the numerical methods used to estimate the uncertainty associated to blade damage are introduced. The hypotheses regarding the input uncertainties are then explained. Finally, in the following subsections the numerical tools used in the required deterministic model evaluations are detailed.

2.1. Stochastic Approach

The stochastic approach exploited in the present work falls into the class of the arbitrary polynomial chaos (aPC) as implemented by Oladyshkin and Novak [13]. This technique falls into the field of the study of aleatory uncertainty, which only accounts for deviations of boundary condition and geometrical parameters. The present approach does not include the contribution of the limits of the numerical approach adopted. The deviation or the effect of such limitation have been considered as negligible. CFD has been validated and run according to best practices, including grid independence study. This approach has the advantage of providing stochastic results (or PDFs) without the need to change the algorithm of the numerical tools employed in the simulations. These kinds of approaches are generally known as non-invasive methods as reviewed by Iaccarino [14] and more detailed in Carnevale [15] and Ahfield [16]. The PDF of a specific quantity of interest is extracted by reproducing a surface response obtained by a certain number of simulations (or deterministic realization). The boundary conditions for these simulations are set to reproduce the PDF representing the aleatory parameter. The process of selecting appropriate boundary conditions is known as sampling. The sampling process is usually obtained by means of selecting the boundary condition using the Monte Carlo method filtered by the proper PDF. The approach as described implies a large number of simulations and it is not reliable for application where CFD solvers are used for each single deterministic prediction. This would require a high computational cost to complete the simulation campaign.

A strategy to overcome this limitation consists of a clever choice of the boundary conditions resulting in a limited number of simulations. The convolution of this boundary conditions is representative of a specific PDF. This approach is known in literature as the probabilistic collocation point (PCM). The PCM are obtained as quadrature points of a linear system built on the basis consisting in a set of polynomials (polynomial chaos, PC). The choice of these polynomials corresponds to make a strong assumption on how the response surface is determined. The surface response will be as the weighted functions corresponding to a specific PDF. Mathematic foundations can be found in Tatang et al. [17]. This particular approach has been successfully applied to CFD simulations in Carnevale et al. [15,18] and Salvadori et al. [19]. The particular approach proposed allows weaker hypothesis to be considered on the PDF of the aleatory parameter. The aPC only demands the existence of a finite number of moments and does not require the complete knowledge or even the existence of a probability density function. This approach has also been employed in Ahlfield et al. [16], where the stochastic behavior physical parameters are characterized by discontinuity and Gibbs phenomena. The aPC extends chaos expansion techniques by employing a global polynomial basis.

Let's consider a generic aleatory variable ξ propagating on a specific output of interest $Y = f(\xi)$, where f is a general unknown stochastic model (or PDF); it can be expressed as a d-order expansion:

$$Y(\xi) \approx \sum_{i=1}^{d} c_i P^{(i)}(\xi) \quad (1)$$

According to the general theory of PCM the characteristic statistical quantities of $Y(\xi)$ can be evaluated by the coefficient c_i, and the momentum and variance are expressed as follows:

$$\mu_Y = c_1,\ \sigma_Y^2 = \sum_{i=1}^{d} c_i \quad (2)$$

The peculiarity of the aPC approach is related to the strategy adopted to determine the orthonormal basis of polynomial $P^{(i)}$. These polynomials have been determined by the moment-based approach detailed in Oladyshkin et al. [13]. Once the aPC, which represents an orthonormal basis, has been identified, the collocation points are obtained by means of a quadrature procedure. Given an aleatory variable $\xi \pm \sigma$ associate with a PDF $f(\xi)$, the more general expression of its quadrature is

$$\int_{-\sigma}^{\sigma} Y(\xi) f(\xi) d\xi = \sum_{k=0}^{d} \omega(\xi_k) P(\xi_k) + R_M(Y) \quad (3)$$

In the previous equation, the left-hand side is the stochastic representation of the aleatory variable ξ associated with the PDF $f(\xi)$. The right-hand side is its expansion on the basis $P(\xi_k)$, where the $\omega(\xi_k)$ is the weighting term (in this context we can consider $\omega(\xi_k) = 1$), $R_M(Y)$ is the remainder approaching zero as d-order of the expansion increases and the collocation points ξ_k are such that the formula $\int_{-\sigma}^{\sigma} Y(\xi) f(\xi) d\xi - \sum_{k=0}^{d} \omega(\xi_k) P(\xi_k) = 0$ is satisfied for the moment $\mu(\xi)$ and the $\mu(\xi) \pm \sigma$.

2.2. Probability Density Functions

The random input variables are introduced in the model using PDFs. Although this is not specifically required by the adopted aPC method, which is on the other hand able to operate on any kind of available data, in the present study PDFs were assumed based on an expert's opinion due to the lack of publicly available information regarding the studied parameters. In fact, the PDFs are based on the assumptions of Bortolotti et al. [20], who also attempt to deal with input uncertainties in aero-servo-elastic wind turbine models. Two beta functions are used for both LE Erosion Factor ε and TE Damage Factor τ. They are appropriately scaled to match the support these variables are defined upon. The values of the PDFs are reassumed in Table 1. The adopted PDFs are assumed as representative of cases where medium-low blade damage is present or of sites with challenging environmental conditions where regular maintenance is performed.

Table 1. Probability density functions for erosion factor andTrailing Edge (TE) damage factor.

Parameter	PDF	α	β	Support
ε	Beta	2.0	6.0	0–10 (%)
τ	Beta	2.0	6.0	0–4 (%)

2.3. Blade-Damage Model

The first stage of the modelling process consists of modelling the blade damage itself. Blade damage is modelled through shape-modification of selected airfoils along the wind turbine's blades. Trailing-edge damage is reproduced by a simple truncation of the airfoil's trailing-edge. The amount of TE truncation with respect to the airfoil's cord is expressed as the above-introduced TE Damage Factor τ. Leading-edge damage is instead modelled through a more complex shape modification. This, which is caused by leading-edge delamination, is based on two main parameters, the maximum erosion depth θ and the chord-wise coverage of the damaged area ε. Both the influence of ε and of θ are studied in this research.

However, in order to quickly estimate the global influence of leading-edge damage with respect to trailing-edge damage and to keep the analysis synthetic with only two random input variables, θ and ε were related through an empiric correlation. This assumption is supported by existing studies, where these two variables seem to be related to each other. However, Gaudern et al. [21] and Sareen et al. [5] found two very different ε-θ curves, as shown in Figure 2. As both curves are found by field examination of the blades and given that there is no clear way of assessing which of the two curves is more accurate for this study, a mean curve is proposed here (also shown in Figure 2). We can now refer only to ε as the aforementioned LE Erosion Factor, as this value now also uniquely determines θ.

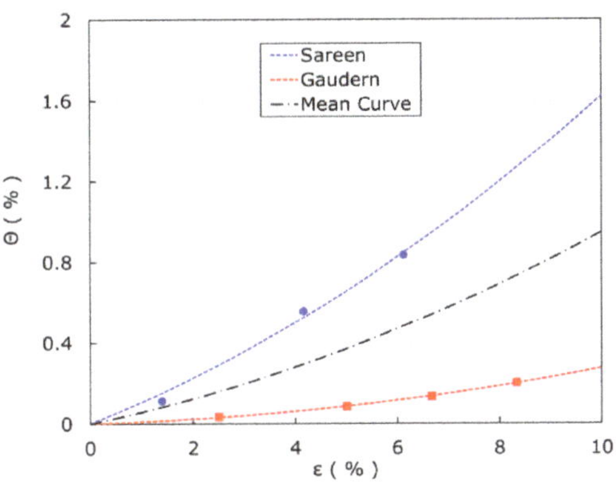

Figure 2. ε-θ correlation in literature [5,21] and proposed correlation in black point-dashed line.

Once a value of ε is selected, the damaged airfoil shape is generated with a purposely developed Matlab® tool. The LE of the airfoil is moved inward by a maximum depth of θ. Similarly to what was done by Schramm et al. [22], the leading-edge was flattened. The height of the flattened area is imposed to be $h = 2\theta$. Damage extends up to ε on the suction side of the airfoil and up to 1.3ε on the pressure side, as done in [5]. This is also motivated by the fact that wind turbine airfoils are designed to operate with a positive angle of attack (AoA), and therefore, the pressure side of the airfoil is more exposed to wear. The depth of delamination at the end of the damaged area is equal to $D_{end} = \theta/3$. The TE and LE damage models are shown in Figure 3. The models are also described in further detail in [10]. The present model is a simplified version of the real LE damage pattern adequate for a parametric study like the present one, which cannot therefore reproduce all the features of a real, three-dimensional damaged blade. The model, however, is in line with the proposals of other authors [22,23] and also qualitatively reproduces the damaged shapes obtained from computational models [8,24], as seen in experiments [5].

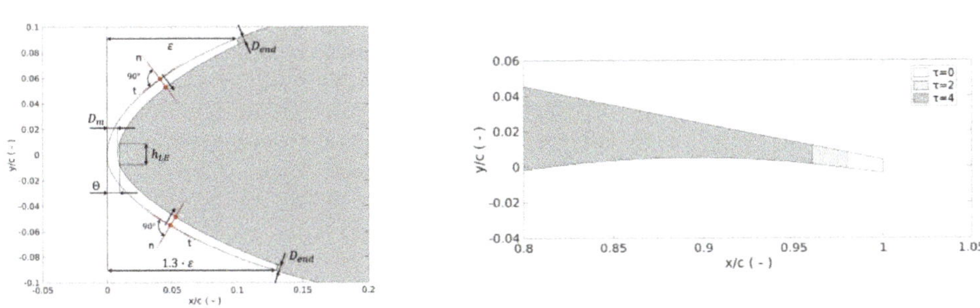

(**a**) (**b**)

Figure 3. Damage modeling. (**a**) Leading-edge (LE) damage; (**b**) TE damage.

2.4. CFD Setup

The lift and drag coefficients of the airfoils are calculated using CFD. The numerical set-up was used by the authors and has been presented in detail in [10]; however, the main parameters will be reassumed herein. The ANSYS® FLUENT® (Version 18.2) solver is used to calculate the 2D polars. A Reynolds-averaged Navier-Stokes (RANS) approach is used. The Navier-Stokes equations are solved in a coupled manner with second order upwind spatial discretization. Turbulence closure is achieved with the k-ω Shear Stress Transport (SST) model. A bullet-shaped computational domain is used, as with this shape open-field conditions can be modelled with only one inlet and one outlet boundary condition. In order to ensure that the boundary conditions do not influence the results, the computational domain is 74 chord lengths long and 40 chord lengths wide, as shown in Figure 4a.

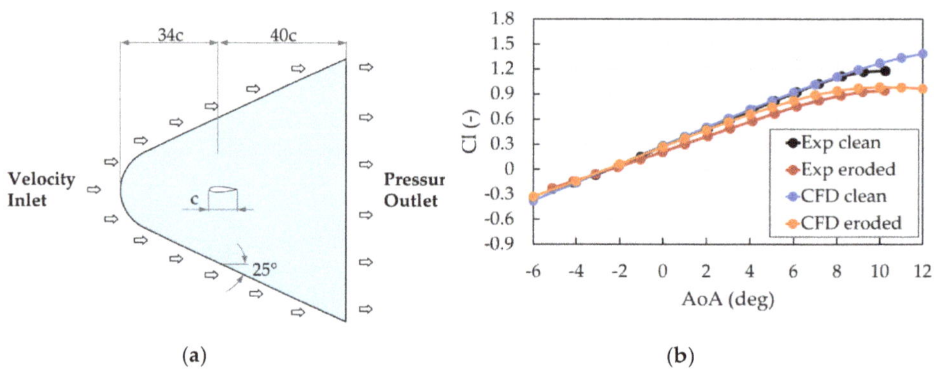

Figure 4. CFD validation: (**a**) Illustration of the adopted computational domain; (**b**)validation of the numerical setup in respect to data from [5] at Re = 1.5×10^6.

An unstructured triangular mesh is used. The airfoil's boundary layer is modelled with a quadrilateral inflation layer from the blade surface. A total amount of 46 prismatic layers are used. To ensure grid independence, three meshes were tested with varying number of elements. A coarse mesh with 1.3×10^5 elements and 500 elements along the airfoil surface, a medium mesh with 2.8×10^5 elements and 650 elements along the airfoil's surface, and a fine mesh with 3.6×10^5 elements and 750 elements along the airfoil's surface. The lift (C_l) and drag (C_d) coefficients are calculated with CFD between 20° and 30° of AoA; values for AoA higher and lower than this are extrapolated using Viterna's method [25]. A total roughness height of 0.4 mm is imposed on the airfoil's nose trough an equivalent sand-grain roughness height, estimated through the simple correlations provided in [26]. This roughness height is selected based on the observations of several authors [5,6,21] and models medium to advanced pitting and gauging of the LE. As the focus of the LE damage model is on advanced stages of damage, a constant value of roughness was considered suitable across all the LE-damaged cases.

It is important to point out that CFD is by its nature deterministic, i.e., the same simulation is expected to give the same results if the same settings are used. In this sense, it does not add any source of uncertainty in the analysis. On the other hand, it is true that using different numerical settings to solve the same test case could lead to different results. On this basis, it is very important that the CFD approach is robust and validated with experiments whenever possible. In the present study, in particular, the numerical set-up was validated with respect to available experimental data from [5]. The clean and eroded data is obtained from the DU96W-180 airfoil that was tested in clean and damaged configurations ("stage 5" erosion in [5]) for a Reynolds number of 1.5×10^6. Figure 4b demonstrates good agreement between the experimental values and CFD predictions, with limited differences that can be attributed to the unspecified wind tunnel turbulence level and to the surface finish of the reference model.

2.5. Aeroelastic Setup

The lift and drag coefficients of the damaged airfoils are used in an aero-servo-elastic model of the DTU 10 MW RWT [11]. The model is developed within NREL's open-source simulation code OpenFAST (NREL, CO, USA) [12]. The aerodynamic module, *AeroDyn* is based on blade element momentum (BEM) theory. As in all BEM codes, the wake is modelled with a series of concentric annuli, upon which a momentum balance is imposed. The blades are modelled trough lift and drag coefficients. Corrections for high induction (Glauert correction), blade tip and root losses, tower shadow, skewed flow and dynamic stall are included [27]. The coefficients of the dynamic stall model are tuned based on the lift, drag and moment coefficients of the damaged airfoils. Blade damage was considered from 70% of the rotor span outwards. The reasoning behind this choice has to do with the fact that the LE damage phenomena considered are mainly related to erosion, which is most influent where the local blade inflow velocities are highest. Other authors also applied damage from 70% of the rotor span outwards [6]. The lift and drag coefficients of the damaged airfoils are applied uniformly to the entire damaged area.

Fully flexible blades and tower are modelled with the structural dynamics module *ElastoDyn*. The modal formulation allows for a fairly accurate computation of the structural dynamics with very low computational cost. The Delft Research Controller (DRC) [28] is used in this study. This open-source baseline controller is able to regulate torque and pitch. Constant-torque operation is selected above rated wind speed. The control parameters are tuned based on the report of [29].

2.6. General DLC Setup

The DTU 10 MW RWT is a state-of-the-art reference rotor, developed in recent years as a benchmark for researchers and industry in the field of wind energy. It features a 178-meter diameter rotor with aerodynamic features like gurney flaps that help this conceptual turbine reach a rated power of 10 MW at a wind speed of 11.4 m/s. The tower height is 119 m and the nominal revolution speed is 9.6 rpm, which equates to a tip speed just shy of 90 m/s. The complete definition of the turbine and all of its parameters can be found in Bak et al. [11]. To estimate the AEP of the turbine, a power-production design load case (DLC) is simulated. This is done through sixty-six 10-minute simulations with wind speeds at a hub height between 4 and 24 m/s. Six turbulent seeds per wind speed are simulated, in compliance with the minimum requirements of the IEC 61400-1 [30]. The wind fields also feature wind shear and misaligned flow with respect to the rotor plane. By simulating several cases, uncertainties regarding atmospheric conditions are dealt with, and their influence is accounted for in this study.

It is important to note that turbulence affects power production and other key turbine figures in a complicated manner, as this depends both on the interaction between the controller and the incoming wind speed and on the complex blade boundary-layer phenomena amongst other things. The interaction between large turbines and the turbulent atmospheric boundary layer is out of the interest of the present study and has been evaluated in detail by Churchfield et al. and Nandi et al. [31,32]. Moreover, as other authors have pointed out when studying a similar multi-MW wind turbine in an aero-servo-elastic modelling framework [20], six turbulent realizations are enough to guarantee good convergence on the AEP statistics.

AEP is calculated using a Rayleigh wind-speed probability density function with a mean of 10 m/s as specified by IEC class IA, which is the design class of the DTU 10MW. The AEP obtained using a Rayleigh distribution with a mean wind speed of 8.5 m/s (corresponding to IEC class IIA) will also be briefly analyzed as this could be more representative of the impact of blade damage on sites with lower mean wind speeds.

3. Results

The aPC resulting collocation points are qualitatively shown in Figure 5 and detailed in Table 2. For each point, the corresponding damaged airfoil geometry is generated and CFD calculations were performed as described in Section 2.3. With the resulting airfoil data, aero-servo-elastic BEM simulations were performed as described in Section 2.4.

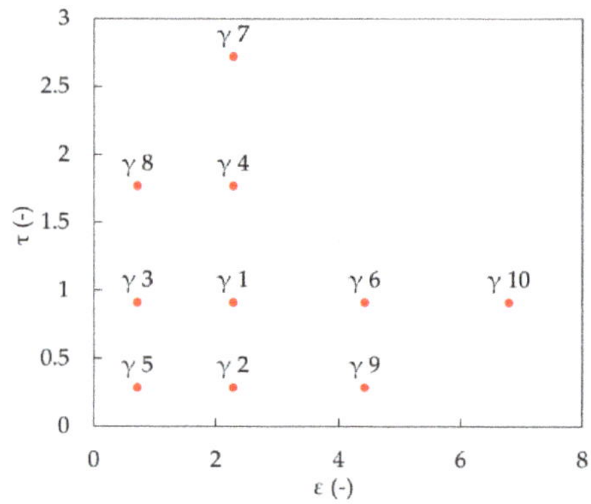

Figure 5. Arbitrary polynomial chaos (aPC) resulting collocation points' plot in ε–τ space.

Table 2. aPC optimal collocation points values.

γ	1	2	3	4	5	6	7	8	9	10
ε	2.2792	2.2792	0.7134	2.2792	0.7134	4.4189	2.2792	0.7134	4.4189	6.7884
τ	0.9118	0.2854	0.9118	1.7677	0.2854	0.9118	2.7157	1.7677	0.2854	0.9118

3.1. Aerodynamic Performance

In this section the aerodynamic performance under uncertainties is discussed. In Figure 6 the mean variation in power coefficient (C_P) with respect to the clean reference turbine is shown. The standard deviation and associated probability contours are also shown. The C_P mean value is lower than the nominal one for all the wind speed bins except for the 4 m/s one. In this wind speed bin, the average gain is about 1%. The reasons that cause such gains are related mainly to the TE damage; however, this gain in performance, while conceptually interesting, is weakened by two factors. First, at 4 m/s the power is about 60 times lower than the nominal one, and thus the effect on the AEP will be minimal. This can be seen clearly in Figure 7. Secondly, there is a high dispersion in the C_P values and therefore the expected value is hard to predict. The high dispersion is due to the extremely different response from the damaged airfoils. Both gain and power losses at this wind speed occur. The time averaged AoA from 30% of the blade span to tip goes from 0° to 5°. This allows some of the damaged airfoils to operate with favourable lift and drag forces with respect to others. More details about this behaviour are given below.

The highest value for the mean decrease in C_P is of −2.6% at 10 m/s. At this wind speed the reduction in C_P can exceed −12%. Moreover, from 8 m/s to 12 m/s, mostly only power losses occur. In this wind speed range, a significant part of the total turbine's energy is produced; therefore, power reductions in this region will eventually lead to a significant reduction in AEP. Finally, for wind speeds higher than 14 m/s, shown in the grey-shadowed region in Figure 6, the damage effects are no longer visible, as from this wind speed onwards a lower pitch-to-feather regulation is able to compensate for the aerodynamic losses.

The power output per wind speed bin is shown in Figure 7. Upon examination of this figure, it is apparent that the blade damage has a greater impact on power output between 8 m/s and 12 m/s, confirming what was seen in the relative trends of Figure 6. At 4 m/s, however, as previously pointed out, the mean power output is only 174 kW, higher than the 172 kW of the nominal case. Due to the little power produced, this difference as well as the high standard deviation of ±7 kW (±4%) are not visible in the plot, further highlighting how such variation has little impact on the overall performance. In order to better understand the global results, each wind speed bin can be examined more in detail.

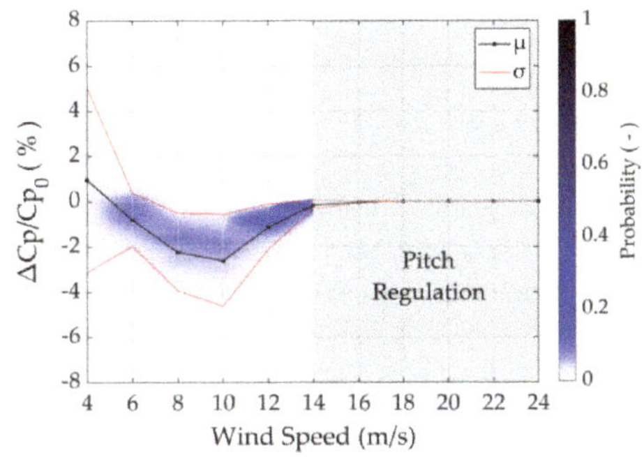

Figure 6. Variation in power coefficient, mean value (μ), standard deviation (σ) and probability.

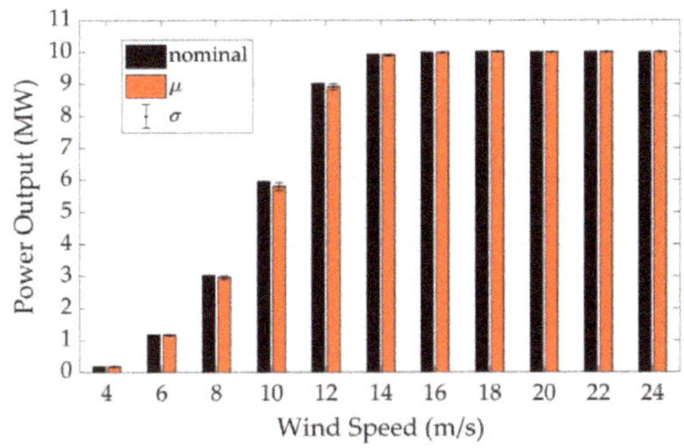

Figure 7. Power output per wind speed bin for nominal and mean damaged (μ) turbine with standard deviation (σ).

The response surfaces reporting the differences in C_P for the wind speed bins that show the most relevant differences are shown in Figure 8. For the wind speed bin of 4 m/s the response surface slightly overestimates the C_P of the nominal geometry. Such behavior is shown in Figure 8a around the $\varepsilon = 0$, $\tau = 0$ point. On the other hand, the response surface prediction gives good results at 8 m/s and 10 m/s where the C_P variation predicted for the nominal geometry is zero as expected.

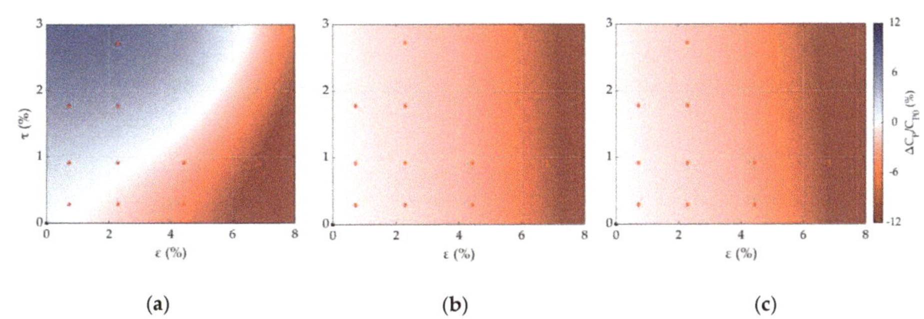

Figure 8. Variation in power coefficient. Response surfaces as contour plots at (**a**) 4 m/s, (**b**) 8 m/s and (**c**) 10 m/s.

In the 4 m/s wind speed bin, an increase in C_P for several combinations of ε and τ can be noted. To explain this unexpected trend, one can consider the collocation point pairs γ 7 & γ 2 (same ε and the highest and the lowest τ, respectively) and γ10 & γ3 (same τ and the highest and the lowest ε, respectively). Therefore, looking at the pair $\gamma2$ & $\gamma7$ the influence of τ is highlighted, while looking at the pair $\gamma10$ & $\gamma3$ the influence of ε is highlighted. Point $\gamma7$ shows the highest increase in C_P (about 10%), while $\gamma2$ shows a mild decrease in C_P, about −1.5%; thus, as shown in Figure 8, power increases as tau increases. The other γ-pair shows the opposite behavior, for $\gamma10$, the power coefficient decreases by 12%, while $\gamma3$ shows an increase in the power coefficient of about 3%, and thus, power decreases as tau decreases. To better understand the trends, the lift and drag coefficients for the FFAW3-241 airfoil (i.e., the airfoil present in the damaged part of the blade) for the four damage levels are shown in Figure 9 with respect to the reference configuration. In general, lift decreases and drag increases for all of the damaged configurations as expected. Focusing on the mean AoA recorded for the various damaged configurations at 4 m/s in Figure 9a, it is clear how the mean AoA increases for all of the damaged cases. This is due to the lower lift of the damaged cases. A new operational equilibrium point in the BEM code is then reached, with a lower induction and thus a higher AoA.

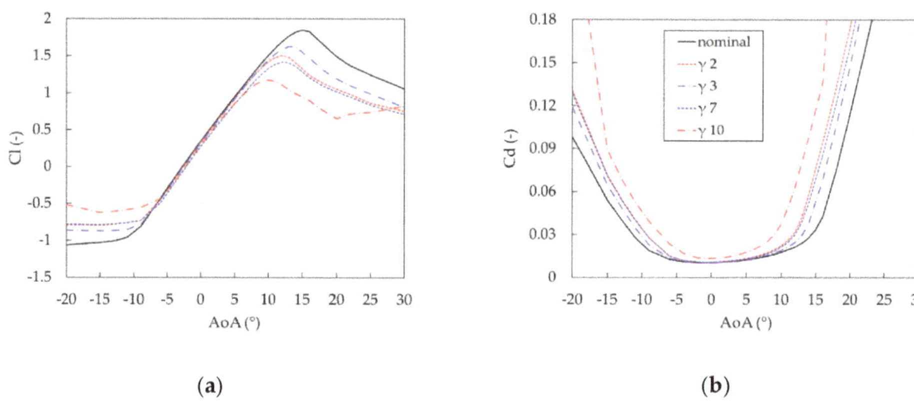

Figure 9. Aerodynamic coefficients for nominal and most significant power coefficients (CPs): (**a**) Lift coefficient; (**b**) drag coefficient.

As a consequence of the increased AoA, lift and drag forces slightly increase and, more importantly, are more tangentially and axially oriented. The new force composition generates more torque and more power for some of the combinations of ε and τ. As shown in Figure 10, the same phenomena are occurring for all the damaged configurations: a change in the lift and drag coefficients leads to a different BEM equilibrium point with different induction and AoA along the entire area of the blade affected by damage. However, increasing ε also significantly increases drag, leading to lower performance and offsetting the benefit of a higher AoA, despite the change of orientation of the forces. For instance, in $\gamma10$, the highest increases in drag are observed, exceeding 30% at an AoA of around 2°.

In Figure 11 the average AoA for the nominal and four damage combinations for all the wind speed bins is shown. For all the damaged combinations, the highest average AoAs are predicted in the 8 m/s and 10 m/s wind speed bins. At 8 m/s mean wind speed the average AoA for the nominal case at 78% blade span is about 6.9°, while the damaged cases work at an even higher AoA due to decreased induction, as previously discussed. In these wind speed bins, there is no power increase in any combination of ε and τ. From the analysis of Figure 9, the higher the AoA, the wider the difference is in lift and drag coefficients. This ultimately leads to the power losses observed in Figures 6–8, with peaks that exceed −10% at 8 m/s and −12% at 10 m/s, respectively. It is also interesting to note that ε is the main cause of performance decrease and has a more pronounced effect than τ. This is due to the fact that LE damage causes a reduction in the stall AoA of the airfoil, which strongly influences high-AoA operation and a more pronounced increase in drag than TE damage.

The probability distributions found from the evaluation of the computed response surfaces at 4 m/s, 8 m/s and 10 m/s are shown in Figure 12. At 4 m/s, the variation in C_P is most affected by uncertainties.

The peak is located at 1% of variation in C_P, but the resulting distribution is fat-tailed. Indeed, the standard deviation is ±4.1% and the probability to lose or gain C_P are about 40% and 60%, respectively. At 8 m/s and 10 m/s, the distributions are strongly asymmetric and have lower standard deviations with respect to the 4m/s case and are equal to ±1.7% and ±2% at 8 m/s and 10 m/s, respectively. In both cases, the probability for a C_P gain is zero and losses always occur. They both have a marked left tail, but a higher dispersion at 10 m/s is found. The probability peak is clearly located on the right of the mean value at −1.5% and −1.7% for 8 m/s and 10 m/s, respectively.

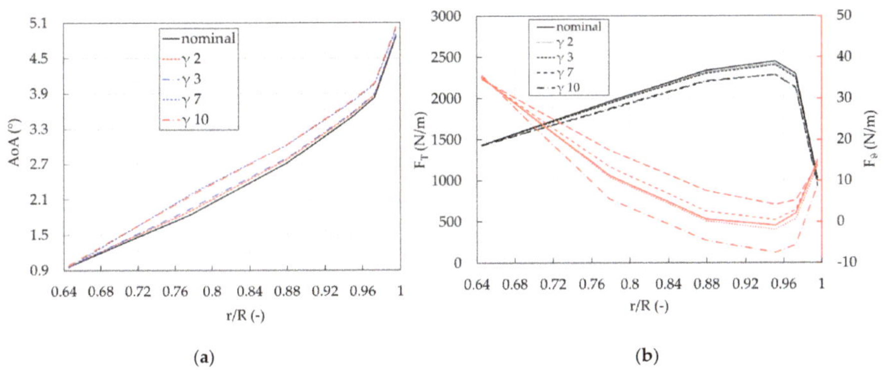

Figure 10. Relevant turbine figures: (**a**) Angle of attack along the outer part of the blade and (**b**) thrust (F_T) and tangential (F_ϑ) for the outer part of the blade at 4 m/s mean wind speed.

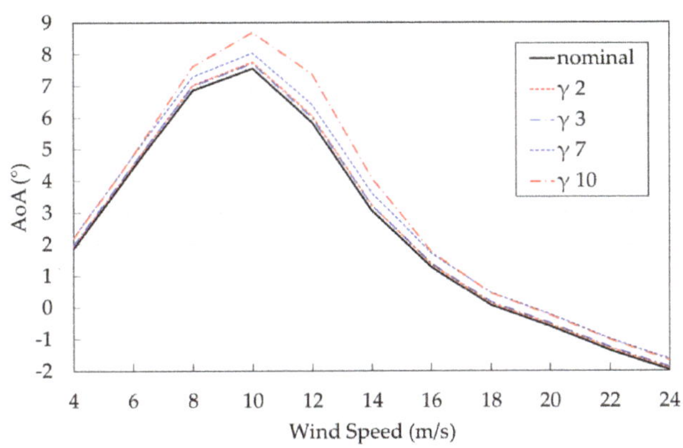

Figure 11. Angle of attack vs. wind speed for nominal and four damaged conditions at 78% blade span.

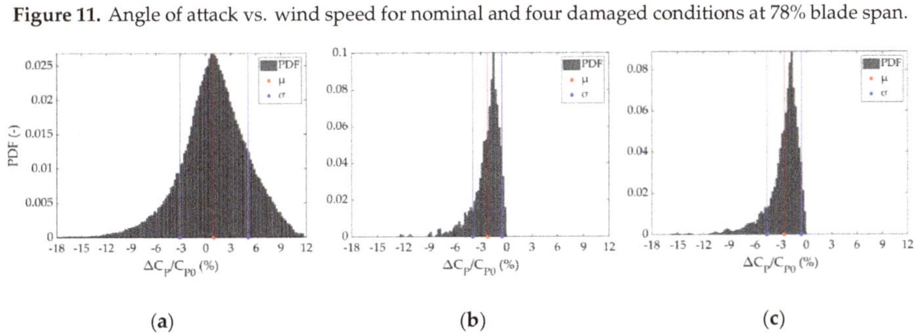

Figure 12. Variation in power coefficient probability density functions (PDFs) with mean value (μ) and standard deviation (σ) at (**a**) 4 m/s, (**b**), 8 m/s and (**c**) 10 m/s.

3.2. Annual Energy Production (AEP)

The uncertainties in AEP estimation are discussed in this section. The AEP was calculated according to IEC 61400-1 standard turbine classes. A Weibull wind speed distribution with shape factor of 2 and average values of 8.5 m/s and 10 m/s were used to model sites of IEC wind class II and IA. In particular, class IA is chosen as this is the design class of the DTU 10 MW RWT and class IIA is chosen as representative of medium wind speed sites, where such a turbine might also be installed. The availability factor was assumed to be 1. This assumption is justified by the fact that relative variations are mainly analyzed in the present study. The variation in AEP for the two wind distributions is shown in Figure 13. Both response surfaces well predict the trends around $\varepsilon = 0$, $\tau = 0$, showing no variation in AEP in that point. The LE erosion, ε, is the main driver for AEP reduction, as decreases are mostly along the ε axis. The trailing edge damage, τ, has a minor influence in AEP, as clearly visible in Figure 13. Moreover, the trailing edge damage contribution seems to be dependent on the erosion level. For instance, if one considers the six combinations of ε and τ (where $\varepsilon = 0, 4, 8$ and $\tau = 0, 3$) for wind class IIA shown in Table 3, the point $\varepsilon = 4$, $\tau = 0$ gives a variation in AEP of −1.87%, while the point $\varepsilon = 4$, $\tau = 3$ gives a variation of −2.24%. Therefore, for $\varepsilon = 4$, the trailing edge damage increases losses by 0.37%. By performing the same consideration for $\varepsilon = 8$, trailing edge damage increases losses by 0.82%. This means that the TE contribution to losses increases as ε increases.

Table 3. Annual Energy Production (AEP) reduction for some of the computed ε-τ combinations.

ε	τ	$\Delta AEP/AEP_0$ (%)
0	0	0.00
0	3	0.00
4	0	−1.87
4	3	−2.24
8	0	−9.69
8	3	−10.51

The highest variation in AEP predicted by the response surface is −10.35% at $\varepsilon = 8$, $\tau = 3$ for class IIA, as seen in Figure 13a. The highest simulated AEP reduction is −6.21% for $\gamma 10$ (class IIA, Figure 13a). For wind class IA, the highest variations in AEP are lower than the ones predicted for class IIA and amount to −8.56% in $\varepsilon = 8$, $\tau = 3$ and −5.02 in $\gamma 10$, as shown in Figure 13b. Such differences are due to the Weibull wind speed PDFs. The probability for the machine to work in the bins range from 8 to 12 m/s, where the highest losses in power occur, are 36% and 31% for IIA and IA, respectively. This difference is the main cause of different variations in AEP for the two classes.

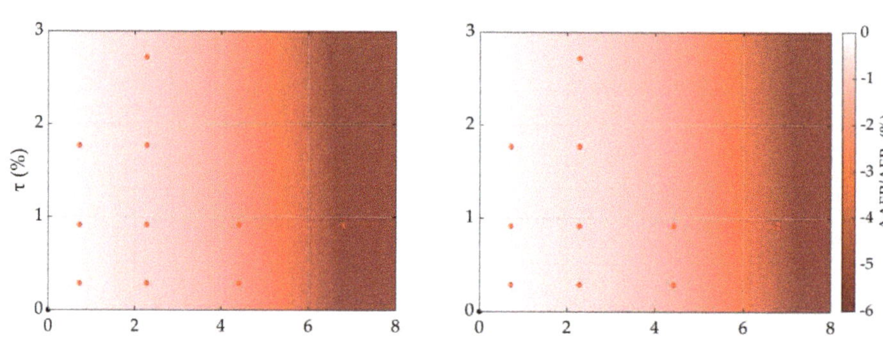

(a) (b)

Figure 13. Variation in AEP. Response surfaces as contours plot for wind classes (**a**) IIA and (**b**) IA.

Finally, we consider the probability distributions AEP variation shown in Figure 14. As is also the case for the previously shown distributions, the PDFs are obtained by sampling the response surfaces 250,000 times. The mean and standard deviations of the PDFs are −1.21% and ±1.04% for class IIA and −0.98% and ±0.84% for class IA. Such mean reductions are indeed significant on a multi-MW scale turbine and are in line with the finding of Eisenberg et al. [33] but seem to be lower than the values indicated by most of the present research [5,6,8]. Both distributions show a clear peak, with the mode of the PDFs below 1% AEP loss in both cases. For both the IEC 61400-1 IA and IIA scenarios, the left tails of the distributions are long, reaching values of 6–8% AEP reductions, coherently with the response surface shown in Figure 13. The probability associated to values of AEP reduction in the order of 3–8%, which most authors indicate, is almost insignificant in the present test case. It is important to stress that these results depend on the assumed PDFs, which are, as discussed, based on published literature and appear reasonable based on the authors' experience. Moreover, as Fiore and Selig have suggested [34], larger turbines seem to be impacted less by LE damage phenomena such as erosion. However, results suggest that the commonly forecasted reductions might be based on heavy-damage scenarios, which, whilst not unrealistic, have low probability of occurrence.

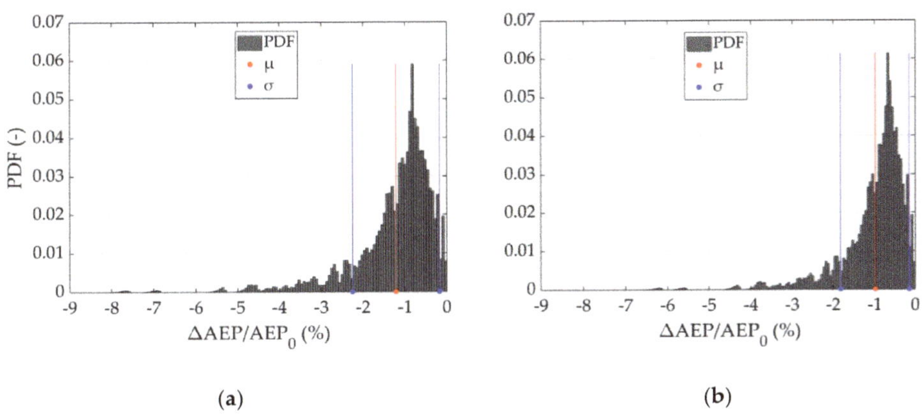

Figure 14. Variation in AEP PDFs for wind classes (**a**) IIA and (**b**) IA.

The wind class IIA shows higher standard deviation and higher left tail length. As previously mentioned, which is due to the fact that in the class IIA scenario the turbine operates at rated power for a shorter period of time with respect to the class IA scenario. In fact, as also pointed out by Eisenberg [33], the turbine's power output does not experience any significant variation for wind speeds above rated when the blades are damaged and therefore the higher the mean wind speed, the lower the variation in AEP. These results clearly depend on the IEC class that was chosen. Lowering the average wind speeds even further (IEC Class III), the turbine is expected to spend less time at rated power, and therefore, AEP losses are expected to further decrease for the present test case. Although low wind speed sites have recently been exploited for wind turbine installation, specially designed machines with low specific power are being installed in such sites, resulting in machines that are able to spend significant time at rated power even in these sites. As noted in [33], a utility-scale machine will spend 40% to 60% of its time at rated power, where blade damage has no effects. In addition, although the main cause of LE erosion is related to the rotational tip velocity, it can be argued that in lower wind speed sites, less transport of abrasive particles will arise, therefore leading to less erosion.

4. Conclusions

This study proposes the use of an uncertainty quantification approach to the modelling of the effects of blade damage on the performance of multi-MW wind turbines. The proposed approach aims at overcoming some of the issues associated with the evaluation of a single test case. In fact, treating blade damage as a random phenomenon, bias due to a specific test case of a combination of blade-damaging factors can be avoided and more general conclusions can be drawn. The entire process is simulated numerically. First, geometric shape modifications are applied to the airfoils that constitute the turbine's blade. Lift and drag coefficients are calculated using CFD. The newly found coefficients are then applied to an aero-servo-elastic model of the wind turbine. Uncertainties are propagated through the model using an arbitrary polynomial chaos method.

Results show that LE damage has the larger influence on power and AEP losses. For the selected test case, TE damage has little impact, except for when the turbine is operating at very low wind speeds, where a slight performance increase is observed due to TE damage. Focusing on absolute values, maximum average power reductions are observed at 8 m/s and 10 m/s mean wind speeds and are of 2.2% and 2.6%, respectively. The most unfavorable damage combinations simulated showed a decrease in AEP of up to over 6%. By looking at the probabilistic framework, however, the configurations with the highest probability of occurring based on the input PDFs show AEP reductions of below 1% in both IEC classes I and IIA. Indeed, mean AEP reductions of just below 1% for class IA and just above 1% for class IIA are estimated. These values, whilst significant, seem to be notably lower than what is commonly forecasted in published literature that, however, is strictly site- or turbine-dependent. It is important to point out that the results of the present study do not indicate that published literature values are unrealistic (even though sometimes a too large span coverage of erosion is considered), however, for the present test case, representative of modern turbine size and design trends, such values seem to have very low probabilities of occurrence. Indeed, AEP decreases exceeding 10% are noted in the present study. Blade damage is an issue that should still be taken very seriously by the industry, due to its structural implications that were not investigated in the present work; however, the impact on AEP does not seem to be as pronounced as early research indicated. A great deal of factors could cause these discrepancies, which could be due to the radial damage extension considered and size and hence the Reynolds number of the turbine, which are not investigated herein and therefore remain an open issue, where additional research would definitely be beneficial. As pointed out by other authors, LE damage seems to have a lower effect on larger wind turbines. Although this is not the focus of the present work, the results of this study, if put into perspective with other published literature that reports higher AEP decreases on smaller turbines, seem to confirm this.

Moreover, as previously discussed, the results strongly depend on the input PDFs. The presented method can be however adapted to different input PDFs, which are hopefully more extensively supported by field data. Nevertheless, the present assumptions can be considered realistic for medium-low damaging environments or for blades where regular maintenance schedules are planned. It is also important to point out that these results are valid strictly only for the present test case. A selection of different study cases might influence the results significantly, as, in the authors' experience, LE damage affects different airfoils to different degrees. Finally, the LE damage model also influences the results. Although the model is calibrated and tested with respect to experimental data and is adequate for the present parametric framework, it is hard, if not impossible, to accurately reproduce small, stochastic features that might influence the sectional efficiency significantly.

In conclusion, even considering these factors, it is apparent that the present statistical approach is able to give designers a better picture of the impact of blade damage.

Author Contributions: Conceptualization, F.P. and A.B.; methodology, F.P., L.C., S.S.; software, F.P., L.C., M.C.; formal analysis, A.B., F.P.; investigation, F.P., L.C.; resources, A.B.; data curation, F.P., L.C., A.B.; writing—original draft preparation, F.P., L.C.; writing—review and editing, A.B., S.S.; visualization, F.P.; supervision, S.S., M.C., A.B.; project administration, A.B. All authors have read and agreed to the published version of the manuscript.

Funding: This research received no external funding.

Conflicts of Interest: The authors declare no conflicts of interest.

Nomenclature

Acronyms	
AEP	Annual energy production, kWh
aPC	Arbitrary polynomial chaos
BEM	Blade element momentum
CFD	Computational fluid dynamics
DLC	Design load case
DRC	Delft research controller
IEC	International electrotechnical commission
LE	Leading edge
PC	Polynomial chaos
PCM	Probabilistic collocation point
PDF	Probability desity function
RANS	Reynolds averaged navier stokes
SST	Shear stress transport
TE	Trailing edge
Latin Letters	
AoA	Angle of attack, deg.
c	Blade chord, m
c_i	Expansion coefficients
Cd	Drag coefficient
Cl	Lift coefficient
C_P	Turbine power coefficient
D_{end}	Delamination depth at the end of damaged area, m
F_T	Thrust force, N/m
F_ϑ	Tangential force, N/m
h	Leading edge flattened area height, m
$P^{(i)}$	Orthogonal polynomials
R_M	Polinomial expansion remainder
Y	Specific output of interest
Greek Letters	
α, β	Beta function's shape parameters
γ	Collocation point
ε	Leading edge erosion factor
Θ	Leading edge erosion depth
μ	Momentum
ξ	Generic aleatory variable
σ	Standard deviation
τ	Trailing edge damage factor
ω	Weighting term

References

1. Rempel, L. Rotor Blade Leading Edge Erosion-Real Life Experiences. *Wind Syst. Mag.* **2012**, *11*, 22–24.
2. Røndgaard, F.-K. February 23, Ersen-; Europe, 2018 Structural Failure Denmark UK WindAction|Siemens Sets Billions: Ørsted must Repair Hundreds of Turbines. Available online: http://www.windaction.org/posts/47883-siemens-sets-billions-orsted-must-repair-hundreds-of-turbines#.XfzepkdKiUl (accessed on 20 December 2019).
3. Régis, D. *Why Inspect Wind Turbine Blades? How to Avoind Major Wind Turbine Blade Damage and Cost.* Sustainable Energy Consulting and Software (3E). 2014. Available online: https://www.3e.eu/wp-content/uploads/2014/06/3E_Why-Wind-Turbine-Blade-Inspections_20140507_final.pdf (accessed on 10 July 2020).
4. Wood, K. Blade Repair: Closing the Maintenance Gap. Available online: https://www.compositesworld.com/articles/blade-repair-closing-the-maintenance-gap (accessed on 9 June 2020).

5. Sareen, A.; Sapre, C.A.; Selig, M.S. Effects of leading edge erosion on wind turbine blade performance: Effects of leading edge erosion. *Wind Energ.* **2014**, *17*, 1531–1542. [CrossRef]
6. Han, W.; Kim, J.; Kim, B. Effects of contamination and erosion at the leading edge of blade tip airfoils on the annual energy production of wind turbines. *Renew. Energy* **2018**, *115*, 817–823. [CrossRef]
7. Jonkman, J.; Butterfield, S.; Musial, W.; Scott, G. *Definition of a 5-MW Reference Wind Turbine for Offshore System Development*; Technical Report NREL/TP-500-38060, 947422; National Renewable Energy Laboratory: Golden, CO, USA, 2009.
8. Castorrini, A.; Corsini, A.; Rispoli, F.; Venturini, P.; Takizawa, K.; Tezduyar, T.E. Computational analysis of performance deterioration of a wind turbine blade strip subjected to environmental erosion. *Comput Mech.* **2019**, *64*, 1133–1153. [CrossRef]
9. Herring, R.; Dyer, K.; Martin, F.; Ward, C. The increasing importance of leading edge erosion and a review of existing protection solutions. *Renew. Sustain. Energy Rev.* **2019**, *115*, 109382. [CrossRef]
10. Papi, F.; Cappugi, L.; Bianchini, A.; Perez-Becker, S. *Numerical Modeling of the Effects of Leading-Edge Erosion and Trailing-Edge Damage on Wind Turbine Loads and Performance*; ASME TurboExpo: London, UK, 2020.
11. Bak, C.; Zahle, F.; Bitsche, R.; Kim, T.; Yde, A.; Henriksen, L.C.; Natarajan, A.; Hansen, M. *Description of the DTU 10 MW Reference Wind Turbine*; DTU Wind Energy: Copenhagen, Denmark, 2013.
12. OpenFAST. Available online: https://github.com/OpenFAST (accessed on 31 October 2019).
13. Oladyshkin, S.; Nowak, W. Data-driven uncertainty quantification using the arbitrary polynomial chaos expansion. *Reliab. Eng. Syst. Saf.* **2012**, *106*, 179–190. [CrossRef]
14. Iaccarino, G. *Quantification of Uncertainty in Flow Simulations Using Probabilistic Methods*; VKI Lecturers Series; von Karman Institute: Rhode-St-Genèse, Belgium, 2009.
15. Carnevale, M.; Montomoli, F.; D'Ammaro, A.; Salvadori, S.; Martelli, F. Uncertainty Quantification: A Stochastic Method for Heat Transfer Prediction Using LES. *J. Turbomach* **2013**, *135*, 1–8. [CrossRef]
16. Ahlfeld, R.; Montomoli, F.; Carnevale, M.; Salvadori, S. Autonomous Uncertainty Quantification for Discontinuous Models Using Multivariate Padé Approximations. *J. Turbomach* **2018**, *140*. [CrossRef]
17. Tatang, M.A.; Pan, W.; Prinn, R.G.; McRae, G.J. An efficient method for parametric uncertainty analysis of numerical geophysical models. *J. Geophys. Res. Atmos.* **1997**, *102*, 21925–21932. [CrossRef]
18. Carnevale, M.; D'Ammaro, A.; Montomoli, F.; Salvadori, S. *Film Cooling and Shock Interaction: An Uncertainty Quantification Analysis with Transonic Flows*; American Society of Mechanical Engineers Digital Collection: New York, NY, USA, 2014.
19. Salvadori, S.; Carnevale, M.; Fanciulli, A.; Montomoli, F. Uncertainty Quantification of Non-Dimensional Parameters for a Film Cooling Configuration in Supersonic Conditions. *Fluids* **2019**, *4*, 155. [CrossRef]
20. Bortolotti, P.; Canet, H.; Bottasso, C.L.; Loganathan, J. Performance of non-intrusive uncertainty quantification in the aeroservoelastic simulation of wind turbines. *Wind Energy Sci.* **2019**, *4*, 397–406. [CrossRef]
21. Gaudern, N. A practical study of the aerodynamic impact of wind turbine blade leading edge erosion. *J. Phys. Conf. Ser.* **2014**, *524*, 012031. [CrossRef]
22. Schramm, M.; Rahimi, H.; Stoevesandt, B.; Tangager, K. The Influence of Eroded Blades on Wind Turbine Performance Using Numerical Simulations. *Energies* **2017**, *10*, 1420. [CrossRef]
23. Cavazzini, A.; Minisci, E.; Campobasso, M.S. *Machine Learning-Aided Assessment of Wind Turbine Energy Losses due to Blade Leading Edge Damage.*; American Society of Mechanical Engineers Digital Collection: New York, NY, USA, 2019; IOWTC2019-7578.
24. Fiore, G.; Selig, M.S. Simulation of Damage Progression on Wind Turbine Blades Subject to Particle Erosion. In Proceedings of the 54th AIAA Aerospace Sciences Meeting; American Institute of Aeronautics and Astronautics, San Diego, CA, USA, 4–8 January 2016.
25. Viterna, L.A.; Janetzke, D.C. *Theoretical and Experimental Power from Large Horizontal-Axis Wind Turbines*; National Aeronautics and Space Administration: Cleveland, OH, USA, 1982.
26. Adams, T.; Grant, C.; Watson, H. A Simple Algorithm to Relate Measured Surface Roughness to Equivalent Sand-grain Roughness. *IJMEM* **2012**, *1*, 66–71. [CrossRef]
27. *Wind Energy: Handbook*; Burton, T., Ed.; J. Wiley: Chichester, UK; New York, NY, USA, 2001; ISBN 978-0-471-48997-9.
28. Mulders, S.; van Wingerden, J. Delft Research Controller: An open-source and community-driven wind turbine baseline controller. *J. Phys. Conf. Ser.* **2018**, *1037*, 032009. [CrossRef]
29. Borg, M. *LIFES50+ Deliverable D1.2: Wind Turbine Models for the Design*; DTU Wind Energy: Risø, Denmark, 2015; p. 29.

30. International Electrotechnical Commission. *Wind Turbines-Part1: Design Requirements*; Technical Standard IEC61400-1:2005; International Electrotechnical Commission: Geneva, Switzerland, 2005.
31. Churchfield, M.J.; Lee, S.; Michalakes, J.; Moriarty, P.J. A numerical study of the effects of atmospheric and wake turbulence on wind turbine dynamics. *J. Turbul.* **2012**, *13*, N14. [CrossRef]
32. Nandi, T.N.; Herrig, A.; Brasseur, J.G. Non-steady wind turbine response to daytime atmospheric turbulence. *Phil. Trans. R. Soc. A* **2017**, *375*, 20160103. [CrossRef] [PubMed]
33. Eisenberg, D.; Laustsen, S.; Stege, J. Wind turbine blade coating leading edge rain erosion model: Development and validation. *Wind Energy* **2018**, *21*, 942–951. [CrossRef]
34. Fiore, G.; Selig, M.S. Simulation of Damage for Wind Turbine Blades Due to Airborne Particles. *Wind Eng.* **2015**, *39*, 399–418. [CrossRef]

energies

MDPI

Article

Fluid–Structure Interaction Simulations of a Wind Gust Impacting on the Blades of a Large Horizontal Axis Wind Turbine

Gilberto Santo [1,*], Mathijs Peeters [2], Wim Van Paepegem [2] and Joris Degroote [1]

[1] Department of Electromechanical, Systems and Metal Engineering, Ghent University, Sint-Pietersnieuwstraat 41, 9000 Ghent, Belgium; joris.degroote@ugent.be

[2] Department of Materials, Textiles and Chemical Engineering, Ghent University, Technologiepark-Zwijnaarde 907, 9052 Zwijnaarde, Belgium; mathijs.peeters@ugent.be (M.P.); wim.vanpaepegem@ugent.be (W.V.P.)

* Correspondence: gilberto.santo@ugent.be

Received: 4 December 2019; Accepted: 15 January 2020; Published: 21 January 2020

check for updates

Abstract: The effect of a wind gust impacting on the blades of a large horizontal-axis wind turbine is analyzed by means of high-fidelity fluid–structure interaction (FSI) simulations. The employed FSI model consisted of a computational fluid dynamics (CFD) model reproducing the velocity stratification of the atmospheric boundary layer (ABL) and a computational structural mechanics (CSM) model loyally reproducing the composite materials of each blade. Two different gust shapes were simulated, and for each of them, two different amplitudes were analyzed. The gusts were chosen to impact the blade when it pointed upwards and was attacked by the highest wind velocity due to the presence of the ABL. The loads and the performance of the impacted blade were studied in detail, analyzing the effect of the different gust shapes and intensities. Also, the deflections of the blade were evaluated and followed during the blade's rotation. The flow patterns over the blade were monitored in order to assess the occurrence and impact of flow separation over the monitored quantities.

Keywords: fluid–structure interaction; wind turbine; atmospheric boundary layer; composite materials; gusts; wind energy

1. Introduction

Renewable energy sources have been gaining more importance in the last few decades as part of the strategies adopted by countries all over the world to limit the use of fossil fuel and fight pollution and global warming. In particular, wind energy has grown rapidly, resulting in the increasing size of modern wind turbines with the objective of reducing the cost of the produced energy [1]. Nevertheless, the adoption of more slender blades has also led to higher deflections during normal operation, and consequently, more interest toward the fluid–structure interaction (FSI) phenomenon in modern wind turbines. Recent works have computed that, while operating in design conditions, the flapwise deflection of a modern horizontal-axis wind turbine (HAWT) blade is in the order of 6–7% of its span [2–4]. Furthermore, their deflections have a sensible impact on the produced power, affecting its oscillation or introducing a reduction up to 6% [2–6].

However, due to the complexities involved, when the FSI of modern wind turbine blades is to be analyzed, simplified models are often adopted for this task. On the aerodynamic side of the problem, among others, blade element momentum theory (BEM) is widely used in the FSI modeling of wind turbines [7–9]. Despite its low computational cost, BEM theory is affected by many limitations, including the need to include tip-loss corrections to account for a blade of finite length [10]. Another class of widely used models are the actuator models, where the blades are represented by lines or surfaces exchanging momentum with the incoming wind flow [11–13]. This strategy has also been used in FSI simulations of a multi-megawatt turbine, coupling it with a structural model based on non-linear beam theory [14].

In order to increase the level of detail in the extracted results, a higher computational cost is required. Computational fluid dynamics (CFD) simulations, performed by solving the Navier–Stokes equations on a computational grid rendering the geometry of the wind turbine, have recently been used in literature within FSI frameworks. In particular, Reynolds-averaged Navier–Stokes (RANS) models are often used to account for turbulence in the atmospheric wind flow [15–19]. This class of turbulence models are computationally cheap when compared to more complex turbulence models, which are also reported in recent literature within FSI simulations of wind turbines [2–4,20,21] but have an extremely high computational cost.

On the structural side of the FSI problem, the complex nature of modern composite blades [22] often leads to the adoption of simplified models. Among others, FSI works relying on multi-body dynamics [21] or using one-dimensional beam elements [14,15,19] can be found, whereas the implementation of detailed finite element models (FEMs), loyally reproducing the composite structure of modern blades, is still limited to only a few works [3,4,6,20] by two distinct research groups.

Despite its importance, the atmospheric boundary layer (ABL), namely the velocity gradient leading to higher wind speeds at higher heights, is very often neglected in FSI simulations of wind turbines. To the best of the authors' knowledge, only a very limited number of works account for it [6,21,23]. These works remark upon the importance of the ABL, highlighting how it induces oscillating loads, power, and deflections on modern blades.

Furthermore, given the aleatory nature of the wind, wind turbines always operate in unsteady and rapidly changing conditions. In particular, local oscillations in the wind speed, referred to as "gusts," lead to the varying performance and structural responses of the blades, which is particularly important in the fatigue life estimation of modern structures. Gusts can be of various shapes, sizes, lengths, and intensities [24], and can be induced by the terrain morphology, as well as by thermal or turbulent effects [25]. Despite their relevance in relation to large HAWTs, the majority of works currently available in the literature focus on vertical-axis wind turbines (VAWTs) or on small HAWTs. Wu et al. [26] carried out 2D RANS simulations to obtain the lift–drag polars of the sections of a VAWT in both steady wind and unsteady (i.e., affected by a gust) wind conditions and used them to obtain the performance of the turbine by means of a BEM-like strategy. Onol and Yesilyurt [27] analyzed a small VAWT by means of a 2D unsteady RANS (U-RANS) model, changing the approaching wind velocity according to the gust time variation prescribed by the IEC-64100 standards for wind turbines. Where 3D simulations are concerned, Bhargav et al. [28] performed CFD simulations of a VAWT while changing the inlet wind speed to follow a sinusoidal function but without considering the FSI.

Regarding to FSI simulations, the work of Timme et al. [29] analyzed the aeroelastic response of a straight wing impacted by a vertical gust. The shape of this gust corresponds to the "$1-\cos$" shape, also prescribed in the IEC-64100 standards. Also operating in an aeroelastic framework, Svacek and Horacek [30] analyzed the response of a flexibly supported airfoil to a vertical wind gust, mimicking the dynamics of a wing by means of 2D U-RANS simulations.

The available literature on the gust aeroelastic response to HAWT is extremely limited. First in time, the work of Younsi et al. [31] featured a BEM code in combination with a structural model reproducing the blade as being homogenously made of an elastic and isotropic material. A single blade was modelled, neglecting the ABL and simulating an extreme gust impacting on the whole blade. More recently, Castellani et al. [32] carried out both experimental and numerical works to investigate the response of a small HAWT (2-m diameter) to a periodical change in the incoming wind speed, aiming at the investigation of an optimal control strategy for such a machine. On the other hand, Ebrahimi and Sekandari [33] investigated the aeroelasticity of a multi-megawatt wind turbine to a sudden change in wind, coupling an unsteady vortex-lattice method (VLM) with a structural model relying on a modal approach. The entire turbine was subjected to a change of velocity magnitude and direction, also accounting for the response of the control systems that react by changing the pitch of the blades and the yaw of the whole machine.

All the aforementioned works on HAWTs neglect the ABL and analyze a wind gust bigger than the analyzed turbine by means of simplified models. The investigated gust impacts on the whole rotor can be counteracted by the turbine control systems [33]. No work about the aeroelastic response of the blades of a large HAWT immersed in the ABL and locally impacted by a wind gust (i.e., soliciting only one blade) was found in the current literature. In this work, for the first time, two high-fidelity models, one for the CFD side and one for the FEM side of the FSI problem, were strongly coupled to dynamically analyze the aeroelasticity of the blade of a 100-m diameter HAWT immersed in the ABL and impacted by wind gusts of different intensities and morphologies. The proposed methodology is believed to be well-suited for advanced engineering applications, and in particular, to analyze the response of modern wind turbines to extreme load cases as part of the design process. This is an added value compared to the available literature, in which the aeroelastic analysis of a wind turbine blade attacked by a wind gust is scarcely reported.

The paper is structured as follows. The CFD model is presented in Section 2.1, the structural FEM model is described in Section 2.2, the coupling strategy is addressed in Section 2.3, followed by the gust model in Section 2.4. Section 3 contains the extracted results and their discussion, and the conclusions are drawn in Section 4.

2. Methodology

In this section, the adopted numerical model is described. It is built analogously to the model in Santo et al. [6]. Therefore, the methodology is shortly summarized here and the reader is referred to Santo et al. [6] for a more extensive description, validation, and sensitivity study of the chosen methods.

2.1. The CFD Model

On the CFD side of the employed FSI model, the domain illustrated in Figure 1 was used.

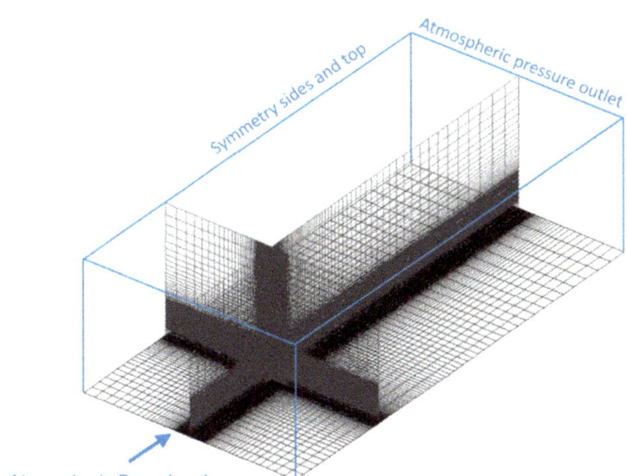

Figure 1. Overview and sections of the CFD computational domain of the background for the ABL with indications of the used boundary conditions.

An overset method (also called Chimera) approach was chosen to handle the rotation and deformation of the blades, resulting in a background grid and several component grids. Every grid was entirely made of hexahedral finite volumes and the state variables were stored in the cell centers. Figure 1 shows the structured grid employed for the background. In its finest region, the cells were cubic and had a 0.275-m edge. In this location, the analyzed HAWT was positioned, being five rotor diameters distant from inlet, symmetry sides, and top surfaces. The atmospheric pressure outlet was placed 15 diameters downstream of the rotor. These distances were chosen according to good practice guidelines for atmospheric flows in urban areas [34]. However, it is reported that remarkably smaller (approximately half) distances are found to be sufficient to minimize boundary influence [35].

A body-fitted hexahedral component mesh was constructed around each blade and around the tower and the nacelle, as depicted in Figure 2. The tower was considered rigid and the geometry of the tower was extracted from Harte and Van Zijl [36], being used for wind turbines of similar size. The hub height was 100 m and the nacelle had a length of approximately 12 m and a section of 5 m × 5 m. The right-hand side of Figure 2 shows different sections of the mesh around each blade, taken at different radial spanwise locations. The mesh on the blade walls was designed to have a y^+ in the log layer (between 30 and 300), while on the border of each body-fitted mesh, a mesh size comparable to the one used in the background was imposed.

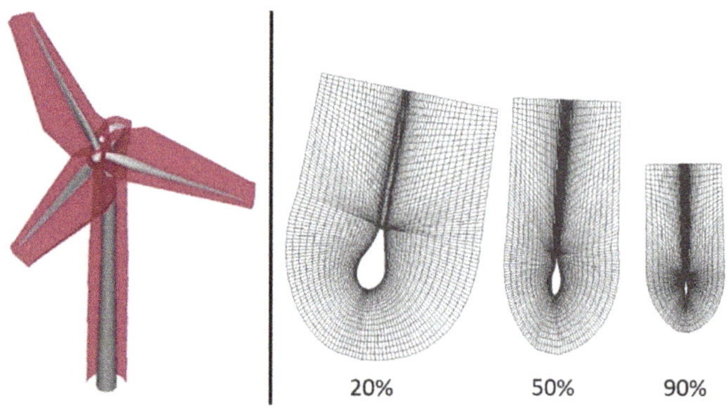

Figure 2. (**left**) Overview of the body-fitted mesh around blades and supporting structures with the overset borders in red, and (**right**) sections of the mesh around each blade, taken at different spanwise positions.

The air flow was modelled as incompressible (air density constant equal to 1.225 kg/m^3) and the k-ε (RANS) turbulence model was used. For this model, Richard and Hoxey [37] developed inlet profiles for the velocity (v), turbulent kinetic energy (k) and its dissipation ratio (ε) as functions of the height from the ground (z). These profiles are obtained as analytical solutions of the transport equations of k and ε and are reported in Equations (1)–(3):

$$u(z) = \frac{v_*}{K} ln\left(\frac{z + z_0}{z_0}\right), \tag{1}$$

$$k = \frac{v_*^2}{\sqrt{C_\mu}}, \tag{2}$$

$$\varepsilon(z) = \frac{v_*^3}{K(z + z_0)}. \tag{3}$$

In these equations, K and C_μ are the von Karman constant (0.4187) and a constant of the k-ε model (0.09), respectively. Moreover, v_* is the friction velocity (a global index of the wind intensity), and z_0 is the aerodynamic roughness length, which provides a measure of how rough the ground wall is [38]. These last two parameters fully define the ABL profiles.

Nevertheless, in order to preserve these profiles while travelling through the computational domain, a modified formulation is necessary for the ground wall, as observed in different works [39,40].

Parente et al. [40,41] have therefore obtained modified wall functions, where the aerodynamic roughness length is directly included in the formulation of the non-dimensional wall distance z^+_{mod} and of the wall function constant E_{mod}:

$$z^+_{mod} = \frac{(z + z_0)u_f \rho}{\mu}, \quad (4)$$

$$E_{mod} = \frac{\mu}{\rho z_0 u_f}. \quad (5)$$

In these equations, ρ and μ are the air density (1.225 kg/m^3) and its viscosity (1.7894×10^{-5} kg/ms), respectively. The modified wall functions were therefore used on the ground wall, while on the walls of the wind turbine, standard wall functions were adopted.

All the investigated load scenarios were carried out at the turbine nominal operating point, where it is expected to operate for most of its life. Therefore, following input from the blade's manufacturer, the rotational speed of the machine was set to 1.445 rad/s and the wind velocity at the hub height was set to 8.5 m/s at the inlet, leading to a tip speed ratio of 8.5 using undisturbed conditions. The axis of rotation was perfectly aligned with the wind flow (i.e., no tilt angle, no yaw misalignment). Consistently, the friction velocity and the aerodynamic roughness length were set equal to 0.671 m/s and 0.5 m, respectively, to achieve the desired wind speed at the hub height. The inlet turbulent kinetic energy was 0.01512 m^2/s^2, leading to a turbulence intensity of 1.3% at the hub height.

The mass and momentum conservation equations were solved together using a pressure-based solver. For the momentum equation, a second-order upwind discretization scheme was adopted, while a first-order implicit scheme was selected for time discretization. All the simulations presented in this work featured the same numerical setup. The entire CFD model was implemented in Fluent 18.1 (Ansys Inc., Canonsburg, PA, USA).

According to the outcome of the mesh and time step sensitivity study carried out in Santo et al. [6], the selected mesh had a total of 55 million cells, 1.9 million belonging to each blade component mesh. The surface of each blade wall was divided into approximately 38,500 faces. The time step size was 0.01812 s, corresponding to 1.5° of rotor rotation per time step. With these settings, the torque coefficient produced by the turbine, considering rigid blades and no gust, was computed to be 0.05221, which compares well with what the manufacturer provides for this operating point (0.0556).

It is remarked that the simulations presented in this work cannot be validated by experimental results, given the difficulty in reproducing the imposed gusts in a controlled way.

2.2. The Structural FEM Model

Each of the three blades was 50-m long and entirely made of composite material. In order to loyally mimic its structure, the blade was divided into approximately 64,000 shell elements with reduced integration. The elements were distributed on the outer mold layer of the blade and on its shear webs and shear caps (Figure 3). In each shell element, a global layup orientation was defined, with respect to which, a ply orientation was defined in each ply of the composite stack in order to fully define the anisotropic properties of the used materials. The generation of the structural mesh is described in Peeters et al. [22] and the comparison of the computed eigenfrequencies with the values provided by the manufacturer is given in Santo et al. [6]. Gravity was also included in the model.

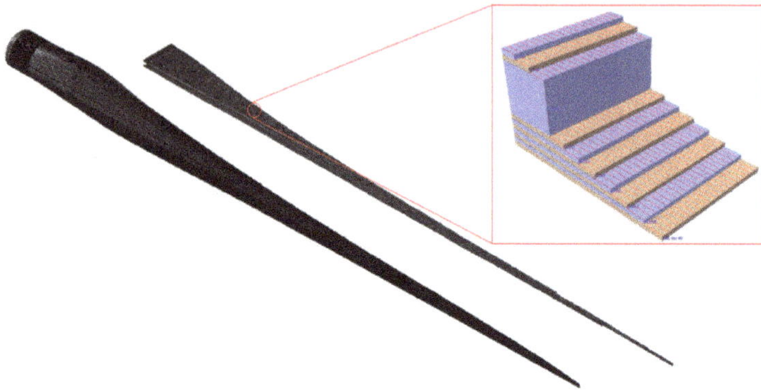

Figure 3. Overview of the structural mesh in the outer shells and in the shear webs of each blade, with a detailed view of the composite layup in one of the 64,000 shell elements.

2.3. FSI Coupling

The CFD and the structural models outlined in the previous sections were coupled by means of Tango, an in-house code developed at Ghent University, resulting in a partitioned FSI simulation [42]. The Gauss–Seidel coupling algorithm was selected and three iterations were enough within each time step to reach a displacement absolute residual of about 5 mm on the fluid–structure interface. The fluid mesh was deformed according to what is prescribed by the structural solver on the fluid–structure interface. Each blade component mesh was deformed by means of a spring-based smoothing method, therefore adopting an arbitrary Lagrangian–Eulerian (ALE) formulation. Compared to other methods present in the literature [43], the adopted methodology can consistently preserve an appropriate y^+ value of the stretched cells in the region adjacent to each wall. This ensured a well-resolved near-wall flow at a reasonable computational cost.

Simulations were performed on 280 cores (10 nodes inter-connected by InfiniBand, each with 2 CPUs of the type 14-core Xeon E5-2680v4, 2.4 GHz) and approximately 10 days were necessary to perform a complete revolution. Starting the FSI simulation from the results of a CFD simulation, approximately 1.2 revolutions were necessary to reach the regime in time. Then, one full rotation was performed and analyzed. During this rotation, the loads, stress, and displacements of each blade could be univocally linked to its azimuth angle. The azimuth angle was set to 0 when the blade was horizontally positioned and in an upward motion. Therefore, a 90° azimuth angle corresponded to the blade pointing upward and 270° to the blade pointing downward.

2.4. Gust Model

Different gust shapes and intensities were imposed to impact one of the three blades. In order to investigate the worst-case scenario, the gusts were chosen to collide on the blade when the blade pointed upward (i.e., at a 90° azimuth angle); in this position, the blade was attacked by the highest wind speed, and therefore, the solicitations acting on it were also the highest. In this work, as in other works on wind gusts [29,44], the gusts were modelled as a local streamwise velocity increase to be superimposed on the local wind velocity field. Using v_g to indicate the gust's additional velocity and v the local wind velocity, the velocity in "gusted" conditions v' can be expressed as:

$$v' = v + v_g. \tag{6}$$

Notice that, at each time instant, all these velocities are functions of the position in the numerical domain. In order to minimize the computational time and numerical dissipation of the gusts [29,44], the gusts were added in the proximity of the turbine when the blade to be hit was positioned at a 60°

azimuth angle, namely 30° and 20 time steps in advance with respect to the impact of the gust on the blade. They affected only a cylindrical region of 25 m in diameter (i.e., half of the blade span) and 12 m in length, whose axis was aligned with the wind direction. The frontal tip of this cylindrical region was positioned at an appropriate axial coordinate (approximately 1 m downstream of the blade tip's axial coordinate) in order to ensure the gust hit the targeted blade. This means that only in this cylindrical region, the gust velocity used in Equation (6) could be written as:

$$v_g = s(a) \cdot f(r) \cdot A_g \neq 0. \tag{7}$$

In Equation (7), s and f are shape functions of the gust, depending respectively on the axial (a) and radial (r) coordinates inside the gusted cylindrical region. In particular, the function s sinusoidally ramped up from 0 to 1 in the first 3 m of the axial length of the cylinder, was kept constant and equal to 1 in the middle, and then sinusoidally ramped down from 1 to 0 in the last 3 m of the axial length. Figure 4 illustrates the initial position and shape of the imposed gusts, as well as the coordinates a and r.

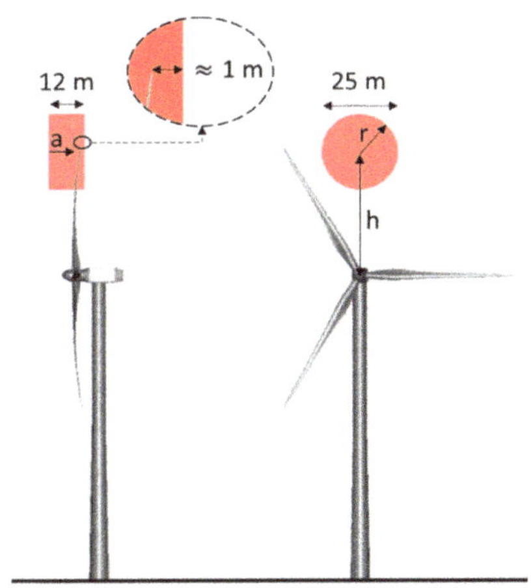

Figure 4. (**left**) Side view with detail of the blade tip axial position relative to the gusted region and (**right**) front view of the initial position and shape of the imposed gusts (in red) with respect to the wind turbine.

For what concerns the gust amplitude A_g, probabilistic analyses of wind gusts have shown that a gust amplitude of 5 m/s has more than an 80% probability of daily occurrence over central Europe when the wind conditions are similar to the ones chosen in this work [45]. Similarly, gusts exceeding this speed are commonly reported in Germany [46]. Also, more intense gusts (18 to 25 m/s) are fairly common in Europe, especially in coastal areas [25,47]. For these reasons, two gust amplitudes A_g were used in this work, namely 5 and 10 m/s.

Lastly, two shape functions f were used in this work. In the first subsection, a novel gust shape function was proposed, imposing a local redistribution of the flow rate and no global change in it, providing a velocity deficit on its border and a velocity increase in its center. This gust shape was conceptually similar to the "extreme operating gust" from the 61400-IEC standards for wind turbines. Subsequently, in the second subsection, a consistently positive gust velocity was used, analogous to the "extreme coherent gust" from the 61400-IEC standards for wind turbines.

3. Results and Discussion

In this section, the results of the performed simulations are presented. In order to analyze the results, the blade was divided into 10 equally spaced sections, as depicted in Figure 5. These sections are addressed as "strips" and numbered according to the notation in the figure.

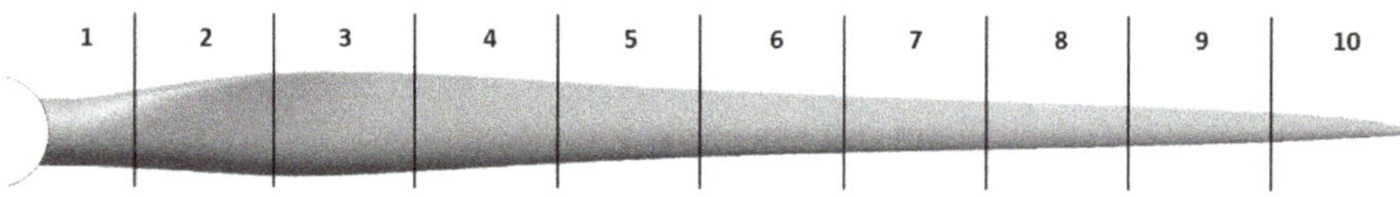

Figure 5. Illustration of the used blade strips.

Furthermore, when presenting the results, the aerodynamic torque (T) and axial force (F, aligned with the axis of rotation) acting on the blades was made non-dimensional by means of Equations (8) and (9), as is normally done in the wind energy community:

$$c_T = \frac{T}{\frac{1}{2}\rho v^2 AR}, \tag{8}$$

$$c_F = \frac{F}{\frac{1}{2}\rho v^2 A}, \tag{9}$$

where ρ is the air density (1.225 kg/m^3), A is the swept area of the rotor, and R is its radius. The velocity v is chosen to be the wind-free stream velocity at the hub height, namely 8.5 m/s.

3.1. Zero Net Flow Rate Gust

In this section, the effect of a gust corresponding to a local redistribution of the available flow rate is investigated. The parameter r indicates the non-dimensional radial coordinate in the gusted cylinder, ranging from 0 (gust center) to 1 (gust border). Imposing C0 and C1 continuity at the border of the cylindrical region ($f(1) = 0$ and $f'(1) = 0$), maximum gust velocity at the center ($f(0) = 1$ and $f'(0) = 0$), and zero net flow rate ($\int_0^{2\pi} \int_0^1 f(r)\ r\ dr\ d\vartheta = 0$), the following polynomial expression is obtained for the radial shape function:

$$f(r) = (1 - 12r^2 + 20r^3 - 9r^4). \tag{10}$$

The obtained function, compared to the "extreme operating gust" from the 61400-IEC standards (used as a function of space, as is done in De Nayer et al. [44]), produced a higher velocity increase, and thus more severe wind conditions, as depicted by Figure 6.

Notice that since this gust shape does not modify the net mass flow rate in the affected area but only redistributes it, it is considered appropriate to be used on inlet boundaries in combination with incompressible solvers, without the necessity to correct the mass flow rate to preserve its steady value [44].

Depending on the radial position where the gust hits the blade, a different effect was found regarding its axial deformation. In order to assess where to hit the blade to obtain the highest effect, several simulations were carried out, positioning the center of the gust at a distance h (Figure 4) equal to 35 m, 40 m, and 45 m starting from the axis. Figure 7 summarizes the results of these simulations.

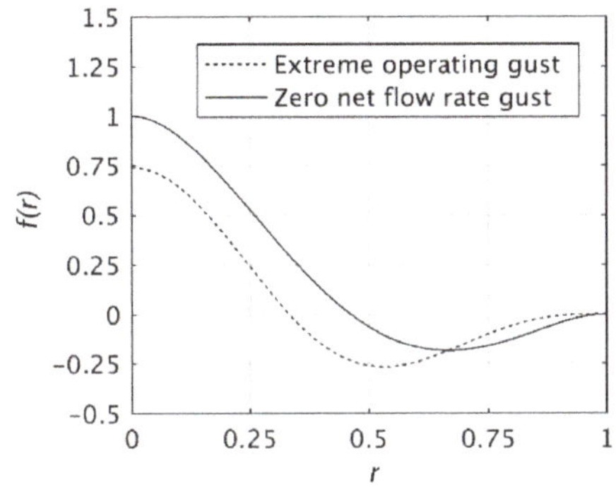

Figure 6. Comparison between the zero net flow rate gust shape function and the "extreme operating gust" from the 61400-IEC standards for wind turbines.

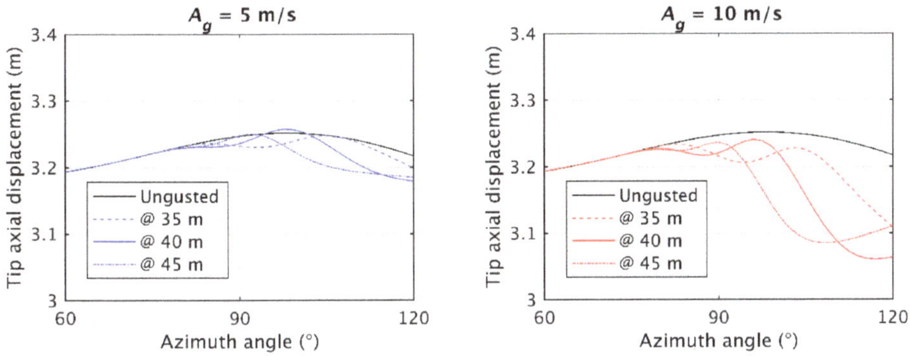

Figure 7. Effect of the radial position of the gust center on the tip axial displacement: (**left**) gust with 5 m/s amplitude and (**right**) 10 m/s amplitude.

Hitting the blade further from its axis of rotation (and thus from its constrained end) increased the lever of the increased axial force. At the same time, due to the tapering of the blade (Figure 5), a smaller area was affected by the pressure increase and thus a smaller axial force increase was obtained. As shown in Figure 7, hitting the blade at 40 m (80%) of its span led to the highest axial deformation, irrespective of the chosen amplitude. For this reason, all the gusts analyzed in the remainder of this work were positioned at a 40 m height from the axis of rotation of the turbine. It was also noticed that, despite the blade always being hit by the gust at a 90° azimuth angle, a higher delay in its peak deflection was obtained when the gust was imposed further from its tip as a result of the higher portion of blade being displaced, and thus, of the higher inertia.

The axial force over the span of the blade was highly influenced by the wind gust. However, the differences on the lower 60% of the blade span (i.e., strips #1 to #6) were negligible, being smaller than 0.77% for $A_g = 5$ m/s and smaller than 1.64% when $A_g = 10$ m/s. Figure 8 shows the axial force evolution over each strip of the outboard 40% span of the blade, as well as the total axial force.

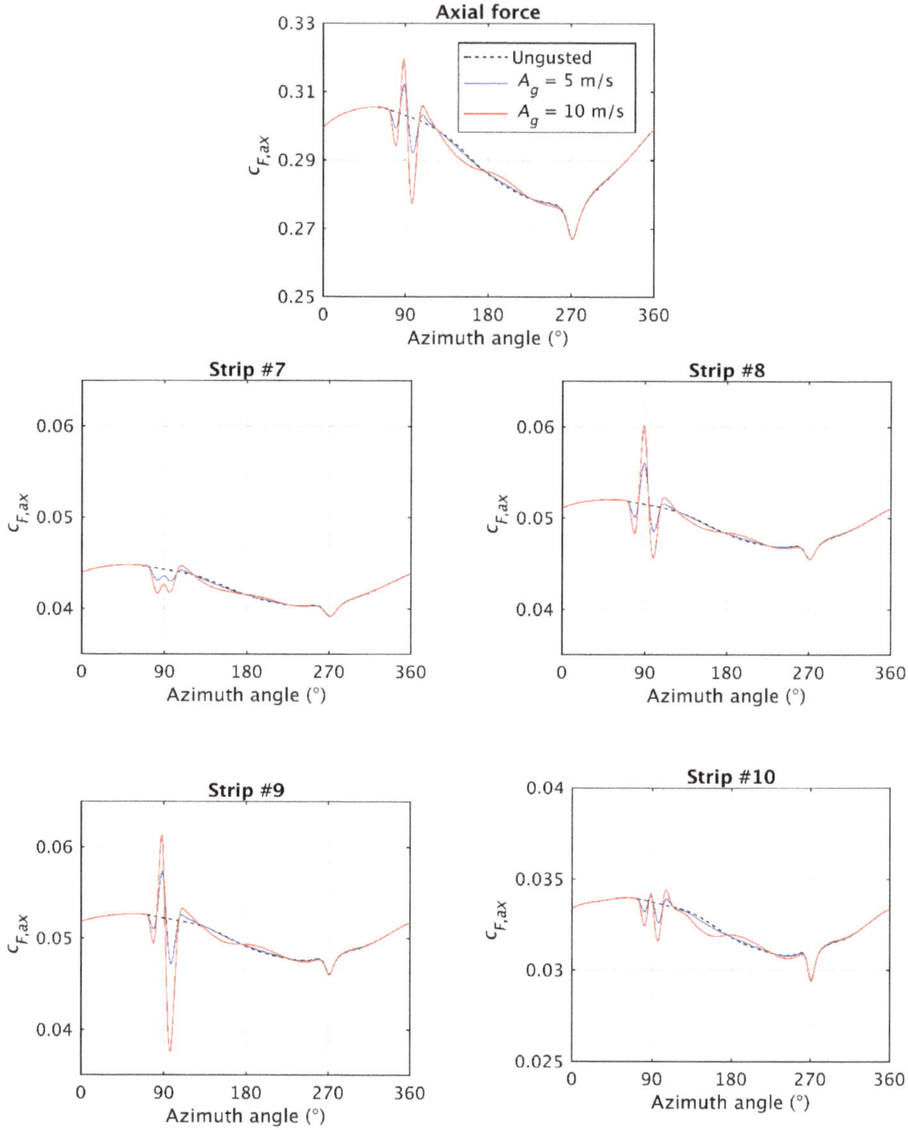

Figure 8. Zero net flow rate gusts: (**top**) total aerodynamic axial force over the blade hit by the gust as a function of its azimuth angle and (**bottom**) axial force contribution of the 4 most outboard strips.

The effect of the ABL is clearly seen in this figure, as well as in the ungusted condition, resulting in a higher loading when the blade pointed upward (90° azimuth angle) and lower loadings when it pointed downward (270° azimuth angle). The strips most sensibly influenced by the gust were #8 and #9 since the center of the gust was positioned exactly between these sections. These strips also provided an important contribution to the total bending moment, as they were located far from the axis of rotation. For both the amplitudes tested, the axial force over each strip resembled the distribution of velocity imposed for the gust, having a positive peak surrounded by two drops. However, it can be noted that the second drop in axial force was larger than the first one, especially for the case with A_g =

10 m/s and on strips #8 and #9. This was due to the occurrence of flow separation in both cases, as illustrated in Figure 9. In this figure, the regions affected by separation were identified by marking the portions of the blade suction side where the tangential component of the wall shear stress was oriented according to the direction of rotation.

Figure 9. Zero net flow rate gusts separation region (in red) over the blade suction side at 90° azimuth angle: (**left**) ungusted case, (**center**) case with A_g = 5 m/s, and (**right**) case with A_g = 10 m/s.

On the root of the blade, a separation region was observed also when no gust was considered. In this region, the blade shape underwent a transition from a cylindrical root to an aerodynamically shaped body. For what concerns the outboard part of the blade, in the case with A_g = 5 m/s, separation occurred only on a limited portion of the blade span and only on a restricted portion of the local chord length. On the other hand, when the gust amplitude was increased to 10 m/s, the separation area grew, expanding both in the spanwise and chordwise directions. This separation region was not reported during the first dip in the axial force. As soon as separation occurred, a sudden drop in the axial force was found. This was most intense on strip #9. As a result, the highest axial force was never reached at a 90° azimuth angle (Figure 8), but always a few degrees earlier. Furthermore, the second drop became longer in time and more intense in the axial force deficit with respect to the first one. This phenomenon strongly influenced the axial tip displacement, as shown in Figure 10. No sensible difference was observed in the tangential displacement because this was strictly related to gravity, as recognized in Santo et al. [6].

In this figure, the impact of the structural inertia was evident. The highest tip displacement was reached with a delay with respect to the axial force. Furthermore, the tip axial displacement immediately started to decrease when the outer border of the gust (where the velocity was decreased) impacted on the blade. Then, when the positive core of the gust hit the blade surface, the displacement started increasing again, and in the case with A_g = 5 m/s, a higher tip displacement was achieved with respect to the ungusted configuration. In the case with A_g = 10 m/s, the higher inertia of the blade due to its initial faster forward movement prevented the blade from reaching high peaks in its tip displacement, even if the axial force was increased, resulting in a maximum displacement lower than the case with A_g = 5 m/s. In addition, when the flow separated (Figure 9), the tip displacement rapidly decreased, preventing the blade from reaching high displacements. This phenomenon was much more intense in the case with $A_g = 10$ m/s since the area affected by the separation was larger (Figure 9), and consequently, the drop in the axial force was also more intense (Figure 8). When the blade moved out of the gust region, the tip showed an oscillatory motion, whose amplitude gradually decreased until the blade reached the (ungusted) regime condition again.

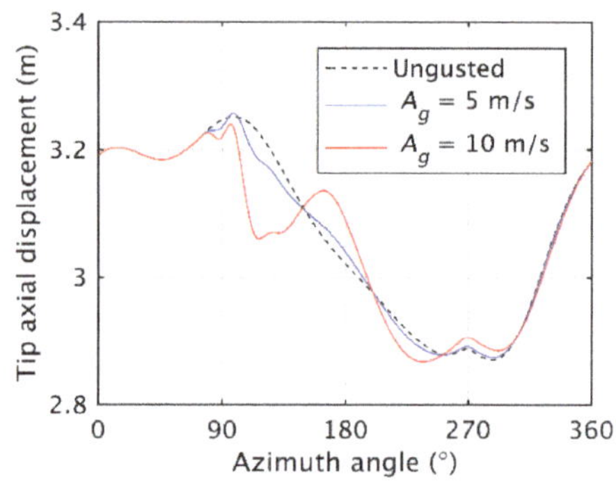

Figure 10. Zero net flow rate gusts: tip axial displacement as a function of its azimuth angle in all analyzed cases.

A slightly different behavior was reported for the torque provided by the blade, as shown in Figure 11. Similarly to the axial force distribution, the torque differences over strips #1 to #6 did not exceed 1.43% for the $A_g = 5$ m/s case and 3.22% for the $A_g = 10$ m/s case.

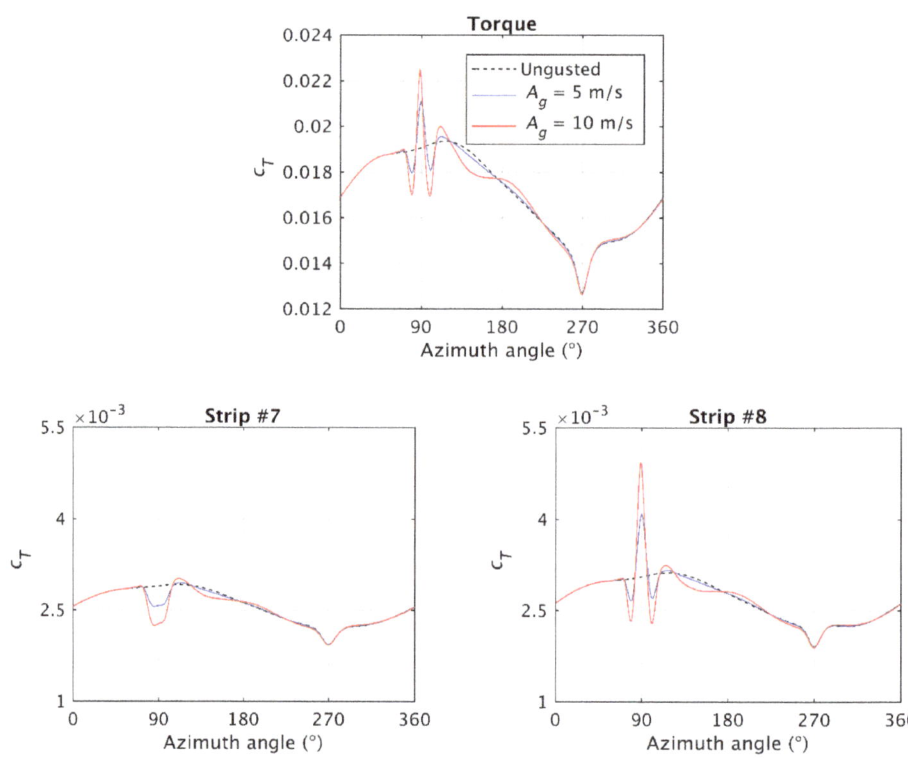

Figure 11. *Cont.*

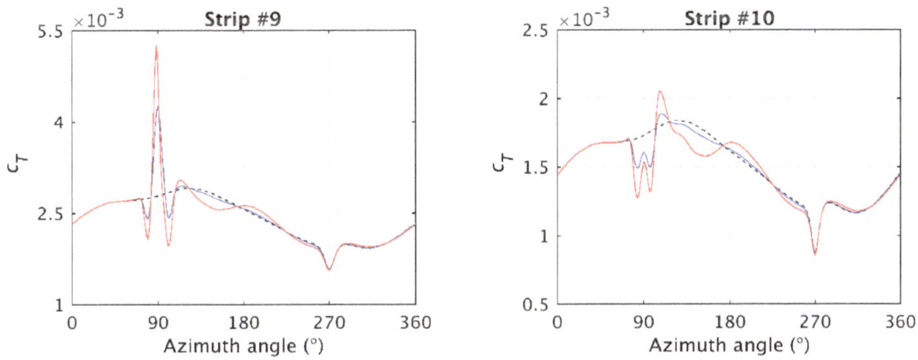

Figure 11. Zero net flow rate gusts: (**top**) total aerodynamic torque over the blade hit by the gust as a function of its azimuth angle and (**bottom**) torque contribution of the four most outboard strips.

Similarly to what was observed for the axial force, the occurrence of separation was reflected in the fact that the highest peak in the torque was never reached at a 90° azimuth angle but always slightly in advance on strips #8 and #9. Differently, the difference in the two drops in torque was much smaller than what was observed for the axial force. This was due to the different effect of separation over the axial and tangential forces. When the flow angle increased, the tangential component of the lift and drag forces also increased, translating into an increase for the tangential force and a decrease for the axial force. At the same time, when separation occurred, the magnitude of lift and drag respectively decreased and increased, leading to a reduction for both the axial and tangential components (assuming that, as is typical, the lift-to-drag ratio was high enough). The two effects compensated for the tangential force and summed up for the axial force, leading to the observed difference in their dynamics. It is also remarked that the tip velocity induced by its transient deformation also sensibly affected the flow angle and the magnitude of the incoming relative velocity, as already observed in Santo et al. [6]. This was reflected in the regions around the 110° azimuth angle, where, on the most outboard strips, a higher torque contribution was found with respect to the ungusted condition as a consequence of the fast forward movement of the blade tip, which increased the incoming flow angle. Lastly, the total torque provided by the machine is provided in Figure 12.

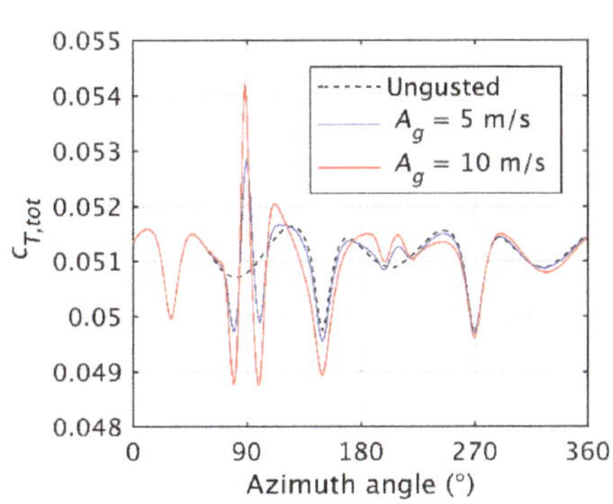

Figure 12. Zero net flow rate gusts: total torque provided by the machine as a function of the azimuth angle of the blade hit by the gust.

The peaks induced by the gust were comparable to the peaks induced by the tower-dam effect. It is also noted that a small effect was observed when the following blade went through the gusted region, i.e., in the azimuthal range around 210°. At this moment, the gust intensity in this region had lowered but had not disappeared.

3.2. The 1 + Cos Gust

In this section, a different gust shape function will be reported on, while the size and position of the gust is the same. The shape function reported on in this paragraph is given in Equation (11), where r indicates the non-dimensional radial coordinate in the gusted cylinder:

$$f(r) = \frac{1}{2}(1 + cos(\pi r)). \quad (11)$$

This gust shape corresponds to the "extreme coherent gust" from the 61400-IEC, as used in similar works [26,29,44]. The present gust shape, when plugged into Equation (7), corresponds to a velocity increase in the whole region affected by the gust, contrary to the gust shape tested in the previous section. As already done for the zero net flow rate gust, two gust amplitudes will be used, namely 5 and 10 m/s.

The total aerodynamic axial force and the contribution of the most outboard strips are illustrated in Figure 13.

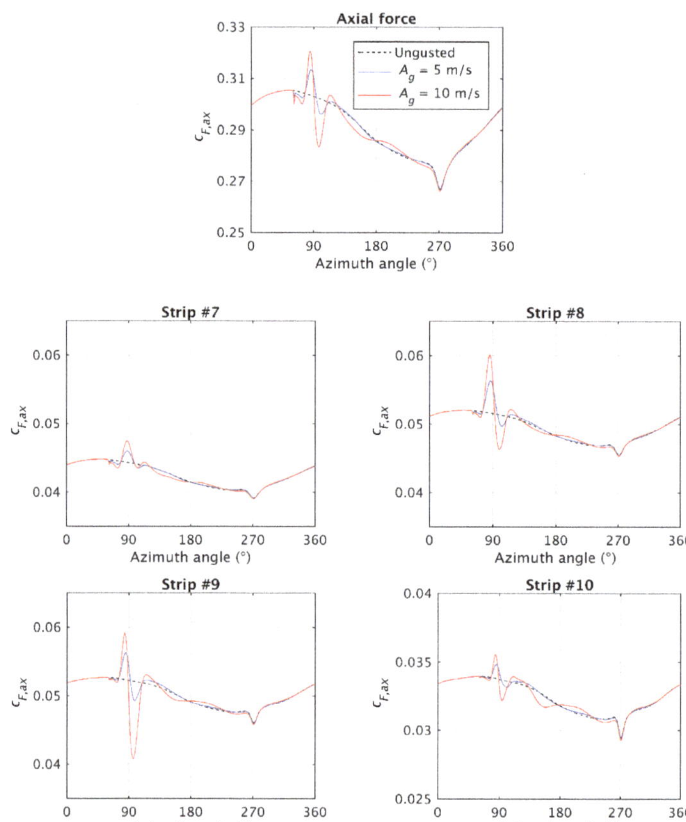

Figure 13. The 1 + cos gusts: (**top**) total aerodynamic axial force over the blade hit by the gust as a function of its azimuth angle and (**bottom**) axial force contribution of the four most outboard strips.

Similarly to what was observed for the zero net flow rate gust, the maximum value in the axial force was always reached in advance of the 90° azimuth angle. Furthermore, a drop in the axial force was observed on each analyzed strip but not due to the gust shape function, indicating, also in this case, the occurrence of a flow separation, as depicted in Figure 14.

Figure 14. The 1 + cos gusts separation region (in red) over the blade suction side at a 90° azimuth angle: (**left**) ungusted case, (**center**) case with A_g = 5 m/s and (**right**) case with A_g = 10 m/s.

Strips #8 and #9 were the most affected by the imposed gusts. Flow separation induced the drops visible in Figure 13. In general, the area affected by the flow separation was broader compared to Figure 9, showing an expansion in both spanwise and chordwise directions. This was due to the more severe gust conditions imposed by the 1 + cos gust. As a result, the drop in axial force was more intense. In particular, when the gust amplitude was increased from 5 to 10 m/s, not only a larger separation region was obtained, but also a lower drop in the axial force. Importantly, the azimuthal range in which the axial force dropped below the ungusted case was wider (especially on strips #8, #9, and #10 in Figure 13) when the gust amplitude was doubled.

The tip axial displacement was also heavily affected by the flow separation, as shown in Figure 15.

It is remarked that, in the case with A_g = 5 m/s, the highest tip axial displacement was obtained. Furthermore, in both cases, separation lowered the tip axial displacement in comparison with the ungusted configuration, acting as a protection mechanism against extreme deflection. This was clearly visible in the case with A_g = 10 m/s, where separation prevented the blade from reaching high deflections and resulted in a maximum deflection lower than the case with A_g = 5 m/s. In the latter case, the effect of the separation on the tip axial displacement was less evident and was restricted only to a marginal influence after a 120° azimuth angle. When the blade surpassed the region affected by the gust, its tip continued oscillating around the ungusted deflection, until it went back into regime conditions by the end of the analyzed full revolution.

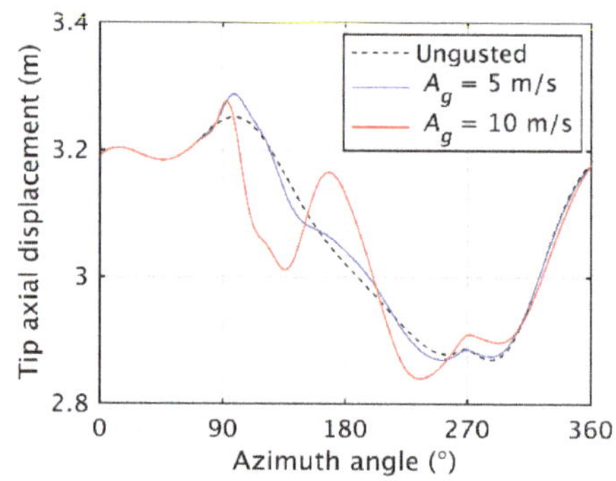

Figure 15. The 1 + cos gusts: tip axial displacement as a function of its azimuth angle in all analyzed cases.

Lastly, the torque provided by the blade hit by the gust is given in Figure 16.

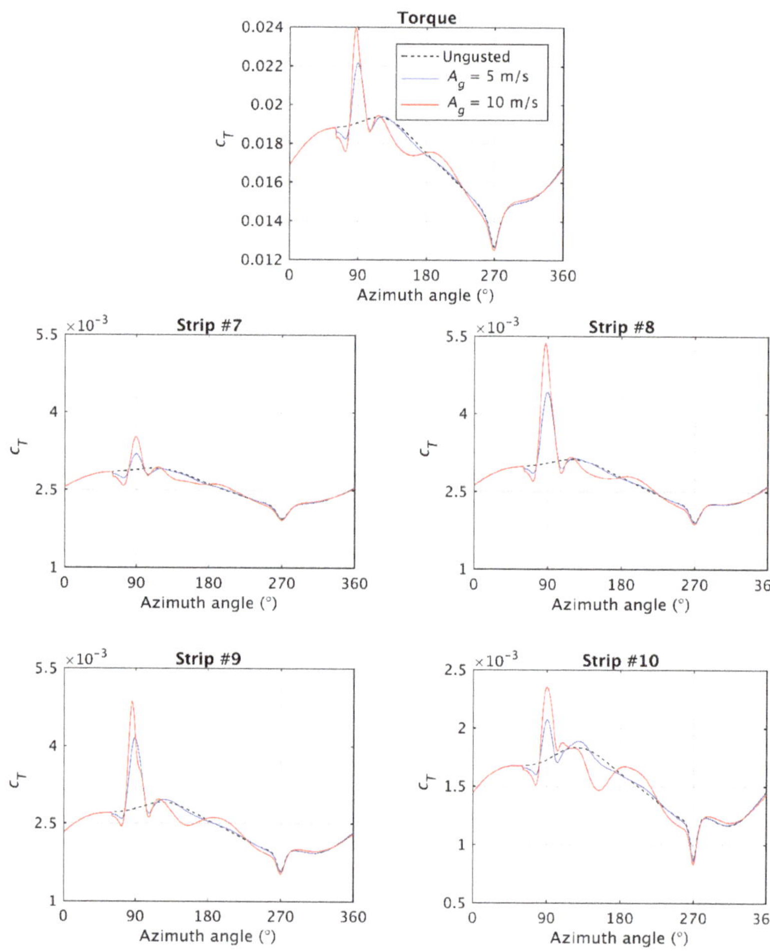

Figure 16. The 1 + cos gusts: (**top**) total aerodynamic torque over the blade hit by the gust as a function of its azimuth angle and (**bottom**) torque contribution of the four most outboard strips.

Similarly to what was explained in the previous section, the effect of separation on the torque was much less intense than on the axial force, resulting in very small drops below the ungusted condition. However, a sudden change in the trend was visible, especially on strips #8 and #9, around an 87° azimuth angle, due to the separation itself, which prevented the maximum torque being reached at a 90° azimuth angle. In the region downstream of the gust (from 110° azimuth angle onwards), the differences observed in the torque, especially in the case with the highest gust amplitude, were attributed to the fast tip movement (Figure 15), which sensibly acted on the incoming flow angle and relative velocity magnitude.

4. Conclusions

In this work, the effect of a wind gust of various shapes and intensities on a modern horizontal axis wind turbine was investigated. A detailed and high-fidelity aeroelastic model was employed, implicitly coupling a computational fluid dynamics (CFD) solver based on an overset technique and a computational structural mechanics (CSM) solver, loyally reproducing the characteristics of the composite material of each blade. The gusts were superimposed on the atmospheric boundary layer.

First, the effect of a wind gust introducing a zero net flow rate was analyzed. This gust was conceptually similar to the "extreme operating gust" from the 61400-IEC standards but introduced a higher velocity increase. Results showed that an initial decrease in the blade loads and displacement was a consequence of the negative velocity increase imposed on the border of the gust. When the positive core of the gust impacted on the blade, the inertia of the structure caused a delay in the tip movement. Furthermore, despite the high peak reached by the aerodynamic axial force on the blade, flow separation over the span affected by the gust prevented the blade from reaching extreme deflections. Increasing the gust intensity, this protective effect was magnified by the broader area affected by the separation.

Subsequently, a gust analogous to the "extreme coherent gust" from the 61400-IEC standards was tested, introducing a consistently positive velocity increase. In this case, the peak tip deflection was shown to be higher than in the previous case as a consequence of the more severe wind conditions. Flow separation was also observed and affected a broader portion of the blade suction side, resulting in a fast reaction of the blade, whose tip underwent a fast axial movement. It was therefore concluded that, for each gust tested, flow separation acted as a protection mechanism and prevented the blade from reaching extreme deflections. For all the tested gusts, different dynamics were observed for the torque and axial force, especially when separation occurred.

Lastly, it can be concluded that the presented methodology allowed for the detailed investigation of the interaction between the blade structural response and the occurring wind gust: such data can be useful in the design stage. Among others, results can be used to better estimate the loading of the blade with respect to the meteorological data about the frequency, size, and intensity of the wind gusts in the site selected for the installation of the analyzed wind turbine.

Author Contributions: Conceptualization, G.S., M.P., W.V.P., and J.D.; methodology, G.S.; software, G.S.; validation, G.S. and J.D.; formal analysis, G.S. and J.D.; investigation, G.S.; resources, J.D.; data curation, G.S. and M.P.; writing—original draft preparation, G.S.; writing—review and editing, G.S., M.P., W.V.P., and J.D.; visualization, G.S.; supervision, J.D. and W.V.P.; project administration, W.V.P. and J.D.; funding acquisition, W.V.P. and J.D. All authors have read and agreed to the published version of the manuscript.

Funding: This work was supported by the Fonds Wetenschappelijk Onderzoek—Vlaanderen (FWO, grant no. G030414N). The computational resources (Stevin Supercomputer Infrastructure) and services used were provided by the VSC (Flemish Supercomputer Center), funded by Ghent University, the Hercules Foundation, and the Flemish Government, department of Economy Science and Innovation (EWI).

Acknowledgments: The authors want to thankfully acknowledge Jan Vierendeels for his contribution to this work.

Conflicts of Interest: The authors declare no conflict of interest.

References

1. Caduff, M.; Huijbregts, M.A.J.; Althaus, H.J.; Koehler, A.; Hellweg, S. Wind power electricity: The bigger the turbine, the greener the electricity? *Environ. Sci. Technol.* **2012**, *46*, 4725–4733. [CrossRef]
2. Bazilevs, Y.; Hsu, M.C.; Akkerman, I.; Wright, S.; Takizawa, K.; Henicke, B.; Spielman, T.; Tezduyar, T.E. 3D simulation of wind turbine rotors at full scale. Part I: Geometry modeling and aerodynamics. *Int. J. Numer. Methods Fluids* **2011**, *65*, 207–235. [CrossRef]
3. Bazilevs, Y.; Hsu, M.C.; Akkerman, I.; Kiendl, J.; Wüchner, R.; Bletzinger, K.U. 3D simulation of wind turbine rotors at full scale. Part II: Fluid–structure interaction modeling with composite blades. *Int. J. Numer. Methods Fluids* **2011**, *65*, 236–253. [CrossRef]
4. Hsu, M.C.; Bazilevs, Y. Fluid-structure interaction modelling of wind turbines: Simulating the full machine. *Comput. Mech.* **2012**, *50*, 821–833. [CrossRef]
5. Liu, X.; Lu, C.; Liang, S.; Godbole, A.; Chen, Y. Vibration-induced aerodynamic loads on large horizontal axis wind turbine blades. *Appl. Energy* **2017**, *185*, 1109–1119. [CrossRef]
6. Santo, G.; Peeters, M.; Van Paepegem, W.; Degroote, J. Dynamic load and stress analysis of a large Horizontal Axis Wind Turbine using full scale fluid-structure interaction simulation. *Renew. Energy* **2019**, *140*, 212–226. [CrossRef]
7. Lee, Y.J.; Jhan, Y.T.; Chung, C.H. Fluid–structure interaction of FRP wind turbine blades under aerodynamic effect. *Compos. Part B Eng.* **2012**, *43*, 2180–2191. [CrossRef]
8. Heinz, J.C.; Sørensen, N.N.; Zahle, F. Fluid–structure interaction computations for geometrically resolved rotor simulations using CFD. *Wind Energy* **2016**, *19*, 2205–2221. [CrossRef]
9. Ahlström, A. Influence of wind turbine flexibility on loads and power production. *Wind Energy* **2006**, *9*, 237–249. [CrossRef]
10. Shen, W.Z.; Mikkelsen, R.; Sorensen, J.N.; Bak, C. Tip loss corrections for wind turbine computations. *Wind Energy* **2005**, *8*, 457–475. [CrossRef]
11. Sørensen, J.N.; Kock, C.W. A model for unsteady rotor aerodynamics. *J. Wind Eng. Ind. Aerodyn.* **1995**, *58*, 259–275. [CrossRef]
12. Shen, W.Z.; Zhu, W.J.; Sørensen, J.N. Actuator line/Navier–Stokes computations for the MEXICO rotor: Comparison with detailed measurements. *Wind Energy* **2012**, *15*, 811–825. [CrossRef]
13. Shen, W.Z.; Zhang, J.H.; Sørensen, J.N. The Actuator Surface Model: A New Navier–Stokes Based Model for Rotor Computations. *J. Sol. Energy Eng.* **2009**, *131*. [CrossRef]
14. Kim, H.; Lee, S.; Son, E. Aerodynamic noise analysis of large horizontal axis wind turbines considering fluid-structure interaction. *Renew. Energy* **2011**. [CrossRef]
15. Yu, D.O.; Kwon, O.J. Predicting wind turbine blade loads and aeroelastic response using a coupled CFD-CSD method. *Renew. Energy* **2014**, *70*, 184–196. [CrossRef]
16. Dai, L.; Zhou, Q.; Zhang, Y.; Yao, S.; Kang, S.; Wang, X. Analysis of wind turbine blades aeroelastic performance under yaw conditions. *J. Wind Eng. Ind. Aerodyn.* **2017**, *171*, 273–287. [CrossRef]
17. MacPhee, D.W.; Beyene, A. Experimental and Fluid Structure Interaction analysis of a morphing wind turbine rotor. *Energy* **2015**, *90*, 1055–1065. [CrossRef]
18. MacPhee, D.W.; Beyene, A. Fluid structure interaction analysis of a morphing vertical axis wind turbine. *J. Fluids Struct.* **2016**, *60*, 143–159. [CrossRef]
19. Wang, L.; Quant, R.; Kolios, A. Fluid structure interaction modelling of horizontal-axis wind turbine blades based on CFD and FEA. *J. Wind Eng. Ind. Aerodyn.* **2016**, *158*, 11–25. [CrossRef]
20. Bazilevs, Y.; Korobenko, A.; Deng, X.; Yan, J. Novel structural modeling and mesh moving techniques for advanced fluid–structure interaction simulation of wind turbines. *Int. J. Numer. Methods Eng.* **2014**. [CrossRef]
21. Li, Y.; Castro, A.M.; Sinokrot, T.; Prescott, W.; Carrica, P.M. Coupled multi-body dynamics and CFD for wind turbine simulation including explicit wind turbulence. *Renew. Energy* **2015**, *76*, 338–361. [CrossRef]
22. Peeters, M.; Santo, G.; Degroote, J.; Van Paepegem, W. High-fidelity finite element models of composite wind turbine blades with shell and solid elements. *Compos. Struct.* **2018**, *200*, 521–531. [CrossRef]
23. Korobenko, A.; Yan, J.; Gohari, S.M.I.; Sarkar, S.; Bazilevs, Y. FSI Simulation of two back-to-back wind turbines in atmospheric boundary layer flow. *Comput. Fluids* **2017**, *158*, 167–175. [CrossRef]

24. Zhou, K.; Cherukuru, N.; Sun, X.; Calhoun, R. Wind Gust Detection and Impact Prediction for Wind Turbines. *Remote Sens.* **2018**, *10*, 514. [CrossRef]
25. Mohr, S.; Kunz, M.; Richter, A.; Ruck, B. Statistical characteristics of convective wind gusts in Germany. *Nat. Hazards Earth Syst. Sci.* **2017**, *17*. [CrossRef]
26. Wu, Z.; Bangga, G.; Cao, Y. Effects of lateral wind gusts on vertical axis wind turbines. *Energy* **2019**, *167*, 1212–1223. [CrossRef]
27. Onol, A.O.; Yesilyurt, S. Effects of wind gusts on a vertical axis wind turbine with high solidity. *J. Wind Eng. Ind. Aerodyn.* **2017**, *162*, 1–11. [CrossRef]
28. Bhargav, M.M.S.R.S.; Kishore, V.R.; Laxman, V. Influence of fluctuating wind conditions on vertical axis wind turbine using a three dimensional CFD model. *J. Wind Eng. Ind. Aerodyn.* **2016**, *158*, 98–108. [CrossRef]
29. Timme, S.; Badcock, K.J.; Da Ronch, A. Gust analysis using CFD derived reduced order models. *J. Fluids Struct.* **2017**, *71*, 116–125. [CrossRef]
30. Svacek, P.; Horacek, J. On mathematical modelling of fluid-structure interactions with nonlinear effects: Finite element approximations of gust response. *J. Comput. Appl. Math.* **2015**, *273*, 394–403. [CrossRef]
31. Younsi, R.; El-Batanony, I.; Tritsch, J.-B.; Naji, H.; Landjerit, B. Dynamic study of a wind turbine blade with horizontal axis. *Eur. J. Mech. A/Solids* **2001**, *20*, 241–252. [CrossRef]
32. Castellani, F.; Astolfi, D.; Becchetti, M.; Berno, F. Experimental and Numerical Analysis of the Dynamical Behavior of a Small Horizontal-Axis Wind Turbine under Unsteady Conditions: Part, I. *Machines* **2018**, *6*, 52. [CrossRef]
33. Ebrahimi, A.; Sekandari, M. Transient response of flexible blade of horizontal axis wind turbines in wind gusts and rapid yaw changes. *Energy* **2018**, *145*, 261–275. [CrossRef]
34. Franke, J.; Hellsten, A.; Schlunzen, K.H.; Carissimo, B. The COST 732 Best Practice Guideline for CFD simulation of flows in the urban environment: A summary. *Int. J. Environ. Pollut.* **2011**, *44*, 419–427. [CrossRef]
35. Sayed, M.; Lutz, T.; Krämer, E. Aerodynamic investigation of flow over a multi-megawatt slender bladed horizontal-axis wind turbine.In Renewable Energies Offshore. In Proceedings of the 1st International Conference on Renewable Energies Offshore (RENEW2014), Lisbon, Portugal, 24–26 November 2014.
36. Harte, R.; Van Zijl, G. Structural stability of concrete wind turbines and solar chimney towers exposed to dynamic wind action. *J. Wind Eng.* **2007**, *95*, 1079–1096. [CrossRef]
37. Richards, P.J.; Hoxey, R.P. Appropriate boundary conditions for computational wind engineering models using the k-ε turbulence model. *J. Wind Eng. Ind. Aerodyn.* **1993**, 145–153. [CrossRef]
38. Wieringa, J. Updating the Davenport roughness classification. *J. Wind Eng. Ind. Aerodyn.* **1992**, *41*, 357–368. [CrossRef]
39. Blocken, B.; Stathopoulos, T.; Carmeliet, J. CFD simulation of the atmospheric boundary layer: Wall function problems. *Atmos. Environ.* **2007**, *41*, 238–252. [CrossRef]
40. Parente, A.; Gorlé, C.; van Beeck, J.; Benocci, C. A Comprehensive Modelling Approach for the Neutral Atmospheric Boundary Layer: Consistent Inflow Conditions, Wall Function and Turbulence Model. *Bound. Layer Meteorol.* **2011**, *140*, 411. [CrossRef]
41. Parente, A.; Gorlé, C.; van Beeck, J.; Benocci, C. Improved k–e model and wall function formulation for the RANS simulation of ABL flows. *J. Wind Eng. Ind. Aerodyn.* **2011**. [CrossRef]
42. Degroote, J. Partitioned simulation of fluid-structure interaction: Coupling black-box solvers with quasi-Newton techniques. *Arch. Comput. Methods Eng.* **2013**, *20*, 185–238. [CrossRef]
43. Ortiz, R.; Casadei, F.; Mouton, S.; Sobry, J.F. Propeller blade debris kinematics: Blade debris trajectory computation with aerodynamic effects using new FSI formulations. *CEAS Aeronaut. J.* **2018**, *9*, 683–694. [CrossRef]
44. De Nayer, G.; Breuer, M.; Perali, P.; Grollman, K. Modelling of wind gusts for large-eddy simulations related to fluid-structure interactions. In *Direct and Large-Eddy Simulation XI*; Springer: Cham, Switzerland, 2019. [CrossRef]
45. Jungo, P.; Goyette, S.; Beniston, M. Daily wind gust speed probabilities over Switzerland according to three types of synoptic circulation. *Int. J. Climatol.* **2002**, *22*, 485–499. [CrossRef]

46. Friederichs, P.; Gober, M.; Bentzien, S.; Lenz, A.; Krampitz, R. A probabilistic analysis of wind gusts using extreme value statistics. *Meteorologische Zeitschrift* **2009**, *18*, 615–629. [CrossRef]
47. Vajda, A.; Tuomenvirta, H.; Juga, I.; Nurmi, P.; Jokinen, P.; Rauhala, J. Severe weather affecting European transport systems: The identification, classification and frequencies of events. *Nat. Hazards* **2014**, *72*, 169–188. [CrossRef]

energies

MDPI

Article

Evaluation of Actuator Disk Model Relative to Actuator Surface Model for Predicting Utility-Scale Wind Turbine Wakes

Zhaobin Li [1] **and Xiaolei Yang** [1,2,*]

1 The State Key Laboratory of Nonlinear Mechanics, Institute of Mechanics, Chinese Academy of Sciences, Beijing 100190, China; zhaobin.li@imech.ac.cn
2 School of Engineering Sciences, University of Chinese Academy of Sciences, Beijing 100049, China
* Correspondence: xyang@imech.ac.cn

Received: 11 May 2020; Accepted: 8 July 2020; Published: 10 July 2020

Abstract: The Actuator Disk (AD) model is widely used in Large-Eddy Simulations (LES) to simulate wind turbine wakes because of its computing efficiency. The capability of the AD model in predicting time-average quantities of wind tunnel-scale turbines has been assessed extensively in the literature. However, its capability in predicting wakes of utility-scale wind turbines especially for the coherent flow structures is not clear yet. In this work, we take the time-averaged statistics and Dynamic Mode Decomposition (DMD) modes computed from a well-validated Actuator Surface (AS) model as references to evaluate the capability of the AD model in predicting the wake of a 2.5 MW utility-scale wind turbine for uniform inflow and fully developed turbulent inflow conditions. For the uniform inflow cases, the predictions from the AD model are significantly different from those from the AS model for the time-averaged velocity, and the turbulence kinetic energy until nine rotor diameters (D) downstream of the turbine. For the turbulent inflow cases, on the other hand, the differences in the time-averaged quantities predicted by the AS and AD models are not significant especially at far wake locations. As for DMD modes, significant differences are observed in terms of dominant frequencies and DMD patterns for both inflows. Moreover, the effects of incoming large eddies, bluff body shear layer instability, and hub vortexes on the coherent flow structures are discussed in this paper.

Keywords: wind turbine wake; actuator disk model; actuator surface model; dynamic mode decomposition; coherent structures; wake meandering

1. Introduction

Nowadays, large wind farms are constructed to respond to the increasing demand of renewable energy. In these wind farms, turbines are installed in cluster to meet the geographical restriction and to reduce the cable and maintenance cost. A turbine may influence its downwind neighbors significantly with the wake effect, leading to a loss of the power production and an increase of the unsteady load on the structure [1]. Therefore, a need to better understand the wake behavior and its influence on the downwind turbines arises.

Understanding turbine wakes in a wind farm is challenging because of its multi-scale nature. For example, the boundary layer on a wind turbine's blade has a thickness of the centimeters, which is orders of magnitudes smaller than the diameter of the rotor ($\approx$100 m) and the thickness of the Atmospheric Boundary Layer (ABL; $\approx$1000 m) [2]. Among others, the difficulty in accurately modeling the flow around the blade of a real wind turbine blade arises both in wind tunnel experiments (due to scale effect [3]) and in numerical simulations (due to the resolution requirement).

Facing this challenge, it is often assumed that the actual geometry of the rotor is of less importance [4] if only the far wake and its influence on downwind turbines are of interest, so that the wind turbine can be approximated by equivalent models. In experiments, the simplest model is the porous disk model, which is broadly used [5,6]. Validations in wind tunnels have demonstrated that porous discs can provide time-averaged wake properties with satisfactory accuracy in the region further than 3.5 rotor diameters downstream, especially when the turbulence of the ABL is concerned [7–9].

In numerical simulations, a series of blade approximated actuator type models representing the turbine blades using equivalent distributed forces have been proposed. The way in which these forces are calculated and distributed distinguishes different models. The simplest is the Actuator Disk (AD) model, which is the numerical equivalence of a porous disk. The thrust on the disk is calculated with one-dimensional momentum theory and is usually distributed uniformly over the rotor swept area with the rotation effects neglected. The Actuator Line (AL) method was proposed to take into account the effects of individual rotating blades [10]. The AL models a wind turbine blade by a rotating line with lift and drag forces determined from tabulated geometric and aerodynamic data of airfoil. To better take into account the geometrical effects of wind turbine blades, the Actuator Surface (AS) method has been proposed, which models a blade as a two dimensional surface with zero thickness [11,12]. Because of its simplicity and computational efficiency, the AD has been widely used in turbine wake simulations especially in farm-scale simulations [13–16]. The capability of the actuator disk model in predicting turbine wakes, especially in the far wake region, has been widely validated in the literature [14,17,18]. However, besides the thrust, experiments revealed that the rotor's rotation also influences both the power output and the wake characteristics significantly [19] and including these rotational effects in the actuator disk model can improve the model's accuracy [20]. In [21], the authors showed that the actuator disk model can reasonably predict the mean velocity profiles but underpredict the turbulence kinetic energy (TKE) for the wake of an axial-flow hydrokinetic turbine. It is noticed that most of the validation studies were focused on time-averaged quantities without probing into the dynamic behavior of turbine wakes, e.g., coherent flow structures and the wake meandering, for which the dataset is difficult to obtain from utility-scale wind turbines [22]. Furthermore, inconsistent results were observed in wind tunnel experiments on the dynamic behavior of turbine wakes when different turbine models were used. For instance, regarding the origin of wake meandering, Medici et al. [3,23] found the wake meandering was related to the bluff body vortex shedding in the experiment of a small scale wind turbine, whereas Espana et al. [6] claimed that the meandering was attributed to the inflow large eddies by carrying out an experiment by representing the turbine with a porous disk. These wind tunnel measurements already make it questionable whether the AD (or the porous disk) model can predict correctly the dynamics of small scale wind turbine wakes. Less is known when applying such a model to utility-scale wind turbines.

To this end, the present study employs simulation results from the well-validated AS model proposed in [11] to examine the capability of the AD model in predicting the dynamic behavior of a utility-scale wind turbine under uniform and fully developed turbulent inflow conditions. Large-Eddy Simulation (LES) is employed for turbulent flow simulations. For both models, exactly the same computational setup is employed. We first compare the time-averaged quantities and then employ the dynamic mode decomposition (DMD) to facilitate the comparison of the most dominant dynamic flow structures and the frequency spectra between the AD and AS models.

The remainder of this paper is structured as follows. Section 2 presents the theory of the AD and the AS models together with a brief description of the LES solver and the DMD method. In Section 3 the simulation setup is provided. Section 4 depicts the simulation results and the DMD analysis in both uniform and ABL conditions. A discussion is provided in Section 5 before the final conclusion in Section 6.

2. Numerical Methods

This section describes the concept of the AD and the AS wind turbine model, the LES flow solver employed in this study, and the dynamic modal decomposition (DMD) employed to analyze the dynamic flow structure from the simulations.

2.1. Wind Turbine Models

2.1.1. Actuator Disk Model

The AD model neglects the geometry detail of individual wind turbine blades. It represents the rotating blades as a fixed 2D porous disk exerting a uniform thrust on the flow, which is the numerical reflection of the perforated disk model usually used in wind tunnel experiments. Neither the rotational effect nor the non-uniform force distribution are considered in the model employed in this work. The axial thrust force f_T per unit area is uniformly distributed over the entire rotor disk surface A and is expressed with the thrust coefficient C_{T} and the inflow velocity V_{ref}:

$$f_T = \frac{1}{2}\rho V_{\mathrm{ref}}^2 C_{\mathrm{T}}. \tag{1}$$

where ρ is the density of air. The reference velocity V_{ref} is defined to be equal to the freestream velocity in uniform inflow condition. In turbulent inflow simulations, the present work approximately calculates V_{ref} by averaging the velocity on a disk of the rotor's size at one diameter in front of the real turbine. The trust coefficient C_{T} remains to be determined according to the turbine operation state. In this work, C_{T} is set to be equal to that of the AS simulations to ensure a fair comparison between the two models.

2.1.2. Actuator Surface Model

The AS model represents the geometry of an individual wind turbine blade with a simplified two dimensional surfaces of zero thickness, which is formed by chords at different radial locations [11,12]. In the actuator model employed in this work, the aerodynamic forces on the surface vary with the radial position and are determined by the tabulated airfoil data in the same ways as the AL model as follows:

$$\mathbf{L} = \frac{1}{2}\rho C_{\mathrm{L}} c |V_{\mathrm{ref}}|^2 \mathbf{e}_{\mathrm{L}} \tag{2}$$

and

$$\mathbf{D} = \frac{1}{2}\rho C_{\mathrm{D}} c |V_{\mathrm{ref}}|^2 \mathbf{e}_{\mathrm{D}}, \tag{3}$$

where $\mathbf{L}$ and $\mathbf{D}$ are the lift and drag force per unit length, ρ is the density of air, c is the chord length, V_{ref} is the flow velocity relative to the rotating blade, $\mathbf{e}_{\mathrm{L}}$ and $\mathbf{e}_{\mathrm{L}}$ are unit directional vectors for lift and drag forces. C_{L} and C_{D} are the lift and the drag coefficients defined in 2D airfoil tables as a function of Reynolds number and the angle of attack. Corrections including the 3D stall delay model of Du and Selig [24] and the tip loss correction of Shen et al. [25,26] are applied.

After calculating $\mathbf{L}$ and $\mathbf{D}$, the force $\mathbf{f}$ per unit area on the surface model is calculated by:

$$\mathbf{f} = (\mathbf{L} + \mathbf{D})/c. \tag{4}$$

The reacting forces exerting by the blade on the air are then distributed to the background Eulerian grid points with a smoothed discrete delta function [27].

2.2. Flow Solver

The turbulent flow is solved with a three dimensional LES code, dubbed as virtual flow simulator (VFS-Wind) [28], in which the governing equations are the filtered Navier–Stokes equations for incompressible flows, which read in the compact tensor form as $(i, j, k, l = 1, 2, 3)$:

$$J\frac{\partial U^i}{\partial \xi^i} = 0, \tag{5}$$

$$\frac{1}{J}\frac{\partial U^j}{\partial t} = \frac{\xi_l^i}{J}\left(-\frac{\partial}{\partial \xi^j}\left(U^j u_l\right) + \frac{\mu}{\rho}\frac{\partial}{\partial \xi^j}\left(\frac{g^{jk}}{J}\frac{\partial u_l}{\partial \xi^k}\right) - \frac{1}{\rho}\frac{\partial}{\partial \xi^j}\left(\frac{\xi_l^j p}{J}\right) - \frac{1}{\rho}\frac{\partial \tau_{lj}}{\partial \xi^j} + f_l\right), \tag{6}$$

where ξ_i is the curvilinear coordinates, $\xi_l^i = \partial\xi_i/\partial x_l$ is the transformation metrics with x_l the Cartesian coordinates, J denotes the Jacobian of the geometric transformation, $U^i = \left(\xi_l^i/J\right) u_l$ is the contravariant volume flux with u_l the velocity in Cartesian coordinates, μ is the dynamic viscosity, ρ is the density, $g^{jk} = \xi_l^j \xi_l^k$ is the components of the contravariant metric tensor, p is the pressure, f_l is the body forces exerted by the actuator models, and τ_{ij} is the sub-grid stress (SGS) resulted from the filtering operation and is modeled with the Smagorinsky SGS model [29] as follows,

$$\tau_{ij} - \frac{1}{3}\tau_{kk}\delta_{ij} = -\mu_t \overline{S_{ij}}, \tag{7}$$

where μ_t is the eddy viscosity and $\overline{S_{ij}}$ is the large-scale strain-rate tensor with $\overline{(\cdot)}$ denoting the grid filtering operator. The eddy viscosity is computed by

$$\mu_t = C_s \Delta^2 |\overline{S}|, \tag{8}$$

where Δ is the filter width, $|\overline{S}| = (2\overline{S_{ij}}\overline{S_{ij}})^{1/2}$ is the magnitude of the strain-rate tensor with C_s the Smagorinsky constant computed via the dynamic procedure of [30].

A second-order accurate central differencing scheme is used for space discretization. The time integration uses the fractional step method [31]. The momentum equation is solved with a matrix-free Newton–Krylov method [32] . The pressure Poisson equation is solved with a Generalized Minimal Residual (GMRES) method with an algebraic multi-grid acceleration [33].

2.3. Dynamic Mode Decomposition

DMD is an equation-free, data-driven method for data analysis and behavior prediction of complex dynamical systems. It was first proposed by Schmid [34] to analyze the high-dimensional fluid dynamics data by decomposing it into coherent spatial structures that oscillate at distinct frequencies. Thanks to its ability both to analyze and to predict, it has gained successes not only in fluid mechanics, but also in other fields, including video processing and finance, where high-dimensional complex dynamic systems are involved [35].

For fluid mechanics applications, the input of DMD analysis is a sequence of snapshots of the flow field. The snapshot $\mathbf{x}_i$ is a column vector of dimension n containing all the interested variables at measure points in the flow field at time $t = t_i$. The snapshots are taken at a fixed time interval Δt. With total m snapshots, the snapshot matrix of the dataset can be written as:

$$\mathbf{X}_1^m = \begin{bmatrix} | & | & & | \\ \mathbf{x}_1 & \mathbf{x}_2 & \ldots & \mathbf{x}_m \\ | & | & & | \end{bmatrix}_{n\times m}. \tag{9}$$

where the sub and sup indexes indicate the starting and the ending index of snapshots. The DMD approximates the dynamic system by a time-independent linear matrix $\mathbf{A}$ representing the temporal evolution from one snapshot to the next, as follows,

$$\mathbf{X}_2^m = \mathbf{A}_{n\times n}\mathbf{X}_1^{m-1}. \tag{10}$$

In practice, the spatial dimension n is usually very large, which makes the direct solution of $\mathbf{A}_{n\times n}$ difficult. Instead, the reduced-order Proper Orthogonal Decomposition (POD) projection matrix $\tilde{\mathbf{A}}_{r\times r}$ with $r \ll n$ is used. r is the number of the most energetic modes kept in the singular value decomposition of $\mathbf{X}_1^{m-1}$, defined as follows,

$$\mathbf{X}_1^{m-1} \approx \mathbf{U}_{n\times r}\mathbf{\Sigma}_{r\times r}\mathbf{V^T}_{r\times(m-1)}, \tag{11}$$

where the left singular vectors in $\mathbf{U}$ are POD modes, the right singular vectors in $\mathbf{V}$ are time coefficient of these modes, and $\mathbf{\Sigma}$ is a diagonal matrix containing the first r largest singular values. $\mathbf{U}$ and $\mathbf{V}$ are orthonormal. $\tilde{\mathbf{A}}$ is defined as

$$\tilde{\mathbf{A}} = \mathbf{U}^T\mathbf{A}\mathbf{U} = \mathbf{U}^T\mathbf{X}_2^m\mathbf{V}\mathbf{\Sigma}^{-1}. \tag{12}$$

The next step is to calculate the eigenvalues and eigenvectors of $\tilde{\mathbf{A}}$,

$$\tilde{\mathbf{A}}\mathbf{W} = \mathbf{W}\mathbf{\Lambda}, \tag{13}$$

where $\mathbf{\Lambda}$ is a diagonal matrix containing r eigenvalues and $\mathbf{W}$ contains r eigenvectors of unit length. The snapshots $\mathbf{x}_k = \mathbf{A}\mathbf{x}_{k-1} = \mathbf{A}^{k-1}\mathbf{x}_1$ can be written as

$$\mathbf{x}_k = \mathbf{U}\mathbf{W}\mathbf{\Lambda}^{k-1}\mathbf{W}^{-1}\mathbf{U}^T\mathbf{x}_1 = \mathbf{\Phi}\mathbf{\Lambda}^{k-1}\mathbf{b}, \tag{14}$$

where $\mathbf{\Phi} = \mathbf{U}\mathbf{W}$ contains the DMD modes (ϕ_i) and $\mathbf{b} = \mathbf{W}^{-1}\mathbf{U}^T\mathbf{x}_1$ contains the amplitudes (b_i) of these modes in the first snapshot. Each mode ϕ_i has a corresponding eigenvalue λ_i in $\mathbf{\Lambda}$. The oscillating frequency (f_i) of mode ϕ_i is equal to

$$f_i = |\frac{\Im(log(\lambda_i))}{2\pi\Delta t}|, \tag{15}$$

and the growth rate g_i is equal to

$$g_i = \frac{\Re(log(\lambda_i))}{\Delta t}. \tag{16}$$

For a stable oscillating system, all the growing rate should be equal to 0. A positive growing rate ($g_i > 0$) indicates that the mode ϕ_i has an amplitude b_i increasing with time; a negative growing rate ($g_i < 0$), in contrast, indicates a mode damping with time.

With the amplitudes b_i, the most energetic DMD modes in the first snapshot can be easily identified and remain the same for other snapshots when the system is stable. In contrast, when growing and damping modes are concerned, the modes should be ordered by the time-averaged amplitudes b_i'. To calculate the time-averaged amplitudes, the eigenvalue-weighted method proposed by Kou and Zhang [36] is selected for its simplicity and computational efficiency among others [37]. The eigenvalue-weighted amplitudes b_i' are calculated as,

$$b_i' = \frac{\sum_{j=1}^{m} |b_i\lambda_i^{j-1}|}{m}. \tag{17}$$

3. Simulation Setup

The AD is compared against the AS models by simulating a three-blade Clipper Liberty 2.5 MW wind turbine located at the EOLOS wind energy research field station in University of Minnesota, USA.

The rotor diameter is $D = 96$ m, the hub height is $z_{\text{hub}} = 80$ m, and the nacelle has a near cuboidal shape with dimensions of 5.3 m × 4.7 m × 5.5 m. The tower as a cylindrical form with a diameter of 3.0 m at the top and 4.1 m at the bottom, respectively. The readers can find more information about this wind turbine in previous works [22,38,39]. Because of proprietary issues, the details of the blade geometry cannot be released in this paper. Interested readers can contact the EOLOS wind energy consortium (Email: eolos@umn.edu, Address: St. Anthony Falls Laboratory, 2 Third Avenue SE, Minneapolis, MN 55414, USA) at the University of Minnesota for these details.

The capability of the employed AS model has been evaluated for different aspects previously. In [40], the employed method was validated against wind tunnel measurements for the time-averaged flow quantities in the wake, such as the velocity deficit and turbulent intensity. Moreover, validations using the same Clipper wind turbine have shown that the AS model is able to predict accurately the near-wake vortex structures as compared with the field measurement using snow-based super-large-scale particle image velocimetry (SLPIV) [41]. Due to these previous validations, in this work, we consider the AS simulation results as references for evaluating the AD model.

Here we present the computational setup for the simulations carried out in this work. In both AD and AS cases, the size of the computational domain is set as $L_x \times L_y \times L_z = 14D \times 7D \times 10D$, where x, y, z represent the stream-wise, the span-wise, and the vertical directions, respectively. The domain is discretized with a Cartesian grid of $N_x \times N_y \times N_z = 281 \times 281 \times 143$. The grid size is uniform in the x, y directions with $\Delta x = D/20$ and $\Delta y = D/40$. In z direction, the mesh is uniform with $\Delta z = D/40$ in $z \in (0, 2D)$ region to resolve the wind turbine wake and the interaction with the ground, and is gradually stretched to the top of the computational domain.

Figure 1 shows the disk and the surface discretized with unstructured triangular surface mesh. Please note that a nacelle model [11] is employed in both AD and AS cases, otherwise there will be a non-realistic jet flow behind the empty rotor center for the AS method. Furthermore, the vortex shedding from the nacelle plays an important role in the wake evolution because of its interaction with the root vortex and the tip shear layer [21]. Although it would be ideal to take the tower into account to have a complete representation of a realistic wind turbine. However, including the tower (diameter = 3.0 m at the top) gives rise to numerical difficulties since the present study (and most numerical studies on the wind turbine's wake as well) employs a grid which is too coarse to resolve the flow details around the tower, and thus complicates the comparison between the AD and AS models. Furthermore, it was shown in [42] that the effect of tower is limited to the near wake region. For these reasons, the tower was not considered and we focus on the differences caused by the two rotor models.

In the AS simulations, the turbine rotates at a fixed tip speed ratio (TSR = $\Omega R/U = 8$, where Ω is the rotor rotational speed, R is the rotor radius and U is instantaneous streamwise velocity averaged over a disk of radius R located $1D$ upstream of the turbine). The thrust is recorded at each time step and then averaged to calculate the thrust coefficient C_T for the AD model. In the AD simulations, the thrust coefficient, which is computed from the corresponding AS simulations, is employed to compute the thrust on disk using Equation (1).

The simulations are conducted with two inflow conditions, i.e., a uniform and a fully developed turbulent inflow. In both cases, the streamwise velocity at the rotor's hub height is $U_{\text{hub}} = 9$ m/s. The Reynolds number based on D and U_{hub} is equal to $Re = DU_{\text{hub}}/\nu = 5.7 \times 10^7$. For the turbulent inflow case, the turbulence density is $\sigma_u/U_{\text{hub}} = 0.08$ at the hub height. The flow at inlet boundary is computed from a precursory LES with a larger computational domain of $L_x' \times L_y' \times L_z' = 62D \times 46D \times 10D$ to capture large scale eddies in the incoming flow. In this inflow generation approach, the velocity fields on a y-z plane are first saved for each time step in the precursory simulation and then applied at the inlet of the turbine simulation. If the mesh and the size of time step employed in the precursory simulation are different from those in the wind turbine simulations, linear interpolations in both space and time are carried out to obtain the inflow velocity for the turbine simulations. Periodical boundary condition is applied in the horizontal directions. The upper boundary condition is the free slip. The wall model based on the logarithmic law is applied on the ground (the roughness length is $z_0 = 5 \times 10^{-3}$ m for the present cases). For the uniform inflow cases, the boundary condition on the lateral walls is the free slip.

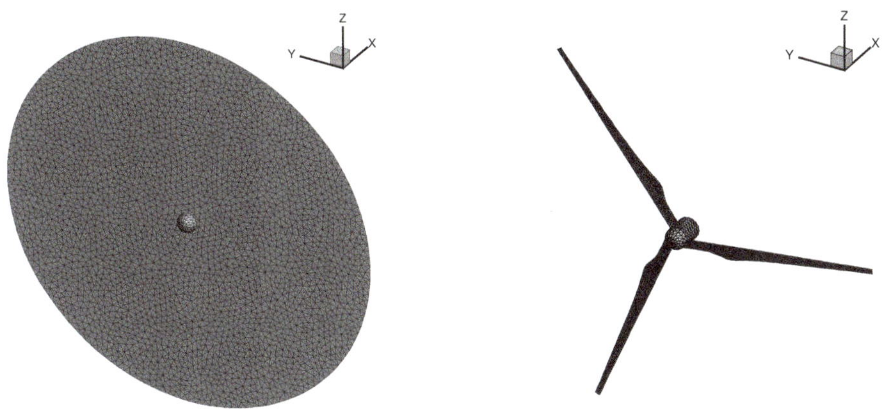

(**a**) (**b**)

Figure 1. The unstructured triangular mesh of wind turbine models. (**a**) AD model. (**b**) AS model.

The simulations use the same fixed time step for both AS and AD cases, which is equal to $1/200$ of the rotor rotational period. The time simulated in the turbulent inflow case for time-averaged quantities is equal to 280 rotor revolutions, which is long enough to take into account the influence of the low-frequency large-scale disturbance in the ABL. For the uniform inflow condition, a shorter simulation of 40 rotor revolutions is carried out, for which large-scale eddies are absent in the inflow.

An extra inflow only simulation with an empty computational domain is carried out with the same setups of the turbulent inflow case to help identify the contribution of turbulent large eddies.

4. Results

In this section, the simulation results are presented. The uniform inflow cases are presented in the first place and are followed by the turbulent inflow simulations. For each inflow condition, we compare instantaneous flow fields, time-averaged flow fields and DMD modes from the AD simulations with those from the AS simulations. In this section, u, v, w denote the instantaneous flow velocity in the streamwise, spanwise and vertical direction, respectively, with U, V, W for the time-averaged values.

4.1. *Uniform Inflow*

4.1.1. Instantaneous Flow Field

Figure 2 depicts the simulated instantaneous velocity fields behind the AD and the AS wind turbine models on the $z = z_{\text{hub}}$ plane. For the streamwise velocity contour in Figure 2a,b, it is found that both wake boundaries are first stable in a small distance behind the turbine and then show fluctuations in the far wake. Inside the wake away from the nacelle, the velocity deficit behind the AD model is more evenly distributed along the radial direction than the AS model, since the tip and the root losses and the radial variation of blade sections are not considered in the AD model. In the hub region, the wake of the nacelle is observed in both models. However, a jet which encompasses the nacelle's wake appears uniquely behind the AS model. Figure 2c,d show the spanwise velocity field. Fluctuations appear behind both wind turbine models. In the near wake region, a Kármán vortex street pattern is observed in the centerline of the near wake due to the nacelle. A significant discrepancy emerges in $2 < x/D < 3$ region, where the AD model's result shows a regularly oscillating pattern on the wake boundary as observed for both streamwise and spanwise velocities (Figure 2a,c). It is also noticed that the spanwise velocity fluctuations in this region are stronger near the wake boundary than those in the wake center. In contrast, the wake behind the AS model does not have this oscillatory boundary, and the vortex shed from the nacelle grows gradually and starts to influence the wake boundary at $x \approx 4D$. In the far wake ($x > 6D$), the spanwise velocity behind the AS model seems to be more energetic than that of the AD model. Quantitative comparisons shall be conducted in the next section to confirm this observation.

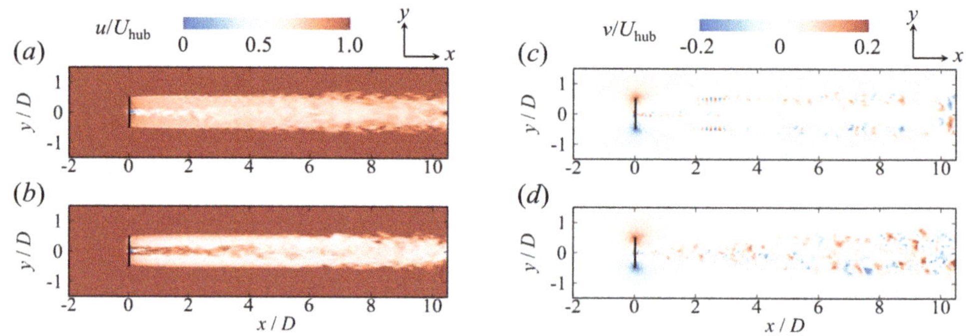

Figure 2. Uniform inflow: contours of the instantaneous velocity field behind the wind turbine on the horizontal plane at hub height. (**a**) streamwise velocity using AD; (**b**) streamwise velocity using AS; (**c**) spanwise velocity using AD; (**d**) spanwise velocity using AS. The solid black line at $x = 0$ illustrates the location and the diameter of the wind turbine.

4.1.2. Time-Averaged Flow Field

In Figure 3 we compare the time-averaged flow field computed from the AD and the AS simulations. As seen in Figure 3a, the mean velocity profiles of both models show an overall agreement in the far wake ($x = 9D$). However, immediately behind the wind turbine, the two velocity profiles differ significantly. The profile of AD is almost uniform (except for the region near the nacelle). In contrast, the profile of AS shows a clear radial variation, which is remarked by a weaker deficit behind the nacelle due to the root loss ($y = 0$) and a smoother transition on the wake boundary due to the tip loss ($y = \pm 0.5D$) at $x = 1D$. In the near wake, the thickness of the shear layer on the wake boundary is smaller for AD, but it grows faster than that in the AS simulation. At $x = 5D$, it is obvious that this transitional region is thicker for the AD than the AS. This faster growing of wake boundary thickness denotes a quicker recovery and expansion of the wake of the AD. By comparison, this transitional region has no remarkable development in the result of AS until $x = 7D$ and expands faster from $7D$ to $9D$. In Figure 3b,c, the turbulence kinetic energy (TKE; k) and the primary Reynolds stress ($<u'v'>$) both indicate that the AD model shows stronger turbulent effects on the wake boundary in the $x < 5D$ region and is surpassed by the AS in the far wake ($x > 7D$). At $x = 9D$ the velocity profile of the two models are in reasonable agreement, while the wake computed by the AS model contains more turbulence kinetic energy and larger Reynolds stress. This result confirms the observation from the instantaneous flow field in Figure 2.

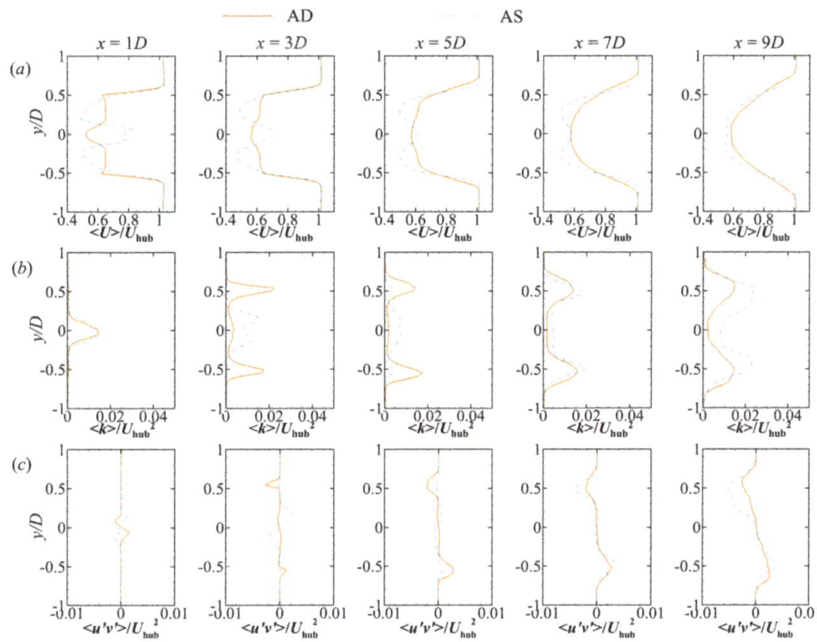

Figure 3. Uniform inflow: horizontal profiles at hub height ($z = z_{\text{hub}}$) of the time-averaged (**a**) streamwise velocity, (**b**) turbulence kinetic energy, and (**c**) the primary Reynolds stress $<u'v'>$ at different downstream location.

4.1.3. DMD Analysis

DMD is conducted to analyze the dynamic coherent structures in the wakes. The velocity field on the horizontal plane at the hub height ($z = z_{\text{hub}}$) is analyzed. Figure 4 depicts the DMD amplitude spectra and the most dominants modes.

In Figure 4a,e, only the modes with Strouhal number less than three ($St = fD/U_{\text{hub}} < 3$) are plotted although the entire spectra are much longer, because the most energetic modes are within this low frequency range. First, these amplitude spectra reveal a significant difference in the energy distribution over the frequencies of the AD and the AS models' wake. The wake from the AS simulations contains generally more energy in the low frequency range below $St < 1.5$ and shows a trend of concentration around $St \approx 0.7$ (marked with ϕ_1 in Figure 4e). However, such a concentration trend does not appear in the spectrum in the AD's case, whose DMD modes are of almost the same amplitudes in $0 < St < 2$ region, except for a distinct peak at $St \approx 1.8$ (marked with ϕ_1 in Figure 4a). Secondly, when comparing the amplitudes between the two cases, it is found that modes of the AS have slightly larger amplitudes than that of the AD, especially in the low frequency region, showing the wake of AS contains more energetic low frequency oscillations. Thirdly, no energy concentration around the vortex shedding frequency of the bluff body ($St \approx 0.168$) is observed in both spectra.

Figure 4b–d,f–h show the three most energetic modes of the AD and the AS cases, respectively. Overall, the spatial scale of the oscillation patterns enlarges as the Strouhal number decreases. However, the results from the AD and the AS models have very different dominant modes. Figure 4b shows the mode of the largest amplitude of the AD case. A spatial energy concentration on the wake boundary around $2D < x < 4D$ is found, which is in agreement with the instantaneous flow field (Figure 2c). Interestingly, no apparent source of disturbance can be traced in the upstream. It suggests that this mode should perhaps be related to the instability of the thin shear layer on the wake boundary that amplifies tiny disturbances in the flow field. It is noticed that this mode dominates for $2D < x < 4D$, but becomes weak in the far field. In contrast, the other two modes of the AD (Figure 4c,d) both stem from the nacelle and dominate the far wake.

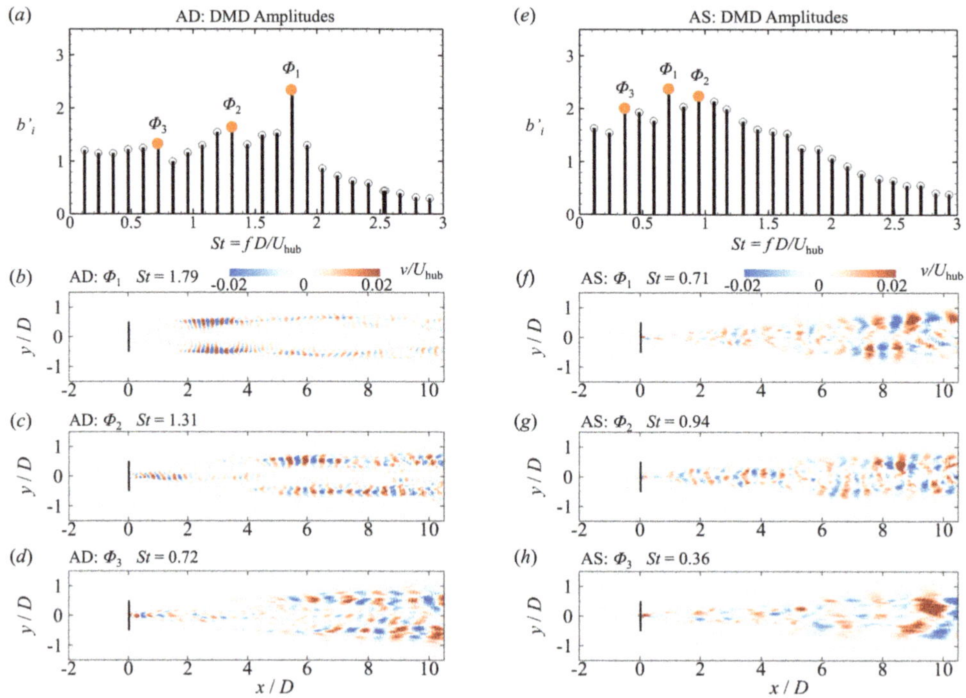

Figure 4. Uniform inflow: Dynamic mode decomposition (DMD) analysis of the velocity field on the horizontal plane at the hub height ($z = z_{hub}$). (**a**,**e**) the eigenvalue-weighted amplitudes of the DMD modes for the AD and the AS models; (**b**–**d**) the largest three DMD modes ordered by amplitude of AD; (**f**–**h**) the largest three DMD modes ordered by amplitude of AS. DMD modes are shown with the spanwise velocity contour.

As to DMD modes in the AS case (Figure 4f–h) all the three modes stem from the upstream nacelle and develop until the far wake. They differ from each other by the frequency and the wave length as shown by the spanwise velocity field. No AS mode similar to ϕ_1 of the AD is found after checking all the modes of the AS (including those not shown in the figure). This observation suggests that for the uniform inflow condition, the wake computed from the AS model is strongly affected by the vortex shedding behind the nacelle, while the AD model predicts a unique mode related to the shear layer instability near the wake boundary, which may not exist in real wind turbine wakes.

4.2. Turbulent Inflow

This section presents the simulation results from the AD and AS cases under the fully developed turbulent inflow condition.

4.2.1. Instantaneous Flow Field

Figure 5 depicts instantaneous velocity fields computed from (a) the AD and (b) the AS models and (c) the inflow case on the $z = z_{hub}$ plane at the same instant. As seen, the wake shapes and the spanwise velocity contours from the AD and AS cases show a reasonably good agreement, except for some differences, such as the jet flow behind the hub of the AS model. Compared with the uniform inflow condition, strong effects of inflow turbulence is observed. For instance, a much stronger spanwise velocity is observed for the turbulent inflow (Figure 5d,e) when compared with that from the uniform inflow (Figure 2c,d). When comparing with the only inflow case (Figure 5f), we can observe an interesting phenomenon, that the patterns of the spanwise velocity computed from the AD and AS model, are very similar to that from the only inflow case. Moreover, it is noticed that both AD and AS models amplify the spanwise velocity fluctuations as the velocity magnitude is larger in the wake than in the inflow. Apart from this, no more information can be extracted with confidence from this instantaneous flow field yet.

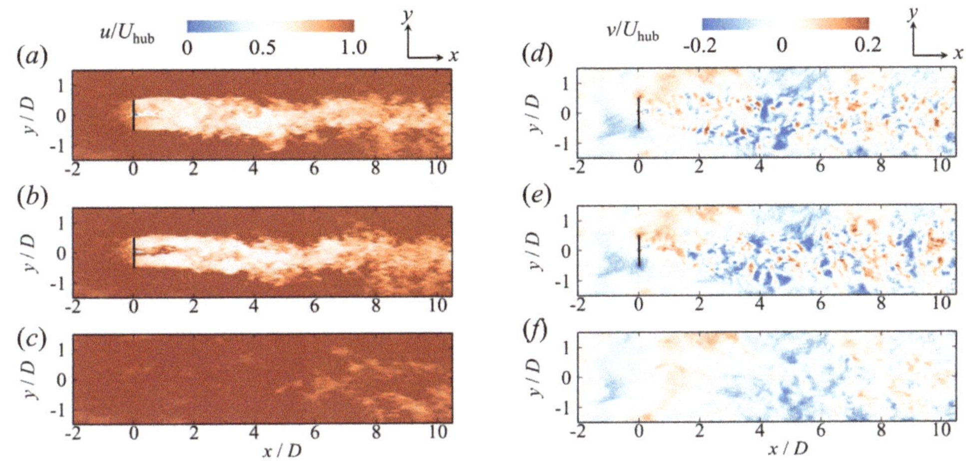

Figure 5. Turbulent inflow: contours of the instantaneous velocity field behind the wind turbine on the horizontal plane at hub height. (**a**) streamwise velocity using AD; (**b**) streamwise velocity using AS; (**d**) spanwise velocity using AD; (**e**) spanwise velocity using AS. The solid black line at $x = 0$ illustrates the location and the diameter of the wind turbine; (**c**,**f**) streamwise and spanwise velocity of the turbulent inflow without wind turbine.

4.2.2. Time-Averaged Flow Field

Figure 6 shows the profiles of time-averaged flow quantities. Before the following detailed analysis, it is clear that the results from the AD model are in an overall good agreement in the far wake ($x > 7D$) for turbulent inflow condition, in contrast to Figure 3 of the uniform inflow condition. For the streamwise velocity, differences between the AD and the AS models are only observed in the near wake region ($x < 3D$) and become insignificant at far wake locations as shown in Figure 6a. The wakes computed by both models slightly skew towards the $-y$ direction due to the insufficient simulation time, which causes the inflow not perfectly symmetric with respect to $y = 0$. The spanwise profiles of TKE are shown in Figure 6b. As seen, the TKE is concentrated near the hub ($y = 0$) and the wake boundary ($y \pm 0.5D$) at $x = 1D$ with larger TKE near the hub and near the boundary for the AS and AD models, respectively. The region of high TKE increases and expands as the wake travels from ($3D < x < 5D$) and the TKE computed by the AS model develops faster and surpasses that computed by the AD after ($x > 3D$). At further turbine downwind locations, the two models show consistent results. Similar trends are observed for the Reynold's stress $<u'v'>$ as shown in Figure 6c. In Figure 6d,e, the vertical distribution of streamwise velocity and the turbulence intensity also confirms the above conclusion that the two models agree well in the far wake and the differences only manifest in the near wake region.

To this extend, the AD model can reasonably predict the time-averaged flow quantities in the far wake for turbulent inflow conditions.

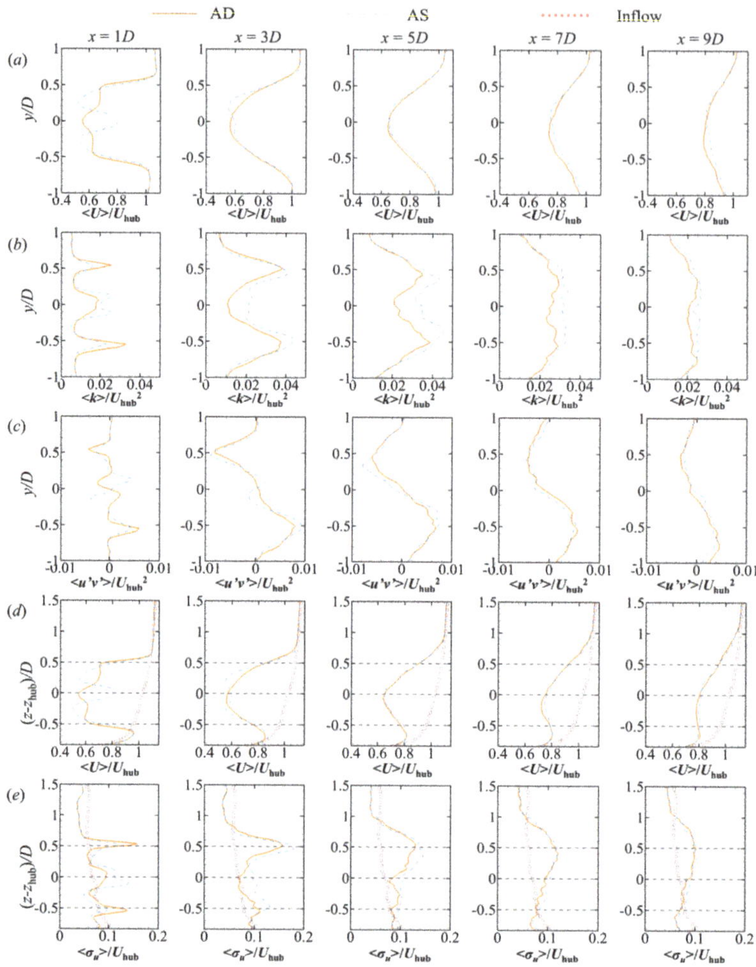

Figure 6. Turbulent inflow: Horizontal profile of time-averaged (**a**) streamwise velocity, (**b**) turbulence kinetic energy, and (**c**) the primary Reynolds stress $<u'v'>$ at different downstream location at hub height ($z = z_{\text{hub}}$); vertical profile of time-averaged (**d**) streamwise velocity and (**e**) turbulence intensity on the plane $y = 0$.

4.2.3. DMD Analysis

Figure 7 depicts the DMD amplitude spectra and the most dominants modes. Figure 7a shows the spectrum computed by the AD model with three distinct peaks remarked. As seen, the Strouhal number of the first peak (ϕ_1) is 0.17, which falls in the range of the bluff body vortex shedding frequencies. This dominant mode, illustrated by Figure 7b, shows that the transversal velocity in the wake alters the direction regularly in a way resembling to the wake behind a circular cylinder [43]. The other two modes of the AS illustrated in Figure 7c,d are at higher frequency. The spatial scale of the oscillation decreases as the Strouhal number increases. These two modes have larger amplitudes on the wake boundary than in the wake center. For the second mode ϕ_2, stronger velocity amplitude is observed in the far wake, whereas ϕ_3 shows a local concentration of energy in $2D < x < 6D$.

On the other hand, the spectrum of the AS case (Figure 7e) shows a dominant peak close to the bluff body vortex shedding frequency at $St = 0.23$, which is the second larger peak in the DMD spectrum with the largest mode at $St = 0.08$. The first mode of the AS case (wavelength $\approx 9D$) is significantly larger as shown in Figure 7f. The second mode from the AS case (Figure 7g), on the other hand, is close to the first mode from the AD case, although the wavelength of ϕ_2 from the AS case is smaller. Moreover, it is observed that the spectrum from the AS case has larger energy at the low frequency range than that of the AD. The two modes of higher frequency (ϕ_2 and ϕ_3) computed from the AS model are both stronger in the far wake than in the near wake with both larger spatial scales when compared with the AD case.

It is also worth noting that the influence of the nacelle on the far wake is less evident compared to the uniform inflow case, that two (the AS case) or three (the AD case) dominant DMD modes are obviously related to the nacelle for the uniform inflow cases, while only the third dominant mode from the AS case (Figure 7h) seems to be magnified by the nacelle for the turbulent inflow cases. Detailed space-time correlation study [44] is needed to further examine the effect of nacelle on wake dynamics under turbulent inflows.

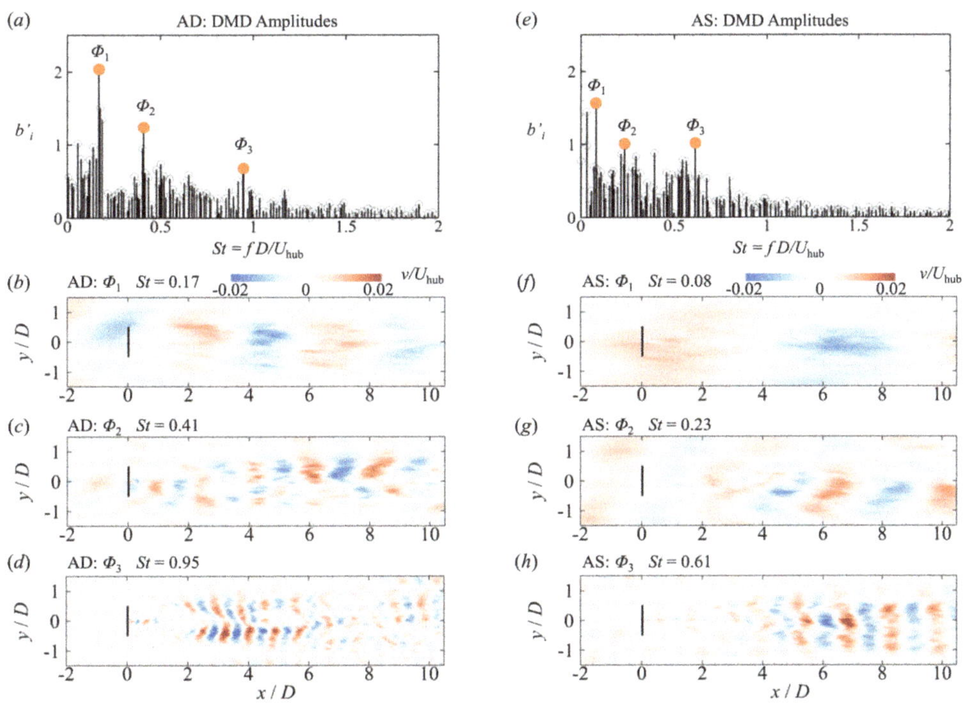

Figure 7. Turbulent inflow: Dynamic mode decomposition (DMD) analysis of the velocity field on the horizontal plane at the hub height ($z = z_{\text{hub}}$). (**a**,**e**) the eigenvalue-weighted amplitudes of the DMD modes for the AD and the AS models; (**b**–**d**) the largest three DMD modes ordered by Strouhal number of AD; (**f**–**h**) the largest three DMD modes ordered by Strouhal number of AS. DMD modes are shown with the spanwise velocity contour.

5. Discussion

The previous section has shown the differences between the wakes computed from the AD and AS models for both uniform and turbulent inflow conditions.

For the uniform inflow cases, external perturbations are excluded and the wakes computed by both models develop on their own properties. The differences are easily identified already on the instantaneous and the time-averaged flow field: the instability near the wake boundary develops earlier for the AD model and results in a faster growth of turbulence near the wake boundary and quicker wake expansion and recovery in the near to intermediate turbine downwind locations. This phenomenon can be explained with the classical linear stability theory of shear flow [45]. According to this theory, the optimal spatial scale of perturbation to destabilize a shear flow is proportional to the transitional thickness so the sharp transition between the freestream flow and the wake of the AD case (as the tip loss effect is neglected) is more sensitive to small perturbations. This sharp transition across the wake boundary and the resultant stronger mixing and expansion in the near wake of the AD model are in accordance with previous wind tunnel experiments of Lignarolo et al. [8,46]. In the DMD analysis, we have compared the two models with the energy spectra and the dominant modes. For both models, a clear influence of the nacelle on the far wake is shown by the DMD modes, which is in agreement with previous numerical studies [21]. However, obvious disparities are also found that the spectrum of the AS case is concentrated in a lower frequency range, whereas that from the AD case has a distinct high frequency dominant mode corresponding to the shear layer instability. Moreover, the wake computed from the AS model has larger coherent structures and oscillates at a lower frequency than that from the AD model, which may also due to the differences of the shear layer near the wake boundary.

For turbulent inflow cases, the velocity deficit recovers faster than the uniform case due to the inflow turbulence, which is in agreement with previous studies [16,47]. In our test cases, the time-averaged streamwise velocity, TKE, and primary Reynold's stresses computed by the AD model reasonably agree with the AS model starting from $x = 7D$. This generally good agreement of the AD and the AS in turbulent inflow condition is in accordance with previous studies [2,17,48]. Furthermore, the DMD analysis shows inflow turbulence shifted the energy spectrum significantly to lower frequency ranges for both models. These findings are in accordance with previous studies where the wake is analyzed using Fourier Transform [6,49]. Compared with the Fourier Transform, the DMD analysis provides additional insightful information about the spatial scale of the coherent structures. These coherent structures reveal that for the turbulent inflow condition, the wakes are dominated by DMD modes of larger scale coherent structures related to the inflow eddies and some DMD modes are enhanced by the hub vortex. Moreover, a mode similar to the bluff-body vortex shedding at Strouhal number $St = 0.17$ is found to be dominant uniquely in the wake behind the AD model, which, on the other hand, happens at $St = 0.23$ for the AS model. The most dominant mode of the AS case appears at lower frequency of $St = 0.08$ and has larger scale coherent flow structures, which shall be related to the passive advection of the wake by the large scale inflow eddies. The origin of these large-scale motions of turbine wakes is often attributed to two different mechanisms, namely the bluff body shear layer instability [23] and the inflow large eddies [6], which convect turbine wake as passive scalars. Recent field measurements [50] and computational studies [49] suggested the co-existence of these two mechanisms. Furthermore, the hub vortex behind the nacelle is shown to have a significant impact on the start and enhancement of wake meandering [21,51]. A recent review on the meandering of turbine wakes can be found in [52]. In this work we observed a complex interaction between the turbine and the inflow eddies: (i) the modes at low frequencies are less affected by the turbine (Figure 7f); (ii) the DMD modes close to the bluff body vortex shedding frequency seem to be enhanced (Figure 7b,g); (iii) some DMD modes seem to be amplified by the hub vortex behind the nacelle (Figure 7c,h); (iv) the instability within the shear layer also seems to be a key factor for some modes (Figure 7d). Although the present work still can not provide a direct answer to the origin of the wake meandering, it suggests that the dynamic structures are different in wakes computed from the AD and the AS models. Due to this difference, the AD model should be used with more attention when the wake dynamics are of interest, e.g., to study the wake meandering, because the AD model can lead to different wake meandering patterns.

6. Conclusions

In this work, we evaluate the capability of an actuator disk model in predicting the wake dynamics of a utility-scale wind turbine by comparing its results with those from an actuator surface model. In the AD model, the same thrust coefficient as that in the AS model is employed. A nacelle model is incorporated into both models. Turbulent flows are simulated using LES with the same mesh and time step for both AD and AS cases. Two inflow conditions are considered, i.e., a uniform inflow and a fully developed turbulent inflow. The wakes computed using the AD model is compared with that from the AS model via the time-averaged field and the DMD analysis. It is found that time-averaged velocity and turbulence kinetic energy computed by the AD model are significantly different from those computed by the AS model until nine turbine rotor diameters downstream for the uniform inflow condition; for fully developed turbulent inflow, the differences between the two models are less significant and agree with each other from seven turbine rotor diameters downstream. The DMD analysis of the uniform inflow cases shows that the vortex shed behind the nacelle triggers the shear layer instabilities on the wake boundary behind both models but of different spatial scales. With a thinner shear layer, the wake predicted by the AD model contains smaller spatial scale oscillations at higher frequency. For the fully developed turbulent cases, the DMD analysis shows that the spectra of both models shift to a lower frequency range and the coherent structures also increase in size. The DMD analysis also reveals significant differences between the two models in the far wake: a bluff-body vortex shedding pattern at $St = 0.17$ appears uniquely in the wake of the AD model as the most dominant DMD mode, whereas the wake computed by the AS model has the most dominant DMD mode of lower energy and at lower frequency $St = 0.08$ which is related to the passive transport by the inflow turbulence large eddies, and with the second dominant mode at a frequency $St = 0.23$ close to the bluff body vortex shedding frequency. It is concluded that the dynamic coherent structures in the wake predicted by the AD model are significantly different from those predicted by the AS models and shall be used with more attention when the dynamics of the wake are of interest. In the present work, the thrust coefficient employed in the AD model is the same as that computed by the AS model. However, since the blade rotation is not modeled in the current AD model, the power coefficient from the AD model is not exactly the same as that from the AS model (the C_P from the AD is approximately 5% to 10% higher). In the current AD model, the thrust coefficient and the power coefficient cannot be specified at the same time. Further studies on how the differences in the power coefficient affect wake evolutions will be carried out using more advanced AD models considering the effect of blade rotation (e.g., [20]).

Author Contributions: Conceptualization, investigation, writing—original draft preparation, Z.L.; conceptualization, methodology, software and writing—review and editing, X.Y. All authors have read and agreed to the published version of the manuscript.

Funding: This work is partially supported by NSFC Basic Science Center Program for "Multiscale Problems in Nonlinear Mechanics" (NO. 11988102).

Conflicts of Interest: The authors declare no conflict of interest.

Abbreviations

The following abbreviations are used in this manuscript:

ABL	Atmospheric Boundary Layer
AD	Actuator Disk
AL	Actuator Line
AS	Actuator Surface
DMD	Dynamic Mode Decomposition
St	Strouhal Number

References

1. Barthelmie, R.J.; Hansen, K.S.; Frandsen, S.T.; Rathmann, O.; Schepers, J.G.; Schlez, W.; Phillips, J.; Rados, K.; Zervos, A.; Politis, E.S.; et al. Modelling and measuring flow and wind turbine wakes in large wind farms offshore. *Wind Energy* **2009**, *12*, 431–444. [CrossRef]
2. Stevens, R.J.; Meneveau, C. Flow structure and turbulence in wind farms. *Annu. Rev. Fluid Mech.* **2017**, *49*, 311–339. [CrossRef]
3. Medici, D.; Alfredsson, P. Measurements on a wind turbine wake: 3D effects and bluff body vortex shedding. *Wind Energy Int. J. Prog. Appl. Wind. Power Convers. Technol.* **2006**, *9*, 219–236. [CrossRef]
4. Vermeer, L.; Sørensen, J.N.; Crespo, A. Wind turbine wake aerodynamics. *Prog. Aerosp. Sci.* **2003**, *39*, 467–510. [CrossRef]
5. Espana, G.; Aubrun, S.; Loyer, S.; Devinant, P. Spatial study of the wake meandering using modelled wind turbines in a wind tunnel. *Wind Energy* **2011**, *14*, 923–937. [CrossRef]
6. Espana, G.; Aubrun, S.; Loyer, S.; Devinant, P. Wind tunnel study of the wake meandering downstream of a modelled wind turbine as an effect of large scale turbulent eddies. *J. Wind. Eng. Ind. Aerodyn.* **2012**, *101*, 24–33. [CrossRef]
7. Aubrun, S.; Loyer, S.; Espana, G.; Hayden, P.; Hancock, P. Experimental Study on the wind turbine wake meandering with the help of a non-rotation simplified model and of a rotating model. In Proceedings of the 49th AIAA Aerospace Sciences Meeting Including the New Horizons Forum and Aerospace Exposition, Orlando, FL, USA, 4–7 January 2011; p. 460.
8. Lignarolo, L.; Ragni, D.; Ferreira, C.S.; Van Bussel, G. *Kinetic Energy Entrainment in Wind Turbine and Actuator Disc Wakes: An Experimental Analysis*; Journal of Physics: Conference Series; IOP Publishing: Bristol, UK, 2014; Volume 524, p. 012163.
9. Camp, E.H.; Cal, R.B. Mean kinetic energy transport and event classification in a model wind turbine array versus an array of porous disks: Energy budget and octant analysis. *Phys. Rev. Fluids* **2016**, *1*, 044404. [CrossRef]
10. Sorensen, J.N.; Shen, W.Z. Numerical Modeling of Wind Turbine Wakes. *J. Fluids Eng. Trans. ASME* **2002**, *124*, 393–399. [CrossRef]
11. Yang, X.; Sotiropoulos, F. A new class of actuator surface models for wind turbines. *Wind Energy* **2018**, *21*, 285–302. [CrossRef]
12. Shen, W.Z.; Zhang, J.H.; Sorensen, J.N. The actuator surface model: A New Navier–Stokes based model for rotor computations. *J. Sol. Energy Eng. Trans. ASME* **2009**, *131*, 011002. [CrossRef]
13. Sanderse, B.; Van der Pijl, S.; Koren, B. Review of computational fluid dynamics for wind turbine wake aerodynamics. *Wind Energy* **2011**, *14*, 799–819. [CrossRef]
14. Castellani, F.; Vignaroli, A. An application of the actuator disc model for wind turbine wakes calculations. *Appl. Energy* **2013**, *101*, 432–440. [CrossRef]
15. Liu, X.; Yan, S.; Mu, Y.; Chen, X.; Shi, S. *CFD and Experimental Studies on Wind Turbines in Complex Terrain by Improved Actuator Disk Method*; Journal of Physics: Conference Series; IOP Publishing: Bristol, UK, 2017; Volume 854, p. 012028.
16. Wu, Y.T.; Lin, C.Y.; Chang, T.J. Effects of inflow turbulence intensity and turbine arrangements on the power generation efficiency of large wind farms. *Wind Energy* **2020**, *23*, 1640–1655. [CrossRef]
17. Stevens, R.J.; Martínez-Tossas, L.A.; Meneveau, C. Comparison of wind farm large eddy simulations using actuator disk and actuator line models with wind tunnel experiments. *Renew. Energy* **2018**, *116*, 470–478. [CrossRef]
18. Yang, X.; Sotiropoulos, F. On the predictive capabilities of LES-actuator disk model in simulating turbulence past wind turbines and farms. In Proceedings of the 2013 American Control Conference, Washington, DC, USA, 17–19 June 2013; pp. 2878–2883.
19. Wu, Y.T.; Lin, C.Y.; Hsu, C.M. An Experimental Investigation of Wake Characteristics and Power Generation Efficiency of a Small Wind Turbine under Different Tip Speed Ratios. *Energies* **2020**, *13*, 2113. [CrossRef]
20. Wu, Y.T.; Porté-Agel, F. Large-eddy simulation of wind-turbine wakes: evaluation of turbine parametrisations. *Bound. Layer Meteorol.* **2011**, *138*, 345–366. [CrossRef]
21. Kang, S.; Yang, X.; Sotiropoulos, F. On the onset of wake meandering for an axial flow turbine in a turbulent open channel flow. *J. Fluid Mech.* **2014**, *744*, 376–403. [CrossRef]

22. Hong, J.; Toloui, M.; Chamorro, L.P.; Guala, M.; Howard, K.; Riley, S.; Tucker, J.; Sotiropoulos, F. Natural snowfall reveals large-scale flow structures in the wake of a 2.5-MW wind turbine. *Nat. Commun.* **2014**, *5*, 4216. [CrossRef]
23. Medici, D.; Alfredsson, P. Measurements behind model wind turbines: Further evidence of wake meandering. *Wind. Energy Int. J. Prog. Appl. Wind. Power Convers. Technol.* **2008**, *11*, 211–217. [CrossRef]
24. Du, Z.; Selig, M. A 3-D stall-delay model for horizontal axis wind turbine performance prediction. In Proceedings of the 1998 ASME Wind Energy Symposium, Reno, NV, USA, 12–15 January 1998; p. 21.
25. Shen, W.Z.; Sørensen, J.N.; Mikkelsen, R. Tip loss correction for actuator/Navier–Stokes computations. *J. Sol. Energy Eng.* **2005**, *127*, 209–213. [CrossRef]
26. Shen, W.Z.; Mikkelsen, R.; Sørensen, J.N.; Bak, C. Tip loss corrections for wind turbine computations. *Wind. Energy Int. J. Prog. Appl. Wind. Power Convers. Technol.* **2005**, *8*, 457–475. [CrossRef]
27. Yang, X.; Zhang, X.; Li, Z.; He, G.W. A smoothing technique for discrete delta functions with application to immersed boundary method in moving boundary simulations. *J. Comput. Phys.* **2009**, *228*, 7821–7836. [CrossRef]
28. Yang, X.; Sotiropoulos, F.; Conzemius, R.J.; Wachtler, J.N.; Strong, M.B. Large-eddy simulation of turbulent flow past wind turbines/farms: The Virtual Wind Simulator (VWiS). *Wind Energy* **2015**, *18*, 2025–2045. [CrossRef]
29. Smagorinsky, J. General circulation experiments with the primitive equations: I. the basic experiment. *Mon. Weather. Rev.* **1963**, *91*, 99–164. [CrossRef]
30. Germano, M.; Piomelli, U.; Moin, P.; Cabot, W.H. A dynamic subgrid-scale eddy viscosity model. *Phys. Fluids Fluid Dyn.* **1991**, *3*, 1760–1765. [CrossRef]
31. Ge, L.; Sotiropoulos, F. A numerical method for solving the 3D unsteady incompressible Navier–Stokes equations in curvilinear domains with complex immersed boundaries. *J. Comput. Phys.* **2007**, *225*, 1782–1809. [CrossRef] [PubMed]
32. Knoll, D.A.; Keyes, D.E. Jacobian-free Newton–Krylov methods: a survey of approaches and applications. *J. Comput. Phys.* **2004**, *193*, 357–397. [CrossRef]
33. Saad, Y. A flexible inner-outer preconditioned GMRES algorithm. *SIAM J. Sci. Comput.* **1993**, *14*, 461–469. [CrossRef]
34. Schmid, P.J. Dynamic mode decomposition of numerical and experimental data. *J. Fluid Mech.* **2010**, *656*, 5–28. [CrossRef]
35. Kutz, J.N. *Data-Driven Modeling & Scientific Computation: Methods for Complex Systems & Big Data*; Oxford University Press: Oxford, UK, 2013.
36. Kou, J.; Zhang, W. An improved criterion to select dominant modes from dynamic mode decomposition. *Eur. J. Mech. B/Fluids* **2017**, *62*, 109–129. [CrossRef]
37. Jovanović, M.R.; Schmid, P.J.; Nichols, J.W. Sparsity-promoting dynamic mode decomposition. *Phys. Fluids* **2014**, *26*, 024103. [CrossRef]
38. Abraham, A.; Dasari, T.; Hong, J. Effect of turbine nacelle and tower on the near wake of a utility-scale wind turbine. *J. Wind. Eng. Ind. Aerodyn.* **2019**, *193*, 103981. [CrossRef]
39. Chamorro, L.P.; Lee, S.J.; Olsen, D.; Milliren, C.; Marr, J.; Arndt, R.; Sotiropoulos, F. Turbulence effects on a full-scale 2.5 MW horizontal-axis wind turbine under neutrally stratified conditions. *Wind Energy* **2015**, *18*, 339–349. [CrossRef]
40. Foti, D.; Yang, X.; Sotiropoulos, F. Similarity of wake meandering for different wind turbine designs for different scales. *J. Fluid Mech.* **2018**, *842*, 5–25. [CrossRef]
41. Yang, X.; Hong, J.; Barone, M.; Sotiropoulos, F. Coherent dynamics in the rotor tip shear layer of utility-scale wind turbines. *J. Fluid Mech.* **2016**, *804*, 90–115. [CrossRef]
42. Schümann, H.; Pierella, F.; Sætran, L. Experimental investigation of wind turbine wakes in the wind tunnel. *Energy Procedia* **2013**, *35*, 285–296. [CrossRef]
43. Kutz, J.N.; Brunton, S.L.; Brunton, B.W.; Proctor, J.L. *Dynamic Mode Decomposition: Data-Driven Modeling of Complex Systems*; SIAM: Philadelphia, PA, USA, 2016.
44. He, G.; Jin, G.; Yang, Y. Space-Time Correlations and Dynamic Coupling in Turbulent Flows. *Annu. Rev. Fluid Mech.* **2017**, *49*, 51–70. [CrossRef]
45. Schmid, P.J.; Henningson, D.S.; Jankowski, D. Stability and transition in shear flows. Applied mathematical sciences, Vol. 142. *Appl. Mech. Rev.* **2002**, *55*, B57–B59. [CrossRef]

46. Lignarolo, L.E.M.; Ragni, D.; Ferreira, C.J.S.; Van Bussel, G.J.W. Experimental comparison of a wind-turbine and of an actuator-disc near wake. *J. Renew. Sustain. Energy* **2016**, *8*, 023301. [CrossRef]
47. Lyu, P.; Chen, W.L.; Li, H.; Shen, L. A numerical study on the development of self-similarity in a wind turbine wake using an improved pseudo-spectral large-eddy simulation solver. *Energies* **2019**, *12*, 643. [CrossRef]
48. Porté-Agel, F.; Wu, Y.T.; Lu, H.; Conzemius, R.J. Large-eddy simulation of atmospheric boundary layer flow through wind turbines and wind farms. *J. Wind. Eng. Ind. Aerodyn.* **2011**, *99*, 154–168. [CrossRef]
49. Yang, X.; Sotiropoulos, F. Wake characteristics of a utility-scale wind turbine under coherent inflow structures and different operating conditions. *Phys. Rev. Fluids* **2019**, *4*, 024604. [CrossRef]
50. Heisel, M.; Hong, J.; Guala, M. The spectral signature of wind turbine wake meandering: A wind tunnel and field-scale study. *Wind Energy* **2018**, *21*, 715–731. [CrossRef]
51. Iungo, G.V.; Viola, F.; Camarri, S.; Porté-Agel, F.; Gallaire, F. Linear stability analysis of wind turbine wakes performed on wind tunnel measurements. *J. Fluid Mech.* **2013**, *737*, 499–526. [CrossRef]
52. Yang, X.; Sotiropoulos, F. A Review on the Meandering of Wind Turbine Wakes. *Energies* **2019**, *12*, 4725. [CrossRef]

energies

Article

An Experimental Study on the Actuator Line Method with Anisotropic Regularization Kernel

Zhe Ma [1], Liping Lei [1,*], Earl Dowell [2] and Pan Zeng [1]

1 Department of Mechanical Engineering, Tsinghua University, Beijing 100084, China; maz14@mails.tsinghua.edu.cn (Z.M.); zengp@mails.tsinghua.edu.cn (P.Z.)

2 Department of Mechanical Engineering & Material Science, Duke University, Durham, NC 27704, USA; earl.dowell@duke.edu

* Correspondence: leilp@tsinghua.edu.cn; Tel.: +86-106-279-4261; Fax: +86-106-2771-279

Received: 19 January 2020; Accepted: 18 February 2020; Published: 21 February 2020

Abstract: Nowadays, actuator line method (ALM) has become the most potential method in wind turbine simulations, especially in wind farm simulations and fluid-structure interaction simulations. The regularization kernel, which was originally introduced to ALM to avoid numerical singularity, has been found to have great influence on rotor torque predictions and wake simulations. This study focuses on the effect of each parameter used in the standard kernel and the anisotropic kernel. To validate the simulation, the torque and the wake characteristics of a model wind turbine were measured. The result shows that the Gaussian width ϵ (for standard kernel) and the parameter in chord length direction ϵ_c (for anisotropic kernel) mainly affect the normal velocity of each blade element when using ALM but have little effect on the tangential velocity calculation. Therefore, these parameters have great influence on the attack angle and rotor torque prediction. The thickness parameter ϵ_t is the main difference between the standard kernel and the anisotropic kernel and it has a strong effect on the wind turbine wakes simulation. When using the anisotropic kernel, the wake structure is clearer and less likely to disperse, which is more consistent with the experimental results. Based on the studies above, a non-uniform mesh is recommended when using the anisotropic regularization kernel. Using a mesh refined in the main flow direction, ALM with anisotropic kernel can predict torque and wake characteristics better while maintaining low computational costs.

Keywords: actuator line method; wind turbine simulation; regularization kernel

---

1. Introduction

Nowadays, the actuator line method (ALM) has been widely used in wind farm simulations due to its capability of wind turbine wakes simulation and its numerical stabilization and low computational cost. This method was developed by Sørensen and Shen [1] in 2002 to overcome the disadvantage of Blade Element Momentum theory (BEM), which cannot simulate the wake characteristics of wind turbines. By combining the computational fluid dynamics (CFD) method and blade element theory, the ALM method avoids the calculation of the boundary layer flow and thus greatly reduces the computational cost compared with resolved CFD approaches. The benefits of low computational cost have two aspects. Firstly, the mesh used in ALM simulation is more regular than the resolved approaches, which means that the orthogonality of the mesh is much better. Therefore, the Large Eddy Simulation (LES) turbulence model, which is more accurate but computationally expensive and sensitive to mesh quality compared with RANS-based turbulence models, can be easily applied to ALM simulations. Combining with the LES model, ALM can make good simulations of velocity field and turbulence field in the wake region [2,3] and it has advantages in wind turbine simulations when the inlet condition is complex, such as the atmospheric boundary layer condition [4,5]. Secondly, due to its low computational cost, ALM can be used in large-scale problems [4,6–8] and can be easily coupled with structural models [9,10]. Therefore, ALM is suitable for wind farm simulations and fluid-structure interaction simulations. In summary, ALM nowadays has become the most potential method in wind turbine simulations, especially in wind farm simulations and fluid-structure interaction simulations.

Regularization kernel was originally introduced to the ALM approach to avoid the numerical singularity [1]. During the ALM approach, the aerodynamic forces of wind turbine are calculated according to the blade element theory and the wind velocity field is calculated by solving the Navier–Stocks equations. Therefore, a regularization kernel must be employed to smoothly apply these aerodynamic forces to the Navier–Stocks equations and a uniform three-dimensional Gaussian function which is suggested by Sørensen and Shen [1] is widely used as the standard regularization kernel.

The regularization kernel also affects the conceptual shape of the wind turbine blade. When using the standard regularization kernel, the conceptual shape of a wind turbine blade will be like a cylinder [11], which is inconsistent with its real shape. Martínez-Tossas et al. [12] proposed a two-dimensional elliptical Gaussian function as the regularization kernel and its direction is based on the global coordinates. Churchfield et al. [13] developed an anisotropic Gaussian function as the regularization kernel whose direction is determined by the local coordinates of each blade element. By using these anisotropic kernels, the shaped of wind turbine blades can be better modeled, which will alleviate the need of tip correction and improve the simulation near the blade tip.

The gaussian width ϵ used in the regularization kernel was found to have great influence on rotor torque predictions, wake characteristics [2,14], and may cause new requirement for the mesh [15,16]. Troldborg states that the value of ϵ should be at least twice the local grid length to avoid numerical oscillation. Martínez-Tossas et al. [12] and Shives et al. [17] suggest that the value of ϵ should be a quarter of the local airfoil chord length. Shives also recommends limiting grid size to a quarter of ϵ. Churchfield et al. [13] states that ϵ should be around 0.035 times the rotor diameter when using the standard regularization kernel. Pankaj et al. [14] developed and tested a series of guidelines for choosing ALM parameters and the results showed that the appropriate ϵ should be determined by a function of blade aspect ratio, grid size and a empirical constant. As for the anisotropic kernel, Martínez-Tossas et al. [12] studied the influence of the chord length parameter ϵ_c and the thickness parameter ϵ_t for two dimensional flow and the result shows that $\epsilon_c \approx 0.4c$ and $\epsilon_t < 0.2c$ are optimal. Churchfield et al. [13] studied the 3-dimensional wake characteristic of NREL (National Renewable Energy Laboratory) phase VI wind turbine with $\epsilon_c = 0.85c$, $\epsilon_t = 0.85t$ and simulations using the anisotropic kernel are more consistent with the experiment than the results of the standard kernel.

However, there is still confusion about the optimal value of the gaussian width and the effect of each parameter used in anisotropic kernel on ALM simulation result is still unclear. The recommended value of the gaussian width from the studies above do not agree with each other. Due to the author's experience, these recommended values of ϵ do not always lead to reliable results. Although the anisotropic kernel was developed in Churchfield's study [13], the parameters used in anisotropic kernel were not systematically studied. Furthermore, different parameters of $\epsilon_c = 0.4c$, $\epsilon_t = 0.2c$ were also used in this study for simulations of NREL 5MW wind turbine and the reason were not explained. Furthermore, the wake effect has great influence on the rotor torque of downstream wind turbines so the wake characteristics is significant in wind farm simulations and must be experimentally validated. In summary, the influence of the gaussian width used in regularization kernel and anisotropic kernel needs further study.

In this study, a new method is developed to measure the three-dimensional velocity field more efficiently and less expensively. Borrowing the idea of a frozen rotor, which is widely used in CFD simulations, the wake characteristics are reconstructed from the simultaneously gathered data of hot-wire anemometer and encoder. This measurement approach not only rebuilds the velocity distribution in a plane (along with or perpendicular to the main flow) but also reconstructs the whole wake region of a wind turbine. In this study, the influence of Gaussian width used in ALM with anisotropic regularization kernel is studied. Validated by the experimental results of power and wake characteristics, the relationship among the parameters of the anisotropic regularization kernel, physical scale of the blade, and mesh grid size are determined. This relationship will be used in further studies of the coupled aeroelastic wake behavior of a wind turbine based on ALM.

2. Method

2.1. Actuator Line Method (ALM)

The actuator line method is realized using OpenFOAM which is an open-source computational fluid software. The Large Eddy Simulation (LES) turbulence model is employed in this study because of its accuracy in wake simulations. The equations are shown as Equations (1) and (2).

$$\frac{\partial \overline{u_i}}{\partial x_i} = 0 \tag{1}$$

$$\rho\frac{\partial \overline{u}_i}{\partial t} + \rho\frac{\partial\left(\overline{u}_i\,\overline{u}_j\right)}{\partial x_j} = -\frac{\partial p}{\partial x_i} + \frac{\partial}{\partial x_j}\left[\mu\left(\frac{\partial \overline{u}_i}{\partial x_j} + \frac{\partial \overline{u}_j}{\partial x_i}\right)\right] + \frac{\partial \tau^s_{ij}}{\partial x_j} + f \tag{2}$$

where $\overline{u}$ is the filtered velocity vector field, p is the scalar field of pressure, μ is a scalar represent the kinematic viscosity, $\tau^s_{ij} = -\rho\left(\overline{u_i u_j} - \overline{u}_i\,\overline{u}_j\right)$ is called the subgrid-scale (SGS) Reynolds stress [18], and the standard Smagorinsky SGS model is employed in this study.

f is the source term which represents the wind turbine blade forces in ALM. When considering the wind turbine blade as a series of blade elements, the force along the blade can be calculated according to Equation (3). Here C_l and C_d are the lift and drag coefficient, respectively. ρ is the density of air, v is the inlet velocity of the blade element, c is the chord length, and L is the length of the each blade element.

$$F_{element} = (F_l, F_d) = (\frac{1}{2}\rho v^2 c C_l L, \frac{1}{2}\rho v^2 c C_d L) \tag{3}$$

The forces calculated by Equation (3) are point forces and a regularization kernel must be employed to avoid a numerical singularity, as shown Equation (4). Traditionally, a uniform three-dimensional Gaussian function is employed as the standard regularization kernel in the actuator line method. ϵ is Gaussian width which adjusts the strength of this regularization kernel.

$$f = \sum F_{element} \otimes \eta_\epsilon \tag{4}$$

$$\eta_\epsilon = \frac{1}{\epsilon^3 \pi^{3/2}}\, e^{-(\frac{r}{\epsilon})^2} \tag{5}$$

However, this uniform function will lead an imprecise approximation of the shape of the wind turbine blade. Although the chord length and the twist angle of the blade elements vary a lot from the root to the tip, the shape of the blade in actuator line model will be like a cylinder because of this uniform function. Furthermore, the uniform smooth function will cause the blade element force to be over concentrated along the chord direction, but more scattered along the thickness direction of the blade element at the meantime. Recently, an anisotropic regularization kernel as shown in Equation (6) was developed to overcome the disadvantages of the standard one.

$$\eta_\epsilon = \frac{1}{\epsilon_c \epsilon_t \epsilon_l \pi^{3/2}} e^{-(\frac{r_c}{\epsilon_c})^2 - (\frac{r_t}{\epsilon_t})^2 - (\frac{r_l}{\epsilon_l})^2} \tag{6}$$

Here r_c, r_t, r_l are the distances between the grid center and the force point in local coordinates of each blade element and ϵ_c, ϵ_t, ϵ_l are the corresponding Gaussian widths. Figure 1 shows the influence range and the strength distribution of these two kernels, the blue curve represents the standard kernel and the red one represents the anisotropic kernel

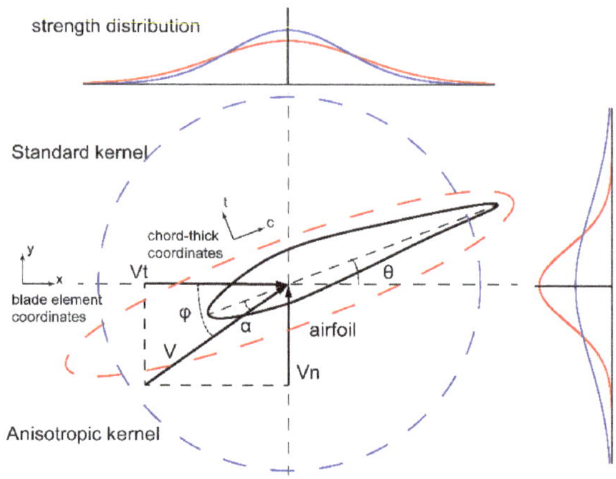

Figure 1. A schematic diagram of the influence range and strength distribution of two types of regularization kernel, the blue curve represents the standard kernel and the red one represents the anisotropic kernel.

2.2. Tip Loss Correction

Tip loss effect was first described by Prandtl who noted that the induced velocity tends to zero exponentially when approaching the blade tip and then the tip loss correction was introduced to BEM to make the simulation more realistic. For ALM, although the relationship between velocity and force is correct, a tip loss correction is suggested by Shen [19] due to the inconsistency between 2D airfoil data and attack angle of the 3D blade. This tip loss correction is employed in this study to compensate for the tip loss effect of wind turbine blade as shown in Equations (7) and (8).

$$F_1 = \frac{2}{\pi}\cos^{-1}[\exp\left(-g\frac{B(R-R_i)}{2R_i sin\phi_i}\right)] \tag{7}$$

$$g = \exp\left(-0.125\left(\frac{B\Omega R}{U_\infty} - 21\right)\right) + 0.1 \tag{8}$$

Here, B is the number of blades, Ω is the angular velocity, ϕ_i is the inflow angle for the ith blade element. R and R_i are the radius of the rotor and the radial position of the ith blade element, respectively.

2.3. Simulation Setup

The dimension and boundary conditions are set up according to the experiment. The inlet velocity is 3.5 m/s. The calculation domain is shown in Figure 2. To get a better result for the wake characteristic, the potential wake region is refined. For the convenience of expression, the calculation domain is subdivided into two regions: the background region and the rotor & wake region. Figure 2 shows the details of the calculation domain. The calculation domain is a little longer than the wind tunnel to avoid the influence of the inlet condition. In summary, the calculation domain is 1.5 m by 1.5 m by 2.8 m and a wake region of 1.0 m by 1.0 m by 1.2 m is refined to get a better simulation of wake characteristics.

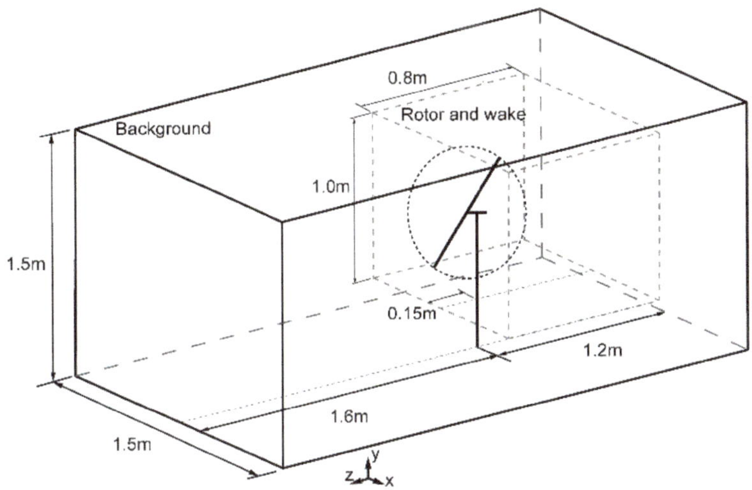

Figure 2. The calculation domain.

2.4. Experimental Setup

In this study, an experimental measurement of torque and wake characteristics was conducted to validate the simulations. Although Particle Image Velocimetry (PIV) technology has been proved to be powerful in wind turbine wake measurement [20–22], there are two restrictions for the PIV technology. When using the PIV system, the main direction of flow must be within the laser plane to guarantee that most of particles do not escape and stay illuminated. On the other hand, the measurement of the velocity field in three dimensions can only be achieved by using two cameras and a special laser generator [23,24]. The cost also constrains the use of PIV system. Due to these limitations of PIV system, hot-wire anemometer is also widely used in wake measurement. The wake experiments using hot-wire anemometer carried out by Schümann et al. [25], Lungo et al. [26], Singh et al. [27], and Dou [28] also made good measurements of wake characteristics. In this study, a new method for wind turbine wake measurement were developed using hot-wire anemometer.

Figure 3 shows the wind tunnel used in this experiment. It is composed of a contraction section, test section, diffuser section, and blower section. The length of the test section is 2.2 m and its cross-sectional dimensions are 1.5 m by 1.5 m. Equipped with three screens and two honeycombs and driven by four 11 kW mixed flow motors, the maximum velocity of the wind tunnel can reach 15 m/s and the turbulence intensity is around 0.5%.

A specifically designed two-blade wind turbine model is used in the experiment. The diameter of the rotor is 0.8 m and an NREL S826 airfoil profile is used all along the span for its high lift-drag ratio and low weight. The chord lengths and twist angles of the blade are shown in Figure 4.

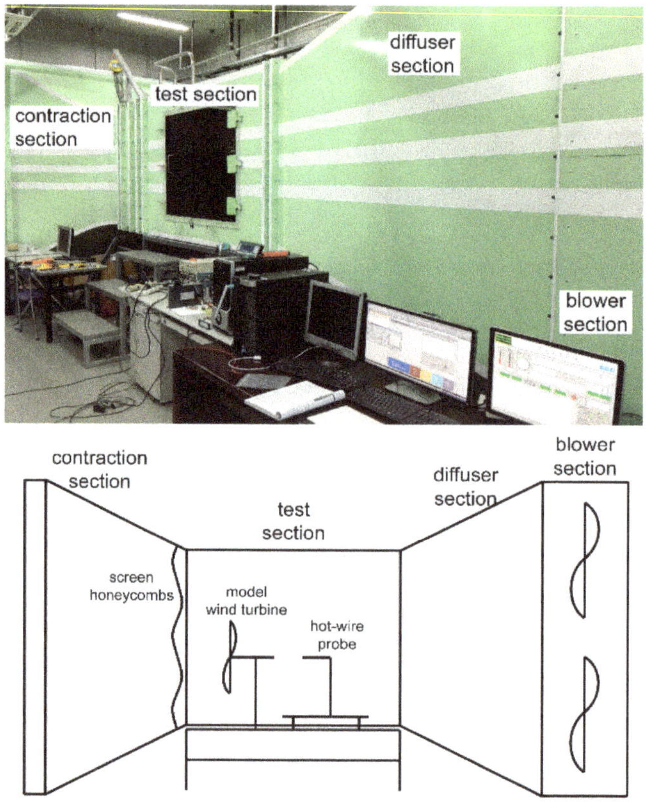

Figure 3. Wind tunnel used in experiment.

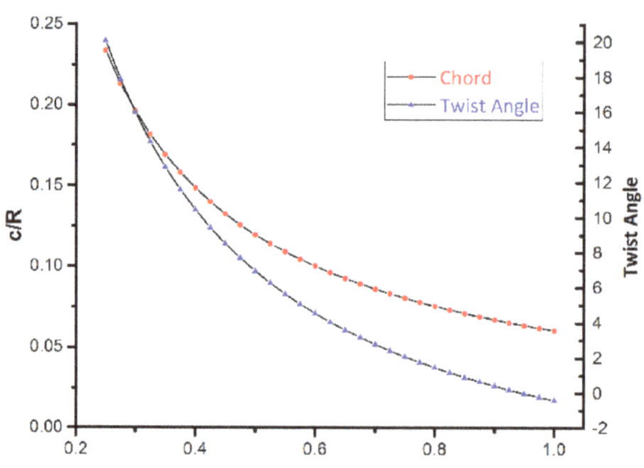

Figure 4. Length and twist angle distribution of the aerodynamics significant part of the blade.

The nacelle is equipped with an encoder of 1000 pulse, a torque sensor of 0.1% precision, and a servo-motor. This nacelle is designed mainly based on Anik's [29] equipment. During the experiment, the rotational speed of the rotor is totally controlled by the servo-motor and the relation between aerodynamic force and the motor force can be determined by the sign of the torque data. The positive sign of the torque data indicates that the rotor is driven by wind and the serves as a load balancing. Furthermore, with this equipment, the friction of the whole system can be measured by a motor-driven experiment without blades. This can help to increase the precision of the experiment. The details of the wind turbine model are shown in Figure 5.

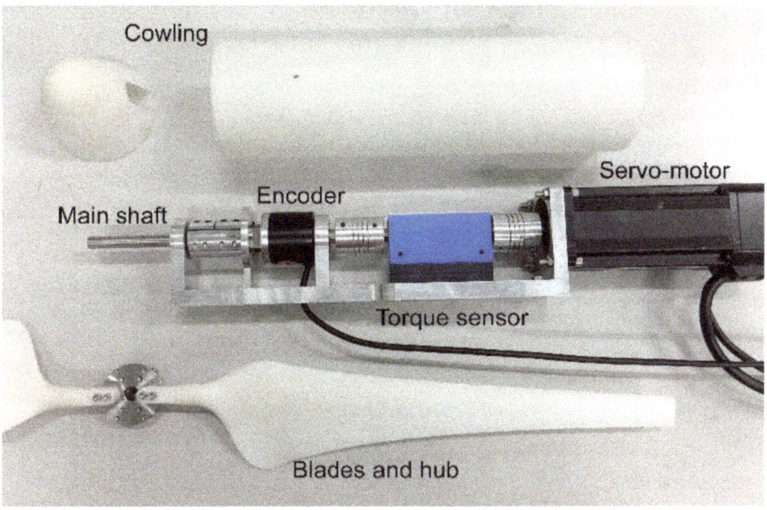

Figure 5. Wind turbine model and measurement equipment.

The inlet wind velocity is set to 3.5 m/s during the whole experiment. However, the rotational speed of the wind turbine model varies from 300 RPM to 650 RPM by 50 RPM. Thus, the tip speed ratio correspondingly varies from 3.6 to 7.8. For each rotational speed, the torque and wake characteristic of the model wind turbine are measured. It should be noted that the dimension and boundary condition of the simulation is strictly based on the experiment.

2.5. Frozen Rotor Method

Although PIV technology has been widely used in wind turbine experiments, it still has some limitations: the measuring plane must along the main flow direction to ensure the particle being illuminated, 2D-PIV can only measure the two-dimensional component of the wind field and 3D-PIV equipment is very expensive.

The frozen rotor method is widely used in CFD simulations for a rotating machine. When considering from the ground coordinates, the fluid field of a rotating machine is a transient problem. However, it can be turned into a static problem by considering form rotating coordinates. In this study, a new measurement method for wind turbine wakes was developed borrowing the idea of frozen rotor. A three dimensional hot-wire anemometer and encoder were employed to measure the velocity field in the wake region. At each measurement point, the wind velocity and the rotor azimuth were collected simultaneously. According to the axial symmetry of the wind turbine flow, this measurement can be considered as the hot-wire probe gathering velocity data around the axis of the rotor while the rotor and fluid field are frozen. Mean velocity was calculated according to the rotor azimuth to avoid the influence of small-scale vortex.

Figure 6 shows the experimental setup of hotwire anemometer probe. In this study, the probe was moved by an auto-controlled platform with the precision of 25 μm to scan a whole plane. Figure 7 shows all the locations of measured points. For each point, the wind velocity and encoder data are collected simultaneously. Using the idea of frozen rotor, with each wind speed marked by a rotor azimuth, the wake characteristic can be reconstructed for the whole wake region. During the experiment, more measurement points were added to the potential tip vortex region, as shown in Figure 7, to get more precise results. By this way, we rebuild the wake region inexpensively but accurately with a hot-wire anemometer system.

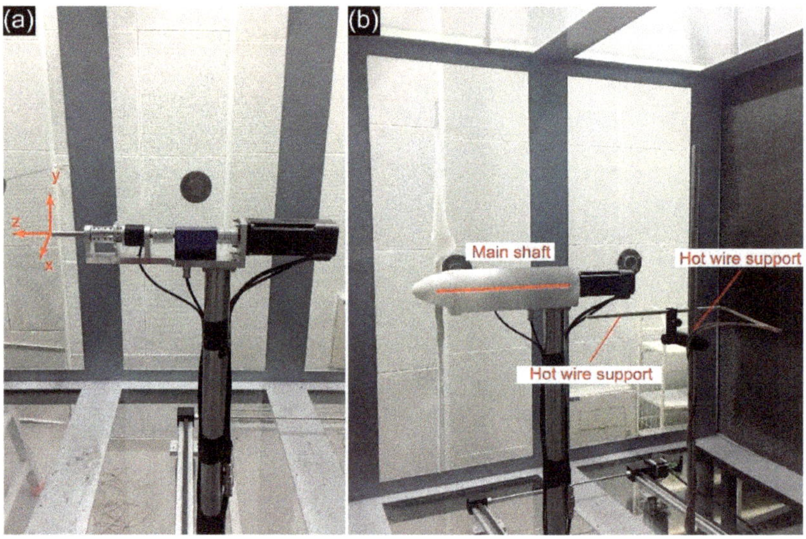

Figure 6. (**a**) Wind turbine model and coordinate system (**b**) hotwire anemometer probe location.

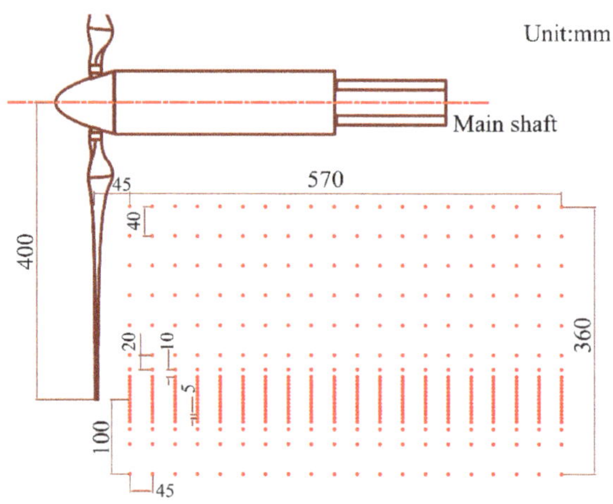

Figure 7. Locations of measurement points.

3. Result

3.1. Mesh Independence

Two levels of the mesh are compared to demonstrate the mesh independence of the simulation results of this study. Table 1 shows comparison of element number and element size between three mesh levels. The element size in the rotor and wake region is designed to be 10 mm in normal mesh and 8 mm in refined mesh. Furthermore, a non-uniform mesh is specially designed to take advantage of the anisotropic kernel. The element size is 10 mm in rotor plane and 6 mm in the main flow direction.

Table 1. The comparison of element number and element size between three mesh levels (Unit: number of elements).

Mesh Level / Region	Background	Rotor and Wake	Total Number	Element Size (Rotor and Wake)
Normal	228,608	800,000	1,112,108	10 mm, uniform
Refined	228,608	1,562,500	1,903,858	8 mm, uniform
Special	228,608	1,340,000	1,673,708	10 mm in rotor plane 6 mm in main flow direction

Figure 8 shows the torque result calculated by different meshes and different ϵ values, which are 16 mm, 20 mm, 24 mm, and 32 mm. Since the Gaussian $\epsilon > 2\Delta_{grid}$ must be guaranteed to avoid numerical oscillation, the torque result calculated with $\epsilon = 16$ mm is only achieved on the refined mesh. All results shown are calculated with the standard regularization kernel.

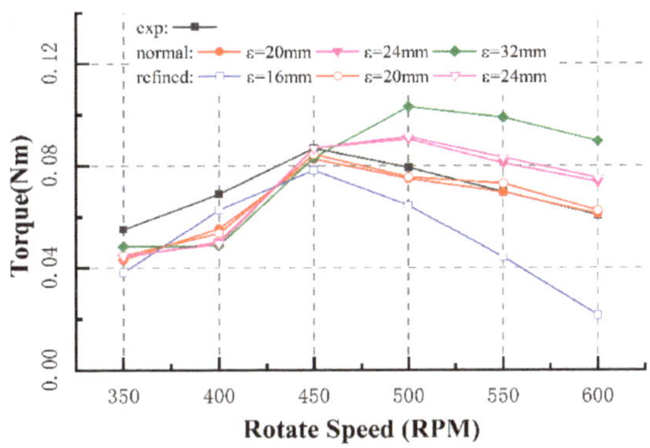

Figure 8. The comparison of torque result with different meshes and different ϵ values. Curves with the same color represent results using different level of mesh but the same Gaussian width.

The torque result shows that the normal mesh gives a similar prediction of the rotor torque compared with the results of the refined mesh when using the same ϵ value. Figure 9 shows the normal velocity (illustrated in Figure 1) of each blade element. It shows that normal mesh gives a similar prediction of the normal velocity compared with the results of the refined mesh. Therefore, it can be concluded that the simulations using a normal mesh are mesh independent.

It should be noticed that $\epsilon = 2\Delta_{grid}$ is not sufficient to guarantee a reliable simulation, because simulations with $\epsilon = 16$ mm and refined mesh give a totally different result compared with simulations with $\epsilon = 20$ mm and normal mesh. Furthermore, a rotating rotor generating torque is a physical phenomenon and it must not relate to the element size. On the other hand, Figure 8 also shows that the torque result is strongly affected by the ϵ value and does not going to converge when the ϵ value grows too much. This phenomenon will be discussed in the next section.

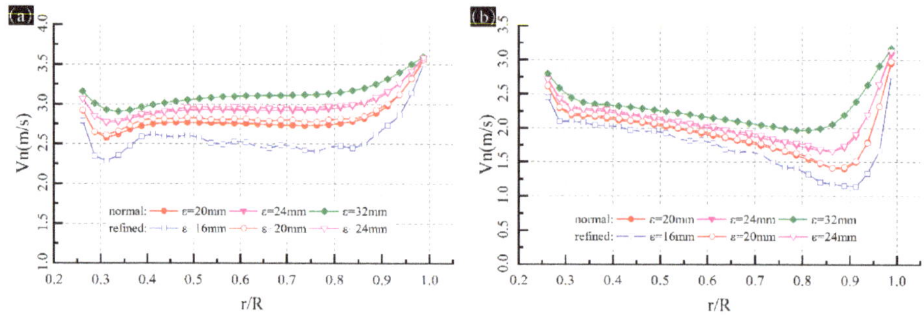

Figure 9. The normal velocity along the blade calculated by two levels of meshes with different Gaussian width when rotating speed is (**a**) 400 RPM, (**b**) 550 RPM. Curves with the same color represent results using the same Gaussian width but different level of mesh.

3.2. The Effect of Gaussian Width

In this section, the simulation results using different ϵ value are compared with the experimental results to study the influence of the Gaussian width. Figure 8 shows the comparison of the torque result between the experiment and simulations. According to the discussion above, the ϵ value is related to the chord length (chord length of blade tip when using standard regularization kernel) to make this study more referential. The lift and drag coefficient data obtained by Sarlak [30] are used in this study and the data were gathered at Re = 100,000, which is a little higher than the Reynolds number of this study. It should be noticed that the cross-section changes from a circle to airfoil S826 in the transition section of the model turbine blade and there are no aerodynamic data for this section and thus the aerodynamic performance of the transition section is neglected in all simulations. It is believed that the difference between the simulation result and the experimental result when rotating speed is 350 and 400 RPM is because of this neglecting. However, when the rotating speed increases, the contribution of the transition section to the aerodynamic load of the whole rotor can be neglected because it is close to the hub and its velocity is low. Nevertheless, the comparison illustrates that the prediction of torque does not converge when the ϵ value grows too much.

Figures 10 and 11 show the velocity and attack angle of each blade element when the rotating speed is 400 RPM or 550 RPM. It shows that the tangential velocity of different cases is almost the same when ϵ value varies and is mainly determined by the rotating speed of the rotor. However, the normal velocity is strongly affected by ϵ and the normal velocity increase with the value of ϵ. This has a significant influence on the attack angle of each blade element as shown in Figure 11 and therefore has a significant influence on the lift and drag force of each blade element. It should be noticed that the influence of the ϵ value on the torque result is not linear. According to the Reynolds number of this study, airfoil S826 gives the best aerodynamic performance when the attack angle is about 8 degree. Therefore, although the simulation with $\epsilon = 0.83c_{tip}$ gives a lower prediction of the normal velocity and the attack angle of each blade element, it gives a higher prediction of the torque, which shows a different trend compared with the results at other rotating speeds.

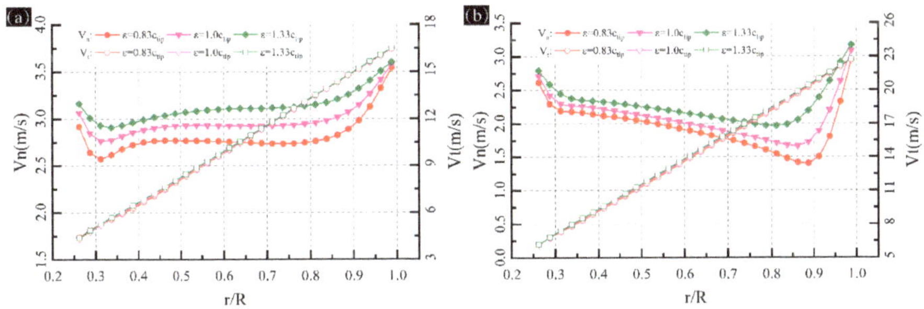

Figure 10. The V_n and V_t of each blade element along the blade when rotating speed is (**a**) 400 RPM, (**b**) 550 RPM.

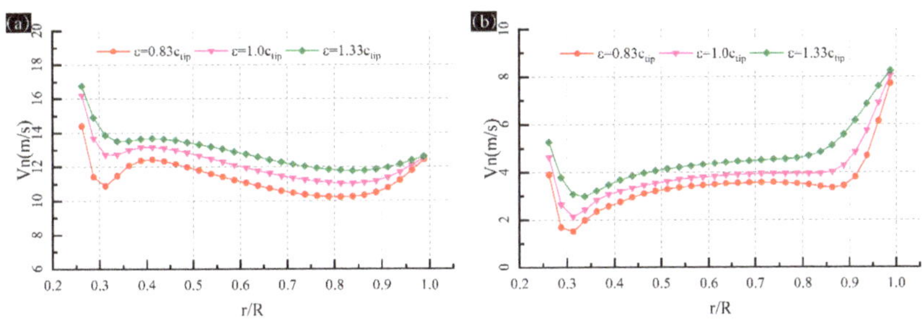

Figure 11. The attack angle of each blade element along the blade when rotating speed is (**a**) 400 RPM, (**b**) 550 RPM.

3.3. The Effect of the Chord Length Gaussian Width

Figure 12 shows the comparison for torque between experiment and simulations using the anisotropic regularization kernel. The best result using the standard regularization kernel is also added to the comparison. This figure also shows that the prediction of torque will not converge with the increment of the ϵc value. The empirical value of ϵ is not suitable for the anisotropic regularization kernel and $\epsilon_c = 1.2c$ shows the best prediction of the torque. However, the torque result is less sensitive to the ϵc value compared with the ϵ value when using standard regularization kernel. The simulation using anisotropic regularization kernel with $\epsilon_c = 1.0c$ also gives a reasonable enough prediction of torque.

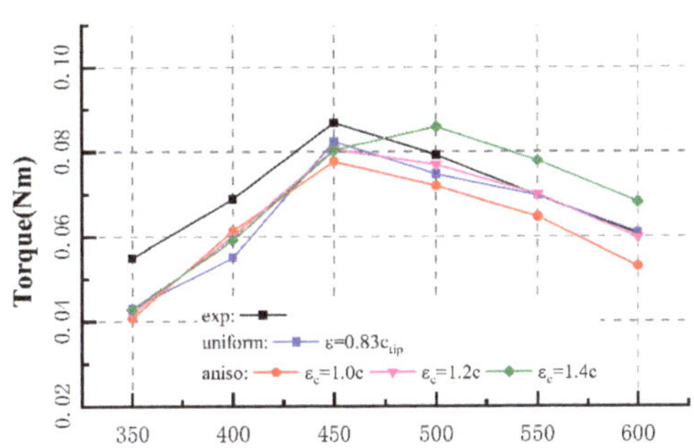

Figure 12. The comparison of torque between experiment, simulation using standard regularization kernel with $\epsilon = 0.83c_{tip}$, and simulations using anisotropic regularization kernel with different ϵ_c values.

The velocity and attack angle results when using anisotropic regularization kernel are different from the standard one. Figure 13 shows the comparison for the velocity between the result using standard regularization kernel with $\epsilon = 0.83c_{\mathrm{tip}}$ and results using anisotropic regularization kernel with different ϵ_c values. It is clear that the tangential velocity of each blade element is less affected by the regularization kernel. However, the normal velocity is strongly affected by the regularization kernel which has been discussed. Compared with the result using anisotropic regularization kernel, the normal velocity is underestimated near the blade root and overestimated near the blade tip when using constant ϵ value. This matches with the previous discussion because the chord length of the blade root is larger than the tip one and a constant ϵ value will mispredict the affect region of a blade element. However, the attack angle of each blade element is not experimentally measured in this study due to the equipment limitation. Further study is needed to make a quantitative conclusion.

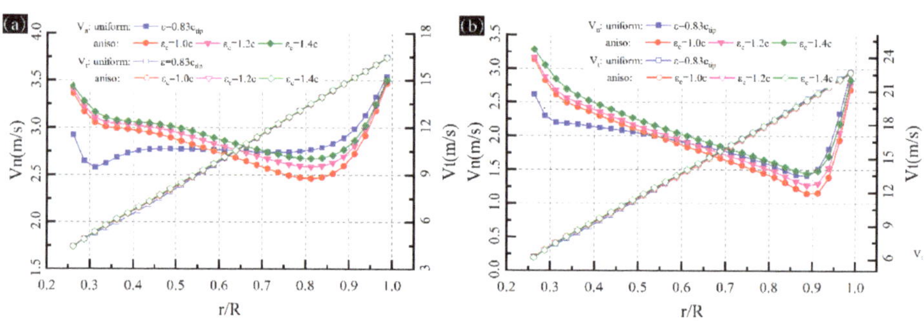

Figure 13. The V_n and V_t of each blade element along the blade when rotating speed is (**a**) 400 RPM, (**b**) 550 RPM.

According to the results above, although the optimal values are different, the effect of ϵ in standard kernel and the effect of ϵ_c in anisotropic is similar. Both ϵ and ϵ_c have little influence on the simulation result of tangential velocity, but will significantly affect the result of V_n and therefore significantly affect the rotor torque result. Figure 14 illustrates the effect region of different ϵ value. With a larger ϵ value, a blade element has a larger effect region in the actuator line method. However, the total value of the force is the same due to the Gaussian function, which means a regularization kernel has a flatter strength distribution when ϵ value is larger. Therefore, the regularization kernel with larger Gaussian width will affect a larger region of the flow and will lead to a higher velocity. This will cause an incorrect prediction of the attack angle of the blade element and has a strong effect on the prediction of lift and drag force.

It should be noticed that the optimal value in this study is different with the study of Martínez-Tossas [13] and Churchfield [12]. Actually, they do not agree with each other. Martínez-Tossas recommended $\epsilon = 0.14c \sim 0.24c$, Churchfield studied NREL 5MW wind turbine with $\epsilon_c = 0.4c$ and studied NREL Phase VI wind turbine with $\epsilon_c = 0.85c$ (here, c is the chord length). The main difference between these studies and this paper is the scale of the wind turbine. It could be inferred that the optimal parameters are related to the Reynolds number and this still needs further study.

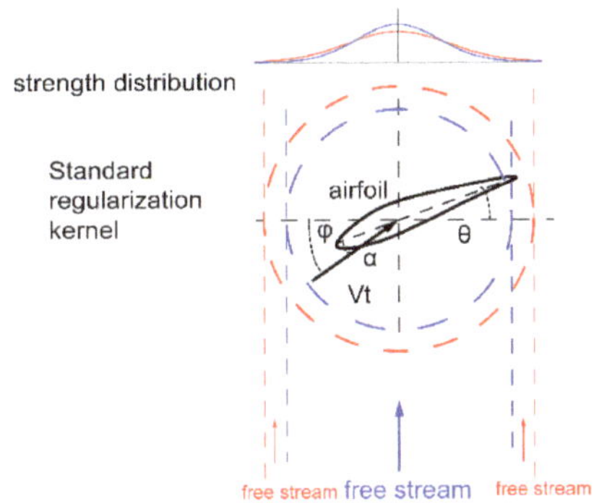

Figure 14. An illustration of the effective region of different ϵ value. The red one represents a regularization kernel with larger Gaussian width.

3.4. The Effect of the Thickness Gaussian Width

Table 2 shows the torque result when using different values of the thickness parameter ϵ_t. Since the thickness of the airfoil is much smaller than the chord length and usually smaller than the limit of $\epsilon > 2\Delta_{grid}$, the ϵt value is usually limited by an absolute value which is related to the grid size. Two different ϵ_t values are compared here. The series results for $\epsilon_t = 20$ mm are calculated on the normal mesh and this value is equal to twice of the grid size. The series results for $\epsilon_t = 9$ mm are calculated on the special mesh and this value is equal to 1.5 times of the grid size in the main flow direction. Table 2 shows that the thickness parameter has little influence on the torque results. Figure 15 shows the velocity component of each blade element and there is only a small difference for V_n which appears at the blade tip. It can be concluded that the thickness parameter ϵ_t has a little influence on the torque prediction of ALM.

Table 2. Torque result when using different ϵ_t value (Unit: Nm).

Speed	Exp	ϵ_c=1.2c, ϵ_t=20 mm	ϵ_c=1.2c, ϵ_t=9 mm
350	0.05504	0.04202	0.04291
400	0.06889	0.06014	0.05979
450	0.08679	0.08384	0.07998
500	0.07922	0.07701	0.07870
550	0.0697	0.07000	0.06982
600	0.06062	0.06228	0.06105

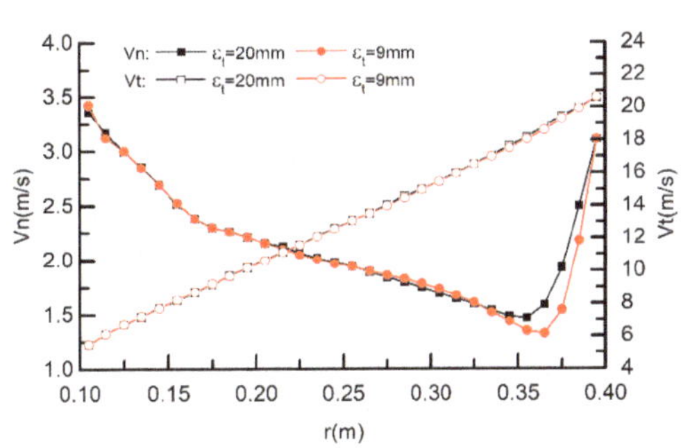

Figure 15. The V_n and V_t of each blade element along the blade when rotating speed is 500 RPM calculated by different ϵ_t value.

3.5. Wake Characteristic

Figure 16 shows the velocity distribution of the plane which is perpendicular to the main flow direction and 45 mm behind the rotor plane. The first row represents the experiment result, the second row represent the result of standard kernel and the third row represents the result of anisotropic kernel. The data area is a ring with outer radius of 500 mm and inner radius of 140 mm. The standard deviation as shown in Equation (9) is used to evaluate the difference between simulation and experimental result. Here, n is the number of sample points, v_{sim} is the simulation result and v_{exp} is the experimental result.

$$E = \sqrt{\frac{\sum_{i=1}^{n}\left(v_{sim} - v_{exp}\right)^2}{n v_{inlet}^2}} \tag{9}$$

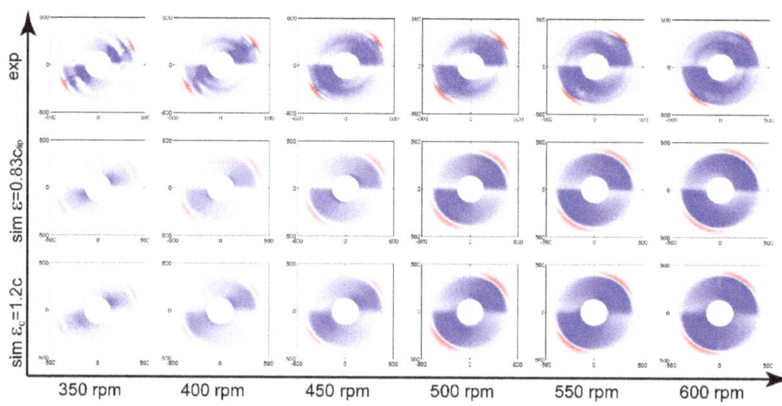

Figure 16. The velocity magnitude contour of the plane (45 mm behind the rotor) perpendicular to the main flow direction. The origin is the main shaft location and the data area is a ring with outer radius of 500 mm and inner radius of 140 mm. The first row represents the experiment result, the second row represent the result of standard kernel and the third row represents the result of anisotropic kernel.

As shown in Figure 17, the simulations show good agreement with the experimental result, the standard deviation between the simulation and experimental result is less than 6%. Furthermore, the standard deviation results illustrate that the ϵ value has little influence on the velocity distribution in the rotor plane.

Figure 18 shows the velocity distribution of the plane along with the main flow. The first row represents the experiment result, the second row represent the result of standard kernel and the third row represents the result of anisotropic kernel. The x coordinate represents the distance from the rotor plane and the y coordinate represents the radius position from the main shaft. The data area is from 45 mm to 615 mm behind the rotor and 140 mm to 500 mm away from the main shaft. The velocity distribution shows that the actuator line method can accurately simulate the pattern of the wake flow of a wind turbine. However, it also shows that the peak velocity of simulations is lower and the wake pattern is flatter compared with the experimental results. Figure 19 shows the standard deviation of velocity between simulations and experimental results. Although the improvement of velocity distribution is not significant, simulations with anisotropic regularization kernel show a more accurate result compared with simulations using the standard kernel.

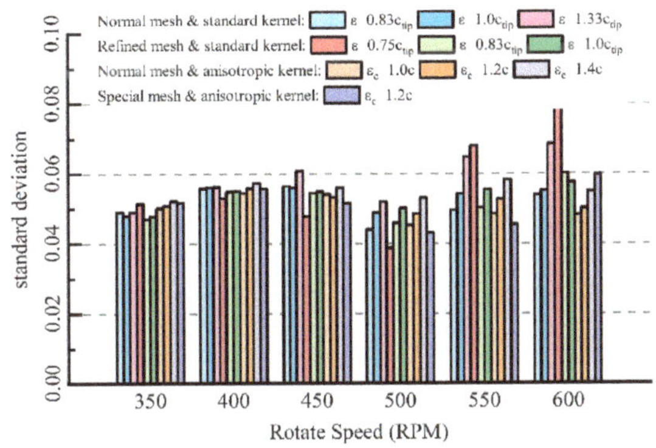

Figure 17. Standard deviation of the velocity magnitude in the plane perpendicular to the main flow direction.

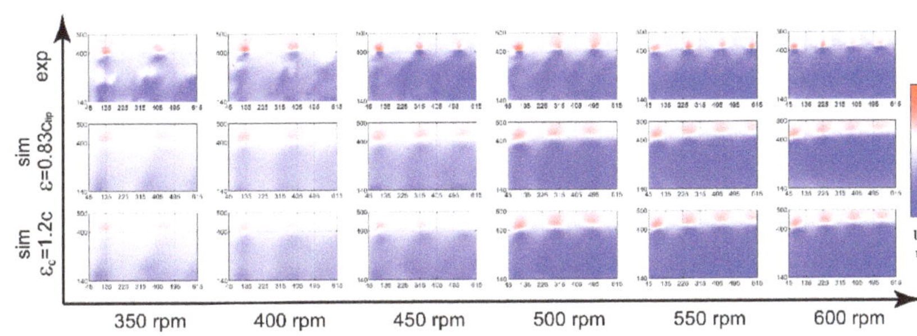

Figure 18. The velocity magnitude contour of the plane along with the main flow. The origin is the rotor center. The data area is from 45 mm to 615 mm behind the rotor and 140 mm to 500 mm away from the main shaft.

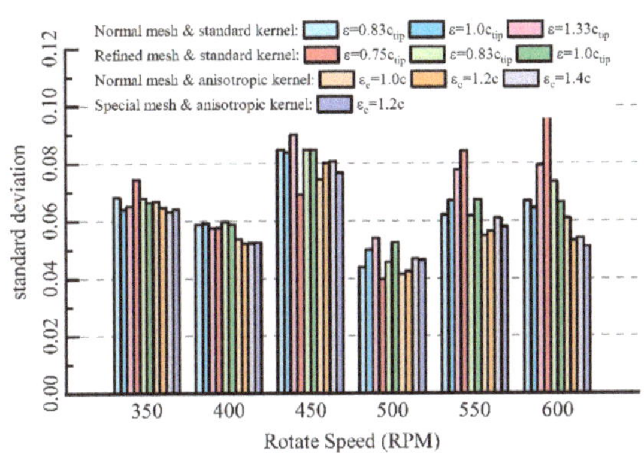

Figure 19. Standard deviation of the velocity magnitude in the plane along with the main flow.

Figure 20 shows the vorticity results. It is clear that the actuator line method gives a flatten prediction of the vorticity distribution because of the simplification of wind turbine blade. The peak vorticity of the experiment is much larger than the simulation result and the region of the high vorticity region is much smaller. However, the actuator line method gives a reliable prediction of the wake pattern. It is difficult to use standard deviation function to evaluate the vorticity results, because both the position and the absolute value must be taken into consideration. Therefore, the correlation coefficient as defined in Equation (10) is used as the evaluation metric.

$$C = \frac{\sum (v_{sim} - \overline{v_{sim}})(v_{exp} - \overline{v_{exp}})}{\sqrt{\sum (v_{sim} - \overline{v_{sim}})^2 \sum (v_{exp} - \overline{v_{exp}})^2}} \quad (10)$$

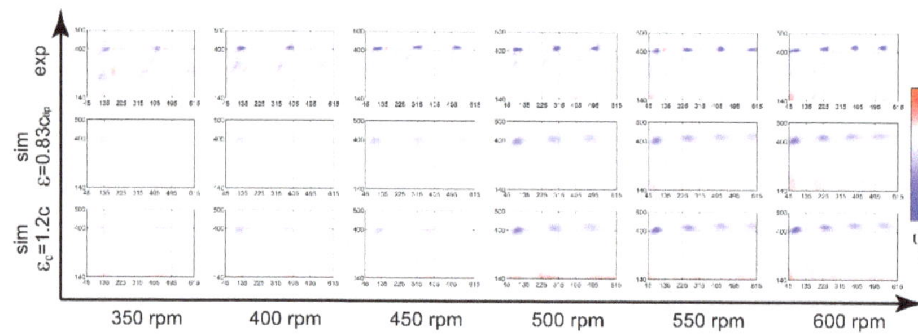

Figure 20. The vorticity magnitude contour of the plane along with the main flow. The origin is the rotor center. The data area is from 45 mm to 615 mm behind the rotor and 140 mm to 500 mm away from the main shaft.

The correlation coefficient neglects the average value difference between two distributions, but focuses on the pattern. In this study, the size of the high vorticity region and its position will significantly affect the correlation coefficient. Figure 21 shows the correlation coefficient between simulations and the experimental results. As the actuator line model predicts a flatter vorticity distribution, the correlation coefficient is not too good. However, simulations with a correlation coefficient higher than 0.5 give a reliable prediction of the position of high vorticity region, which means these simulations give a reliable prediction of the wake pattern. Furthermore, the correlation coefficient is significantly improved when using the anisotropic regularization kernel, especially when using the special mesh at the same time. As the main difference between the standard kernel and the anisotropic one is the thickness parameter, it can be concluded that the thickness parameter has a significant influence on the wake pattern prediction.

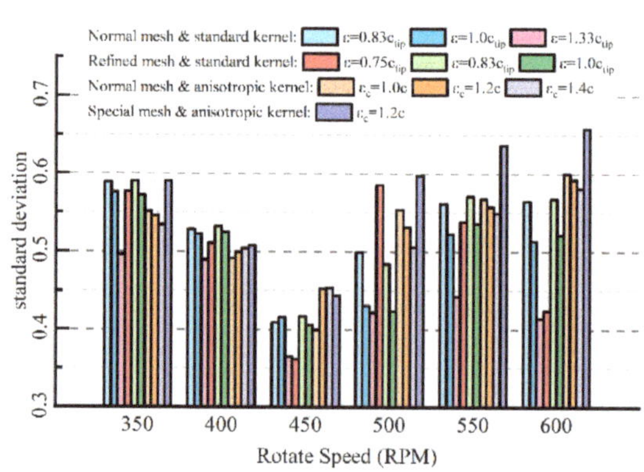

Figure 21. The comparison of the correlation coefficient.

It should be noticed that the number of elements of the special mesh is lower than the refined mesh, but the correlation coefficient of the simulations using the special mesh is much higher than the simulations using refined mesh. These simulations obviously take the advantages of the anisotropic regularization kernel and significantly improve the performance of the actuator line model. A special mesh with refinement in the main flow direction together with the anisotropic regularization kernel will give a more accurate and lower computational cost simulation of the wind turbine.

4. Conclusions

In this paper, the mesh and parameters of the actuator line method with standard and anisotropic regularization kernel are studied. An experiment of the torque and wake characteristic was carried out, using hot-wire anemometer and borrowing the idea of frozen rotor method, to evaluate the simulation results. The relationship between Gaussian width ϵ, element size, attack angle of blade elements, and the simulation result of torque, and wake characteristic are discussed. The conclusion is as follows:

1. Gaussian width ϵ will strongly affect the torque result during actuator line simulations and it does not converge when ϵ becomes larger. Larger ϵ value will cause a higher prediction of the normal velocity of each blade element, but has little effect on the tangential velocity. The influence of the ϵ value on the attack angle is the main reason for its effect on the torque prediction.
2. In this study, $\epsilon = 0.83c_{\text{tip}}$ for standard regularization kernel and $\epsilon_c = 1.2c$ for anisotropic kernel can guarantee a reliable torque result. However, according to the state-of-art studies, the optimal value for ϵ varies with the scale of wind turbine. It can be inferred that the suitable parameters are related to the Reynolds number.
3. The thickness parameter ϵ_t has little influence on the torque prediction. However, the thickness parameter significantly affects the prediction of the wake characteristics. The anisotropic regularization kernel will improve the performance of the actuator line model in wake simulations.
4. Borrowing the idea of frozen rotor method, this study developed a reliable method to measure the wind turbine wakes. The wake characteristics were reconstructed by simultaneously gathered velocity data and rotor azimuth.
5. A special mesh with refinement in the main flow direction will take advantages of the anisotropic regularization kernel. Using a mesh refined in the main flow direction, ALM with anisotropic kernel can predict torque and wake characteristics better while maintaining low computational costs.

Author Contributions: Conceptualization, Z.M.; Formal analysis, Z.M.; Supervision, L.L. and P.Z.; Writing—original draft, Z.M.; Writing—review & editing, L.L., E.D. and P.Z. All authors have read and agreed to the published version of the manuscript.

Funding: This study was funded by National Natural Science Foundation of China (No.51875305) and Short-term Visiting Foundation of Tsinghua University (No.2018020).

Conflicts of Interest: The authors declare no conflict of interest.

Nomenclature

Variables

B	number of wind turbine blades
U_∞	free stream velocity [m/s]
Ω	rotor speed [rad/s]
R	rotor radius [m]
R_i	radial position of *i*th blade element
ϕ_i	inflow angle for *i*th blade element
η_ϵ	regularization kernel
ϵ	Gaussian width for standard regularization kernel [m]
ϵ_c	chord length Gaussian width for anisotropic regularization kernel [m]
ϵ_t	thickness Gaussian width for anisotropic regularization kernel [m]
ϵ_l	length Gaussian width for anisotropic regularization kernel [m]

ρ	air density [kg/m^3]
F_l	lift force [N]
F_d	drag force [N]
c	chord length [m]
L	length of blade element [m]
v	local velocity on blade element [m/s]
C_l	lift coefficient
C_d	drag coefficient
r	distance from the center of regularization kernel [m]
r_c	projection of r on the chord length direction of blade element [m]
r_t	projection of r on the thickness direction of blade element [m]
r_l	projection of r on the length direction of blade element [m]
E	standard deviation
C	correlation coefficient

References

1. Sorensen, J.N.; Shen, W.Z. Numerical modeling of wind turbine wakes. *J. Fluid. Eng.-Trans. ASME* **2002**, *124*, 393–399. [CrossRef]
2. Martinez-Tossas, L.A.; Churchfield, M.J.; Leonardi, S. Large eddy simulations of the flow past wind turbines: Actuator line and disk modeling. *Wind Energy* **2015**, *18*, 1047–1060. [CrossRef]
3. Mikkelsen, R.; Sørensen, J.N.; Øye, S.; Troldborg, N. Analysis of power enhancement for a row of wind turbines using the actuator line technique. *J. Phys. Conf. Ser.* **2007**, *75*, 012044. [CrossRef]
4. Lee, S.; Churchfield, M.J.; Moriarty, P.J.; Jonkman, J.; Michalakes, J. A Numerical Study of Atmospheric and Wake Turbulence Impacts on Wind Turbine Fatigue Loadings. *J. Solar Energy Eng.-Trans. ASME* **2013**, *135*. [CrossRef]
5. Storey, R.C.; Cater, J.E.; Norris, S.E. Large eddy simulation of turbine loading and performance in a wind farm. *Renew. Energy* **2016**, *95*, 31–42. [CrossRef]
6. Na, J.S.; Koo, E.; Munoz-Esparza, D.; Jin, E.K.; Linn, R.; Lee, J.S. Turbulent kinetics of a large wind farm and their impact in the neutral boundary layer. *Energy* **2016**, *95*, 79–90. [CrossRef]
7. Na, J.S.; Koo, E.; Jin, E.K.; Linn, R.; Ko, S.C.; Muñoz-Esparza, D.; Lee, J.S. Large-eddy simulations of wind-farm wake characteristics associated with a low-level jet. *Wind Energy* **2018**, *21*, 163–173. [CrossRef]
8. Zhong, H.M.; Du, P.G.; Tang, F.N.; Wang, L. Lagrangian dynamic large-eddy simulation of wind turbine near wakes combined with an actuator line method. *Appl. Energy* **2015**, *144*, 224–233. [CrossRef]
9. Ma, Z.; Zeng, P.; Lei, L.P. Analysis of the coupled aeroelastic wake behavior of wind turbine. *J. Fluids Struct.* **2019**, *84*, 466–484. [CrossRef]
10. Meng, H.; Lien, F.S.; Li, L. Elastic actuator line modelling for wake-induced fatigue analysis of horizontal axis wind turbine blade. *Renew. Energy* **2018**, *116*, 423–437. [CrossRef]
11. Sorensen, J.N.; Mikkelsen, R.F.; Henningson, D.S.; Ivanell, S.; Sarmast, S.; Andersen, S.J. Simulation of wind turbine wakes using the actuator line technique. *Philos. Trans. A Math. Phys. Eng. Sci.* **2015**, *373*, 20140071. [CrossRef] [PubMed]
12. Martinez-Tossas, L.A.; Churchfield, M.J.; Meneveau, C. Optimal smoothing length scale for actuator line models of wind turbine blades based on Gaussian body force distribution. *Wind Energy* **2017**, *20*, 1083–1096. [CrossRef]
13. Churchfield, M.J.; Schreck, S.J.; Martinez, L.A.; Meneveau, C.; Spalart, P.R. An advanced actuator line method for wind energy applications and beyond. In Proceedings of the 35th Wind Energy Symposium, Grapevine, TX, USA, 9–13 January 2017; p. 1998.
14. Jha, P.K.; Churchfield, M.J.; Moriarty, P.J.; Schmitz, S. Guidelines for Volume Force Distributions Within Actuator Line Modeling of Wind Turbines on Large-Eddy Simulation-Type Grids. *J. Solar Energy Eng. Trans. ASME* **2014**, *136*. [CrossRef]
15. Troldborg, N. Actuator Line Modeling of Wind Turbine Wakes. Doctoral Thesis, Technical University of Denmark, Lyngby, Denmark, 2009.
16. Troldborg, N.; Sorensen, J.N.; Mikkelsen, R.; Sorensen, N.N. A simple atmospheric boundary layer model applied to large eddy simulations of wind turbine wakes. *Wind Energy* **2014**, *17*, 657–669. [CrossRef]

17. Shives, M.; Crawford, C. Mesh and load distribution requirements for actuator line CFD simulations. *Wind Energy* **2013**, *16*, 1183–1196. [CrossRef]
18. Wu, Y.T.; Porte-Agel, F. Large-Eddy Simulation of Wind-Turbine Wakes: Evaluation of Turbine Parametrisations. *Bound.-Layer Meteorol.* **2011**, *138*, 345–366. [CrossRef]
19. Shen, W.Z.; Sørensen, J.N.; Mikkelsen, R. Tip Loss Correction for Actuator/Navier–Stokes Computations. *J. Solar Energy Eng.* **2005**, *127*, 209–213. [CrossRef]
20. Snel, H.; Schepers, J.; Montgomerie, B. The MEXICO project (Model Experiments in Controlled Conditions): The database and first results of data processing and interpretation. *J. Phys. Conf. Ser.* **2007**, *75*, 012014. [CrossRef]
21. Hong, J.; Guala, M.; Chamorro, L.; Sotiropoulos, F. Probing wind-turbine/atmosphere interactions at utility scale: Novel insights from the EOLOS wind energy research station. *J. Phys. Conf. Ser.* **2014**, *524*, 012001. [CrossRef]
22. Hong, J.; Toloui, M.; Chamorro, L.P.; Guala, M.; Howard, K.; Riley, S.; Tucker, J.; Sotiropoulos, F. Natural snowfall reveals large-scale flow structures in the wake of a 2.5-MW wind turbine. *Nat. Commun.* **2014**, *5*, 4216. [CrossRef]
23. Hu, H.; Wei, T.; Wang, Z. An Experimental Study on the Wake Characteristics of Dual-Rotor Wind Turbines by Using a Stereoscopic PIV Technique. In Proceedings of the 34th AIAA Applied Aerodynamics Conference, Washington, DC, USA, 13–17 June 2016; p. 3128.
24. Wang, Z.Y.; Ozbay, A.; Tian, W.; Hu, H. An experimental study on the aerodynamic performances and wake characteristics of an innovative dual-rotor wind turbine. *Energy* **2018**, *147*, 94–109. [CrossRef]
25. Schumann, H.; Pierella, F.; Saetran, L. Experimental investigation of wind turbine wakes in the wind tunnel. *Energy Procedia* **2013**, *35*, 285–296. [CrossRef]
26. Iungo, G.V.; Viola, F.; Camarri, S.; Porte-Agel, F.; Gallaire, F. Linear stability analysis of wind turbine wakes performed on wind tunnel measurements. *J. Fluid Mech.* **2013**, *737*, 499–526. [CrossRef]
27. Singh, A.; Howard, K.B.; Guala, M. On the homogenization of turbulent flow structures in the wake of a model wind turbine. *Phys. Fluids* **2014**, *26*, 025103. [CrossRef]
28. Dou, B.Z.; Guala, M.; Lei, L.; Zeng, P. Experimental investigation of the performance and wake effect of a small-scale wind turbine in a wind tunnel. *Energy* **2019**, *166*, 819–833. [CrossRef]
29. Anik, E.; Abdulrahim, A.; Ostovan, Y.; Mercan, B.; Uzol, O. Active control of the tip vortex: An experimental investigation on the performance characteristics of a model turbine. *J. Phys. Conf. Ser.* **2014**, *524*, 012098. [CrossRef]
30. Sarlak, H.; Mikkelsen, R.; Sarmast, S.; Sørensen, J.N. Aerodynamic behaviour of NREL S826 airfoil at Re= 100,000. *J. Phys. Conf. Ser.* **2014**, *524*, 012027. [CrossRef]

energies

MDPI

Article

Implementation and Validation of an Advanced Wind Energy Controller in Aero-Servo-Elastic Simulations Using the Lifting Line Free Vortex Wake Model

Sebastian Perez-Becker *, David Marten, Christian Navid Nayeri and Christian Oliver Paschereit

Chair of Fluid Dynamics, Hermann Föttinger Institute, Technische Universität Berlin, Müller-Breslau-Str. 8, 10623 Berlin, Germany; david.marten@tu-berlin.de (D.M.); christian.nayeri@tu-berlin.de (C.N.N.); oliver.paschereit@tu-berlin.de (C.O.P.)
* Correspondence: s.perez-becker@tu-berlin.de

Abstract: Accurate and reproducible aeroelastic load calculations are indispensable for designing modern multi-MW wind turbines. They are also essential for assessing the load reduction capabilities of advanced wind turbine control strategies. In this paper, we contribute to this topic by introducing the TUB Controller, an advanced open-source wind turbine controller capable of performing full load calculations. It is compatible with the aeroelastic software QBlade, which features a lifting line free vortex wake aerodynamic model. The paper describes in detail the controller and includes a validation study against an established open-source controller from the literature. Both controllers show comparable performance with our chosen metrics. Furthermore, we analyze the advanced load reduction capabilities of the individual pitch control strategy included in the TUB Controller. Turbulent wind simulations with the DTU 10MW Reference Wind Turbine featuring the individual pitch control strategy show a decrease in the out-of-plane and torsional blade root bending moment fatigue loads of 14% and 9.4% respectively compared to a baseline controller.

Keywords: wind energy; wind turbine control; load mitigation; individual pitch control; lifting line free vortex wake; vortex methods

Citation: Perez-Becker, S.; Marten, D.; Nayeri, C.N.; Paschereit, C.O. Implementation and Validation of an Advanced Wind Energy Controller in Aero-Servo-Elastic Simulations Using the Lifting Line Free Vortex Wake Model. *Energies* **2021**, *14*, 783. https://doi.org/10.3390/en14030783

Academic Editor: Robert Castilla
Received: 6 January 2021
Accepted: 28 January 2021
Published: 2 February 2021

1. Introduction

Over the past years, wind turbines have increased significantly in size in an effort to reduce the cost of wind energy and make it a competitive source of energy. So far, this has been a successful approach. An example of this success can be seen in the increased installation numbers of new wind turbines in Germany [1]. Yet increasing the turbine size also has its drawbacks. A larger rotor will be subjected to higher aerodynamic and gravitational loads. For example, the aerodynamic bending moments and the moments due to the self-weight of the blade scale as the third and fourth power of the rotor diameter respectively ([2], (pp. 97–123)). As a consequence, the structure of the turbine components such as rotor blades has to be stiffer, which requires more or stronger material. This leads to an increase in the cost of energy. A way of counteracting the load increase seen in larger wind turbines is through the use of advanced wind turbine controllers. This allows for less material use in the design of the different components and hence results in a decrease of the cost of energy.

The most common actuator used for load alleviation is the blade pitch actuator. This comes from it being already used in the power regulation strategy of modern turbines. Many advanced load alleviation strategies using pitch actuators have been proposed. One of the best-known ones is the Individual Pitch Control (IPC) strategy [3], which commonly relies on the out-of-plane Blade Root Bending Moments (BRBM) of the individual blades as input signals. It has been used in combination with different input sensors—such as inflow sensors [4,5]—and also in combination with different actuators—such as active trailing edge flaps [6–9]. Other studies have studied advanced model-based [10] or adaptive [11] controllers for load alleviation. A controller that uses neural networks as part of its architecture has also been recently proposed in [12].

Oftentimes, the results of these studies are difficult to compare and reproduce. This is partly because many research groups use self-developed baseline controller strategies as a basis for load reduction comparative studies. The source code of these controllers is rarely available. Alternatively, studies that use the NREL 5 MW Reference Wind Turbine (RWT) [13] also use the baseline controller available in the model definition. This is an older controller—based on [14,15]—that offers limited functionality and is inconvenient to use with other turbine models. This is because the controller parameters are hard-coded in the source file and the controller has to be recompiled every time the parameters change.

As part of the UpWind Project, a more advanced baseline controller was developed which allows for better power tracking and smoother transitions between the operating regions. These features lead to better energy capture and to reduced loading compared to the NREL 5 MW controller. Many of the aforementioned features are also implemented in the Basic DTU Wind Energy Controller [16].

In recent years, the wind energy community has started to address the problem of reproducible controller results by introducing several modern, open-source reference wind turbine controllers. Mulders and van Wingerden publish the Delft Research Controller (DRC) [17], which is expanded by Abbas et al. into NREL's Reference Open-Source Controller (ROSCO) [18]. Meng et al. also extend the Basic DTU Wind Controller to form the DTU Wind Energy Controller (DTUWEC), which includes advanced industrial features [19].

All of the cited reference controllers use classical controller architectures such as PID controllers. As discussed in ([20], (pp. 506–518)), there are several reasons for this choice. An important one is that with classical controller architectures, controller stability can be guaranteed. This is of utmost importance since modern large wind turbines are expected to operate reliably under all circumstances with minimum supervision. Another reason is that individual features can be added to the controller without the need to re-calibrate the whole controller. Also, the integration of a supervisory controller is much more straightforward when classical controller architectures are used. Other control techniques such as neural networks are powerful techniques that could potentially be used to solve specific control objectives. They are therefore well suited to be part of advanced controller features. They are however less appropriate as candidates for reference baseline controllers. Because of their black-box model nature, it is difficult to guarantee their stability, add specific features or integrate them with supervisory controllers.

Having a reference controller is one aspect of accurate and comparable load reduction estimation. In order to have a complete picture of the design loads of different wind turbine components, we need to simulate the wind turbine in realistic aeroelastic scenarios that often include controller faults or other unforeseen events. Current industry standards prescribe a large number of aeroelastic simulations of the complete turbine [21]. These are grouped together into Design Load Case (DLC) groups, each considering a particular scenario in the wind turbine's design life. Most of the aforementioned studies only include a selection of DLC groups, such as the power production DLC group. While being a good approximation for certain component's loads, using a selection of DLCs will not give a complete picture of the load reduction capabilities of the different turbine controllers. This is in part because many of the available wind turbine controllers do not feature a supervisory controller. The latter oversees the pitch and torque controller and reacts to unforeseen events, shutting down if threshold values of certain signals are passed in order to ensure the structural integrity of the turbine components. Only a full load calculation according to industry standards will give an accurate estimate of the load reduction capabilities of advanced controller strategies.

Another important aspect for accurate load estimation is the use of appropriate models in the aeroelastic simulations. Current aeroelastic codes mostly rely on the Blade Element

Momentum (BEM) aerodynamic model to calculate aerodynamic loads ([20], (pp. 57–66)). BEM models are attractive because they are computationally inexpensive. Yet in order to capture the more challenging unsteady aerodynamic phenomena present in DLCs, BEM models require a series of engineering corrections. These corrections have been developed and tested so that they work for a wide range of operating conditions. If the wind turbine operates in a condition outside this range, then the BEM corrections could introduce inaccuracies and overestimate the aerodynamic loading on the turbine. Examples of this include turbine operation in extreme yawed conditions or inflow conditions that are inhomogeneous across the rotor. This could arise if the turbine is operating in the partial wake of another turbine, in sheared and/or turbulent inflow due to the Earth's boundary layer or if there is a large difference in the individual pitch angles of the blades (e.g., with pitch actuator faults) [22–25].

Vortex methods such as the Lifting Line Free Vortex Wake (LLFVW) aerodynamic model have a higher order representation of the unsteady aerodynamic phenomena and are capable of modeling these with far fewer assumptions than BEM models [26]. This is because in the LLFVW method the wake is explicitly modeled. In [27] the authors show that there are significant differences in fatigue and extreme loading if the aeroelastic simulations are performed with a BEM aerodynamic model compared to a LLFVW model, with the BEM model predicting increased fatigue loads. If we wish to accurately asses the load reduction potential of advanced control strategies based on individual pitch action, we require an accurate representation of the local aerodynamic effects that occur on each blade. Since the advanced controller action will be individual on each blade, the resulting induction field will be non-homogeneous. This in turn requires a higher-order aerodynamic model such as the LLFVW model to accurately estimate the effect of the resulting aerodynamic loads.

In this study, we address the aforementioned issues by introducing the TUB Controller (TUBCon). It is an open-source reference wind turbine controller with advanced load reduction strategies that features a complete supervisory controller. It can therefore be used to perform a full load calculation so that an accurate load picture of the turbine loads is obtained. Furthermore, this controller is fully compatible with the aeroelastic software QBlade [28]—which features the LLFVW aerodynamic model. In combination, QBlade and TUBCon can be used to accurately calculate wind turbine loads and also evaluate the performance of advanced load reduction strategies. We give an example of this by analyzing the load reduction capabilities of the well-known IPC strategy in power production with turbulent wind conditions. This paper is structured as follows: In Section 3 we present and fully describe TUBCon, including its advanced load reduction strategy and supervisory controller. In Section 4, we validate TUBCon in steady and turbulent wind conditions by comparing its performance to an established wind turbine controller from the literature. In Section 5 we analyze TUBCon's advanced load reduction capabilities in turbulent wind conditions by simulating a reference wind turbine using the baseline and IPC variants of TUBCon. Conclusions are drawn in Section 6.

2. Methods

We chose the DTU 10 MW RWT as the turbine model for this study as it is representative of the new generation of wind turbines and has been used in several research studies. The complete description of the turbine can be found in [29].

As explained in [25], large rotors such as the one from the DTU 10MW RWT have similar scales to the scales from turbulent wind found in Earth's boundary layer. As a consequence, the wind field that the turbine rotor sees will be much less homogeneous than the wind field seen by a smaller turbine rotor. BEM codes typically average the induction factor across a blade annulus, thus leading to inaccurate load predictions. In addition, large blade deflections that occur in modern large blades may introduce radial flow, thus violating the assumptions of many BEM models. Both of the aforementioned issues are addressed with higher order aerodynamic models such as the LLFVW model found in TU Berlin's aeroelastic simulation tool QBlade. We did the testing and simulation of the TUB Controller presented in this study using QBlade.

2.1. QBlade

To model the unsteady aerodynamics, QBlade uses the Lifting Line Free Vortex Wake (LLFVW) method [28]. Here, the blade aerodynamic forces are evaluated on a blade element basis using standard airfoil polar data. The wake is modeled with vortex line elements. These are shed at the blade's trailing edge during every time step and then undergo free convection behind the rotor. Vortex methods can model the wake with far less assumptions and engineering corrections compared to BEM methods. Especially when the wind turbine is subject to unsteady inflow or varying blade loads, the LLFVW method increases the accuracy compared to BEM methods [27]. To model the dynamic stall of the blade elements, QBlade uses the ATEFlap unsteady aerodynamic model [30], modified so that it excludes contribution of the wake in the attached flow region [31].

Regarding the structural model, QBlade uses the open-source multi-physics library CHRONO [32]. It features a multi-body representation of the turbine which includes Euler-Bernoulli beam elements in a co-rotational formulation. With it, QBlade is able to accurately simulate blade deflections including the blade torsion, which has a significant influence on the blade loads. A detailed comparison between QBlade and OpenFAST can be found in [27].

3. Description of the TUB Controller

This section describes the TUB Controller (TUBCon). A graphical representation of the controller architecture is shown in Figure 1.

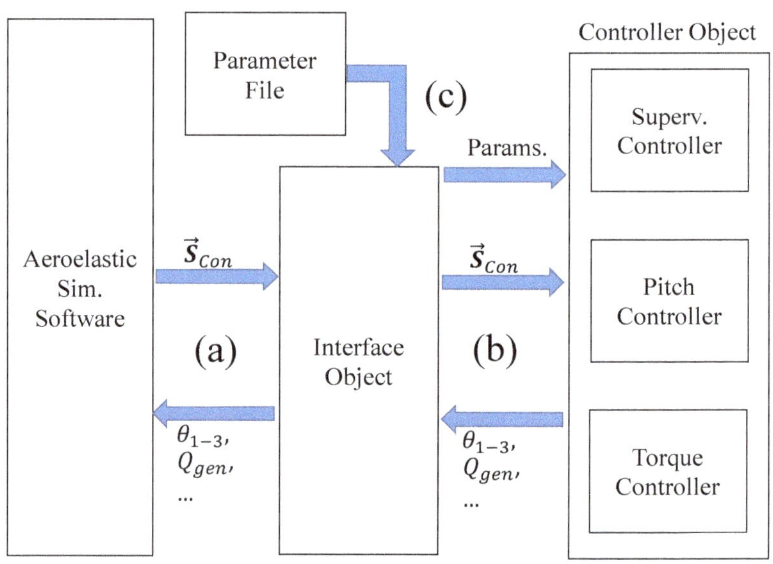

Figure 1. Graphical representation of the controller architecture. The communication with the aeroelastic software is done in two stages using an interface and a controller object. The controller parameters are passed through an external parameter file. $\vec{S}_{Con}$ stands for controller signals.

The controller code is written in C++ in an object-oriented manner. It features a two-stage interaction with the aeroelastic simulation software. The first stage is an interface object that handles the data exchange with the aeroelastic code (Figure 1(a)). The current supported interfaces are the conventional Bladed-Style communication via a swap-array and a modified swap-array communication specially designed for interacting with QBlade.

This makes TUBCon compatible with common aeroelastic simulation codes including OpenFAST, Bladed and QBlade. The modular nature of the interface object makes it easy to extend the compatibility of TUBCon with other aeroelastic codes.

The interface object then communicates with the turbine controller in a second stage Figure 1(b). The individual controller components are designed in Simulink and compiled to a C++ object using the Simulink Coder package. The resulting object has an initialization function, a step function and a termination function. These are called by the main routine at the appropriate simulation times. Such an architecture allows us to rapidly develop and test new controller features that can be incorporated to the compiled controller library. The controller is parameterized externally using an appropriate parameter file (Figure 1(c)). This XML file is read and the controller parameters set as part of the initialization routine.

The controller object is adapted from the Basic DTU Wind Energy Controller [16] and shares many of its capabilities. It features a state-of-the-art pitch and torque controller that enable the turbine to operate reliably under unsteady turbulent wind conditions. Although the description of the torque and collective pitch controller is given in [16], it is also included here for the sake of completeness.

Figure 2 shows a schematic overview of the controller.

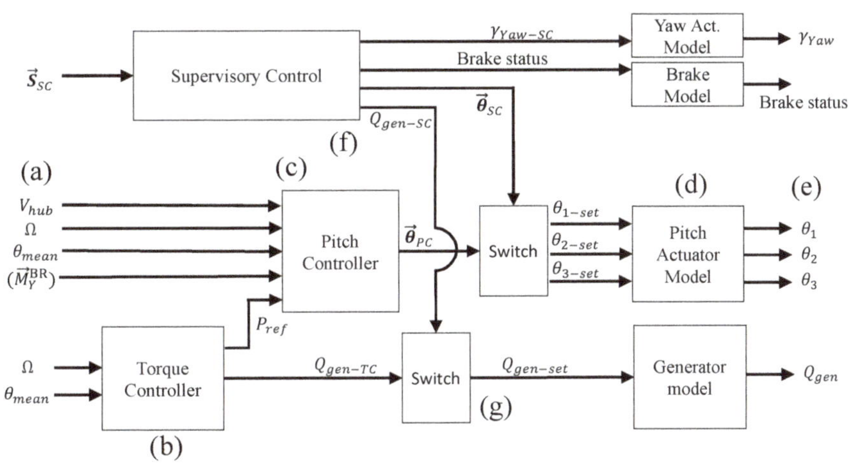

Figure 2. Overview of the controller object. It is comprised of three major modules: the pitch controller, the torque controller and the supervisory control. The latter is able to override the former two with the use of switches. The setpoint signals are passed to the respective actuator model to calculate the actuation signal.

Both the pitch and torque controllers share the rotor speed Ω and the mean pitch angle θ_{mean} as input sensors (Figure 2(a)). The output of the torque controller is the generator set-point torque $Q_{gen-set}$ and the reference power P_{ref}. The pitch controller uses the aforementioned Ω and θ_{mean} as well as P_{ref} and the measured wind speed at hub height V_{hub} to determine the set-point angles of all three blades ($\theta_{1-set}, \theta_{2-set}$ and θ_{3-set}) (Figure 2(c)). If the IPC strategy is enabled, then the pitch controller also needs the out-of-plane BRBM of all blades ($\vec{M}_Y^{BR}$). The controller includes actuator models for the pitch and yaw actuators as well as the brakes and generator to model the dynamics of the different actuators (Figure 2(d)). The set-point signal of each controller is passed to the respective actuator model to calculate the actual signals passed back to the aeroelastic code (Figure 2(e)). In the following sections, the individual parts of the controller will be further explained.

Figure 2 also shows the supervisory control. It is included to allow the controller to perform full load calculations. It will be explained in Section 3.3 and is included in Figure 2(f),(g) to show the location where the supervisory control overrides the signals of the individual controllers.

The pitch and torque controllers explained in the following sections operate using the low speed shaft side of the drive train. That is, they assume that the generator speed and torque are identical to the low speed shaft speed and torque. For geared turbines, the gear box ratio G is applied to convert the high speed shaft quantities into low speed shaft quantities and back.

3.1. Torque Controller

The torque controller is divided into two main parts: the partial and the full load regimes. There is also the switching logic to change between both regimes. The input for the controller is the rotor speed Ω and the mean pitch angle θ_{mean}. The latter is used for the switching logic between the two load regimes. The output of the controller is the generator torque Q_{gen} and the reference power P_{ref} for the pitch controller. The reference power is simply the product of the instantaneous generator torque and the rotor speed: $P_{ref}(t) = Q_{gen}(t) \cdot \Omega(t)$.

The variable speed controller is a PID controller with two different rotor speed set-points. The controller is saturated with a maximum and a minimum limit so that Ω is adjusted to follow an optimum tip speed ratio in the partial load regime.

As a first step, the rotor speed Ω is low-pass filtered with a second order low-pass filter to exclude unwanted high frequency dynamics. The function of the low-pass filter is given by Appendix A Equation (A2). The speed error e_{TC} is calculated using the difference of the low-passed rotor speed Ω_{LP} and the current speed set-point Ω_{set}: $e_{TC} = \Omega_{LP} - \Omega_{set}$. The speed set-point is defined as

$$\Omega_{set} = \begin{cases} \Omega_{min} & \text{if } \Omega_{LP} < \frac{\Omega_{rated}+\Omega_{min}}{2} \\ \Omega_{rated} & \text{if } \Omega_{LP} \geq \frac{\Omega_{rated}+\Omega_{min}}{2}, \end{cases} \tag{1}$$

where Ω_{rated} and Ω_{min} represent the rated and minimum rotor speed. The generator torque is computed using the equation

$$Q_{PID}(t) = k_{P-TC} \cdot e_{TC}(t) + k_{I-TC} \cdot \int_0^t e_{TC}(\tau) d\tau + k_{D-TC} \cdot \frac{de_{TC}(t)}{dt}. \tag{2}$$

Here, k_{P-TC}, k_{I-TC} and k_{D-TC} represent the proportional, the integral and the differential gain of the PID controller. Because the values of the generator torque are saturated, each time step, the integral term of Q_{PID} is recalculated as

$$Q_I(t) = k_{I-TC} \cdot \int_0^t e_{TC}(\tau) d\tau = Q_{PID}(t) - k_{P-TC} \cdot e_{TC}(t) - k_{D-TC} \cdot \frac{de_{TC}(t)}{dt}. \tag{3}$$

This helps to avoid windup of Q_I and enables the controller to react quickly if the required torque signal changes from a previously saturated value.

3.1.1. Partial Load Regime

In order for the torque controller to enforce an optimum tip speed ratio in the partial load regime, the generator torque is saturated with upper and lower limits: Q^P_{max} and Q^P_{min}. Between Ω_{min} and Ω_{rated}, the limits are identical and follow the optimal torque speed curve of the turbine:

$$Q^P_{max} = Q^P_{min} = K_{opt} \cdot \Omega^2_{LP} \tag{4}$$

$$= \frac{\pi \rho R^5 C_{p-opt}}{2\lambda^3_{opt}} \cdot \Omega^2_{LP}. \tag{5}$$

In the last equation, ρ represents the air density, R the rotor radius and C_{p-opt} the power coefficient at optimum tip speed ratio λ_{opt}.

At two given rotor speed ranges, $[\Omega_{min-a}, \Omega_{min-b}]$ and $[\Omega_{max-a}, \Omega_{max-b}]$, the torque limits open up to allow the PID controller to keep the rotor speed at the required set-point. The switching logic for Q^P_{min} is given by:

$$Q^P_{min} = \begin{cases} \min\left(K_{opt} \cdot \Omega^2_{LP} \cdot \sigma(\Omega_{min-a}, \Omega_{min-b}, \Omega_{LP}), K_{opt} \cdot \Omega^2_{max-a}\right) & \text{if } K_{opt} \cdot \Omega^2_{max-a} \leq Q^F_{ref} \\ Q^F_{ref} & \text{if } K_{opt} \cdot \Omega^2_{max-a} > Q^F_{ref}. \end{cases} \quad (6)$$

In the above equation, Q^F_{ref} represents the generator torque in full load regime and $\sigma(\cdot)$ represents a smooth switching function between the two limits Ω_{min-a} and Ω_{min-b}. The switch function is given by Equations (A5) and (A6). Complementary to Equation (6), Q^P_{max} follows the switching logic:

$$Q^P_{max} = \max\Big((1 - \sigma_{max}) \cdot K_{opt} \cdot \Omega^2_{LP} + \sigma_{max} \cdot Q^F_{ref}, K_{opt} \cdot \Omega^2_{min-b}\Big). \quad (7)$$

The above equation uses the symbol σ_{max} to denote the switch function $\sigma(\Omega_{max-a}, \Omega_{max-b}, \Omega_{LP})$.

Figure 3 shows the torque limits as a function of Ω.

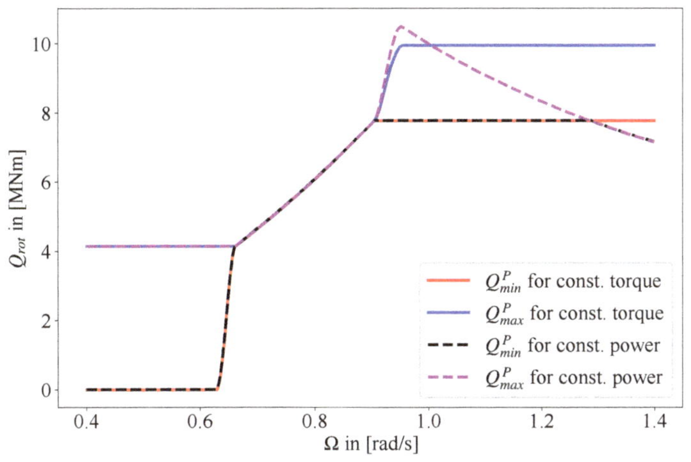

Figure 3. Limits for torque controller at the partial load regime. Limits are shown for the constant power and constant torque strategies.

Below the minimum rotor speed, Q^P_{min} is kept at 0 kNm to allow the rotor to gain enough rotational speed at low wind speeds. Once the rotor speed reaches Ω_{min-a} (95% of Ω_{min} in the figure), Q^P_{min} is increased and equaled to $K_{opt} \cdot \Omega^2$. This enforces the optimal torque-speed curve of the wind turbine below the rated speed. Once Ω gets close to Ω_{max-a} (95% of Ω_{rated}), Q^P_{max} opens up to allow the PID controller to regulate the rotor speed. The parameters $\Omega_{min-a}, \Omega_{min-b}, \Omega_{max-a}$ and Ω_{max-b} are externally defined by the user. Figure 3 includes the behavior of the torque limits for the two torque control strategies in the full load regime: constant power and constant torque. These are explained in the next section.

3.1.2. Full Load Regime

The full load regime of the controller uses the same PID controller for the controller torque, but now the torque limits Q^P_{min} and Q^P_{max} take the same value Q^F_{ref}. The control strategy of the torque controller in the full load regime can be either constant power or constant torque. Q^F_{ref} takes different values depending on the strategy. These are:

$$Q^F_{ref} = \begin{cases} \frac{P_{rated}}{\Omega_{rated}} & \text{for constant torque} \\ \frac{P_{rated}}{\Omega} & \text{for constant power.} \end{cases} \tag{8}$$

While Q^F_{ref} is constant for the constant torque strategy, it is inversely proportional to the (unfiltered) rotor speed signal for the constant power strategy. This explains the $1/\Omega$ behavior of Q^P_{min} and Q^P_{max} for the constant power strategy in Figure 3.

3.1.3. Switching between Regimes

The controller switches between partial and full load regimes depending on the first order low-pass filtered switch variable $\sigma_{\theta-LP}$. σ_θ is defined using Equation (A5) as $\sigma(\theta_{f1}, \theta_{f2}, \theta_{mean})$. The limits θ_{f1} and θ_{f2} are user-defined parameters. For this study, the limits were defined as $\theta_{f1} = \theta_{f2} = \theta_{min} + 0.5°$. In this configuration, the switching function σ_θ behaves like a step function. The variable θ_{min} is the minimum pitch angle of the pitch controller, defined in Section 3.2. The equation of the first order low-passed filter is given by (A1).

Including the switching between regimes and the torque strategy in the above rated regime, the full torque limits take the form

$$Q_{min} = (1 - \sigma_{\theta-LP}) \cdot Q^P_{min} + \sigma_{\theta-LP} \cdot Q^F_{ref} \tag{9}$$

$$Q_{max} = (1 - \sigma_{\theta-LP}) \cdot Q^P_{max} + \sigma_{\theta-LP} \cdot Q^F_{ref}. \tag{10}$$

3.1.4. Drivetrain Damper

The torque controller also includes a drivetrain damper to damp unwanted oscillation of the turbine shaft due to e.g., excitation of its torsional eigenfrequency. The drivetrain damper is implemented as a proportional term k_{dmp} to the band-passed rotor speed signal Ω_{BPF}:

$$Q_{dmp} = k_{dmp} \cdot \Omega_{BPF}. \tag{11}$$

The equation of the band-pass filter is given in Equation (A3).

Q_{dmp} is added to Q_{PID} to obtain the total output torque of the controller:

$$Q_{gen}(t) = Q_{dmp}(t) + Q_{PID}(t). \tag{12}$$

3.1.5. Disable for Low Rotor Speeds

An additional feature of the controller is a switch-off mechanism for low rotor speeds. The rotor speed signal Ω is filtered using a low pass filter (see Equation (A2)) and compared to a fraction of the minimum rotor speed: $k_{\text{off}} \cdot \Omega_{Min}$. If the filtered rotor speed signal is below this value, then the torque controller shuts down by setting $Q_{max} = Q_{min} = 0$ Nm. Note that if the turbine rotor speeds up again and the (filtered) rotor signal exceeds the aforementioned limit, then the torque controller is activated again and functions as described above.

3.2. Pitch Controller

The pitch controller in TUBCon features two strategies. The first is the Collective Pitch Control (CPC) strategy for power regulation. The second is the advanced Individual Pitch Control (IPC) strategy for additional load reduction. The CPC strategy is always active while the IPC strategy can be enabled by the user through the external parameter file.

3.2.1. Collective Pitch Control

The CPC controller combines two PID controllers that react to the rotor speed error and the power error. The inputs for the pitch controller are the rotor speed Ω, the mean pitch angle θ_{mean}, the reference power from the torque controller P_{ref} and the measured wind speed at hub height V_{hub} (see Figure 2(a)).

The rotor speed signal is filtered using a second order low-pass filter as described by Equation (A2). Both the pitch and the torque controller share the same filter frequency and damping ratio to calculate Ω_{LP}. The speed error e_{Ω} is calculated as the difference between the low pass filtered rotor speed Ω_{LP} and the rated rotor speed Ω_{rated}. The error e_{Ω} is then passed through a notch filter to filter out the drivetrain eigenfrequency. The equation of a notch filter is given in Equation (A4).

In parallel, the power error e_P is calculated as the difference between P_{ref} and P_{rated} and passed through a notch filter (Equation (A4)) to obtain e_{P-NF}. The proportional, integral and differential terms of the pitch controller are calculated as follows:

$$\theta_P(t) = \frac{1}{2}(k_{P-\Omega} \cdot e_{\Omega-NF}(t) + k_{P-P} \cdot e_{P-NF}(t)), \tag{13}$$

$$\theta_I(t) = \frac{1}{2}\left(\int_0^t (k_{I-\Omega} \cdot e_{\Omega-NF}(\tau) + k_{I-P} \cdot e_{P-NF}(\tau))d\tau\right), \tag{14}$$

$$\theta_D(t) = k_{D-\Omega} \cdot \frac{de_{\Omega-NF}(t)}{dt}. \tag{15}$$

Here, k_{P-X}, k_{I-X} and k_{D-X} are the proportional, integral and derivative constants of each PID controller. The subscript Ω denotes that the PID constants are applied to $e_{\Omega-NF}$. Likewise, the subscript P denotes the affiliation of the constants to e_{P-NF}. Note that k_{D-P} is always set to zero.

Additionally, the pitch signal is gain-scheduled with two factors. The first one accounts for the non-linear effect that larger blade pitch angles have on the aerodynamic torque. The second factor increases sensitivity of the pitch controller to large speed excursions in order to limit them. The total gain schedule is calculated via

$$\eta = \eta_\theta \cdot \eta_\Omega = \left(\frac{1}{1 + \frac{\theta_{LP}}{K_1} + \frac{\theta_{LP}^2}{K_2}}\right) \cdot \left(1 + \frac{e_\Omega^2}{(\Omega_2 - \Omega_{rated})^2}\right). \tag{16}$$

In this equation, θ_{LP} is the low-pass filtered mean pitch signal θ_{mean}. K_1, K_2 and Ω_2 are parameters given by the user. To calculate θ_{LP}, a first order filter as described in Equation (A1) is used.

The pitch angle signal is limited by the parameters θ_{max} and θ_{min}. In addition, θ_{min} can be modified by V_{Hub} using a look-up table given by the user. In this case, the measured wind speed at hub height is also low-passed using a first order low pass filter (Equation (A1)).

In order for the pitch controller to react quickly if the required pitch signal changes suddenly from a saturated value, an anti-windup scheme like the one described in Equation (3) is used for θ_I.

The total collective pitch angle signal of the pitch controller is therefore given by

$$\theta_{PC} = \begin{cases} \eta \cdot (\theta_P + \theta_I + \theta_D) & \text{if } \theta_{min} \leq \theta_{PC} \leq \theta_{max} \\ \theta_{min}(V_{hub}) & \text{if } \theta_{PC} < \theta_{min}(V_{hub}) \\ \theta_{max} & \text{if } \theta_{PC} > \theta_{max}. \end{cases} \tag{17}$$

Additionally, the pitch rate is also limited by two user defined parameters: $\dot{\theta}_{max}$ and $\dot{\theta}_{min}$.

The power dependency of θ_{PC} will make the pitch controller raise the pitch angle set point when P_{ref} from the torque controller is close to P_{rated}. This in turn helps trigger the switching procedure from partial to full load regime of the torque controller as explained in Section 3.1.3.

3.2.2. Individual Pitch Control

One of the most common advanced pitch control strategies in wind energy research is the Individual Pitch Control (IPC) strategy [3,33]. Several studies have used this strategy as a comparison to other advanced pitch controller strategies [4,11,34]. It has also been used as a comparison or a complementary strategy for trailing edge flap controllers [7,9,10].

Figure 4 shows the graphical representation of the IPC strategy implemented in TUBCon.

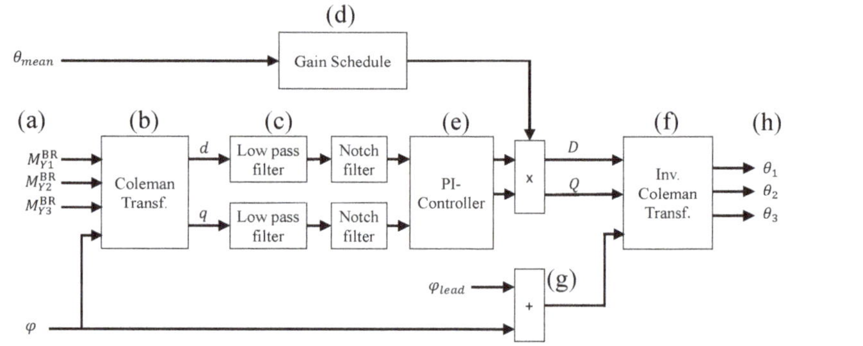

Figure 4. Graphical representation of the Individual Pitch Control (IPC) strategy implemented in the controller. The Blade Root Bending Moments (BRBM) signals are transformed into a non-rotating coordinate system and filtered using a low-pass and a notch filter. The filtered signals are then used in a PI controller to calculate the control signals. These signals are transformed back into the rotating frame of reference to be used as input for the individual pitch angles.

The IPC uses as input signals the out-of-plane BRBM of the three blades ($M^{BR}_{Y1}, M^{BR}_{Y2}, M^{BR}_{Y3}$) as well as the mean pitch angle θ_{mean} and the rotor azimuth angle φ (Figure 4(a)). The out-of-plane BRBMs are transformed to the direct and quadrature axes using the once-per-revolution or 1P-Coleman transformation (Figure 4(b)). The 1P-Coleman transformation for a three bladed rotor takes the form

$$\begin{pmatrix} d \\ q \end{pmatrix} = \frac{2}{3}\begin{pmatrix} \cos(\varphi) & \cos(\varphi+\frac{2\pi}{3}) & \cos(\varphi+\frac{4\pi}{3}) \\ \sin(\varphi) & \sin(\varphi+\frac{2\pi}{3}) & \sin(\varphi+\frac{4\pi}{3}) \end{pmatrix}\begin{pmatrix} M^{BR}_{Y1} \\ M^{BR}_{Y2} \\ M^{BR}_{Y3} \end{pmatrix}, \tag{18}$$

where d and q are the quantities expressed in the direct and quadrature axes respectively. The physical interpretation of these quantities is the rotor tilt and yaw moment in the non-rotating coordinate system.

These rotor moments are then passed through a second order low-pass filter (Equation (A2)) and a notch filter (Equation (A4)) Figure 4(c) to filter out unwanted high frequency content as well as the 1P component of the load signals in the non-rotating frame of reference. This step is needed as a 1P oscillation in the non-rotating frame of reference can lead to an increased 3P excitation of the turbine [35]. Following a concept presented in [9], a gain schedule is implemented using θ_{mean} to adapt the individual pitching action to the wind speed (Figure 4(d)). The gain schedule has the same mathematical expression as η_θ in Equation (16) but uses different parameter values. A PI controller is implemented with a zero set-point to reduce the rotor tilt and yaw moments (Figure 4(e)).

The control demand is transformed back to the demands of the individual pitch angles using the inverse 1P-Coleman transformation (Figure 4(f)). The inverse transformation is given by

$$\begin{pmatrix} \theta_1 \\ \theta_2 \\ \theta_3 \end{pmatrix} = \begin{pmatrix} \cos(\varphi') & \sin(\varphi') \\ \cos(\varphi'+\frac{2\pi}{3}) & \sin(\varphi'+\frac{2\pi}{3}) \\ \cos(\varphi'+\frac{4\pi}{3}) & \sin(\varphi'+\frac{4\pi}{3}) \end{pmatrix}\begin{pmatrix} D \\ Q \end{pmatrix}, \tag{19}$$

where the quantities D and Q represent the control signals in the non-rotating coordinate system and θ_1, θ_2 and θ_3 the individual pitch angles for each blade (Figure 4(h)). The inverse Coleman transformation uses a modified azimuth angle φ'. Figure 4(g) shows that $\varphi' = \varphi + \varphi_{lead}$, where φ_{lead} is the lead angle. This constant angle helps decoupling the rotor yaw and tilt moments in the non-rotating frame of reference so that they can be treated as independent systems. As explained in [4], certain factors such as blade stiffness, collective pitch angle and pitch actuator response time introduce a dependency between the two rotor moments. Using the lead angle is a straightforward way to counter this problem.

It is possible to generalize this strategy to other frequencies by using nP Coleman transformations. In this case the rotor angle φ and the respective shift for each blade in Equations (18) and (19) are replaced by n-times their value. Using for example an IPC-like strategy with a 2P Coleman transform helps to reduce the 3P asymmetrical loads in the yaw bearing of the turbine [10].

The IPC pitch angles are limited to a maximum and minimum value ($\theta_{max-\mathrm{IPC}}$ and $\theta_{min-\mathrm{IPC}}$) given by the user. Since the PI controller determines the quantities D and Q in the non-rotating coordinate system, the limits are transformed into this coordinate system:

$$\frac{\theta_{min-\mathrm{IPC}}}{\sqrt{2}} \leq D, Q \leq \frac{\theta_{max-\mathrm{IPC}}}{\sqrt{2}}. \tag{20}$$

As with the CPC strategy, an anti-windup procedure equivalent to Equation (3) is used to limit the increase of the integral part of the controller.

The pitch angle signal of the IPC strategy is added to the collective pitch signal to obtain the final pitch angle signal. The use of this strategy increases the pitch activity significantly. In order to limit this increase in pitch activity, the control strategy is phased out in the partial load regime. Bergami and Gaunaa show in [36] that the majority of fatigue loading that can be alleviated with this strategy occurs in the above-rated region. Phasing out the IPC strategy helps limit the pitch activity and optimizes energy capture by keeping the pitch angle constant in the below-rated region. The phasing out is done by multiplying the IPC PI-constants with a switch function (Equation (A5)) based on θ_{mean}. The switch limits used in this study are $x_0 = 0.5°$ and $x_1 = 3°$.

3.3. Supervisory Control

The supervisory control of TUBCon has two main functions. The first one is to detect whenever some security limit (e.g., generator over-speed) has been tripped and enforce an appropriate maneuver (e.g., normal stop). The second function is to trigger specific events. These can be controller fault events (as required by current industry standards) or stop maneuvers (if a specific alarm has been triggered).

The control signals from the supervisory control will always override the signals from the pitch and torque controllers. This is shown in Figure 2 with the switch for each controller. In this figure, $\vec{\theta}_{SC}$ represents the pitch signal from the supervisory control that overrides the signal from the pitch controller. In the same manner, Q_{gen-SC} overrides the generator torque from the torque controller. In addition, the supervisory controller can activate the brake and the yaw actuator of the turbine.

The internal logic of the supervisory control is based on [37] and shown here in Figure 5. It features three control blocks. The first one checks if any turbine sensor is outside its allowable range and triggers an alarm should this happen (Figure 5(a)). The turbine sensors are bundled into the vector $\vec{S}_{SC}$ in Figure 5 for the sake of clarity (see Section 3.3.1 for a full description of the considered sensors). Depending on the situation, the alarm can have several levels. An example of this could be the tripping of the first over-speed limit and the tripping of the second over-speed limit, both alarms depending on the sensor Ω.

The second block manages the alarms that the first block triggered and sets an appropriate reaction maneuver (Figure 5(b)). Keeping the example above, if the alarm block detected a first over-speed trigger, the managing block would request a normal stop of the turbine. If a second over-speed trigger is detected, then the manager block would request an emergency stop procedure. The latter clearly has a higher priority than the former. Both stop procedures have a higher priority than the normal control action of the controller.

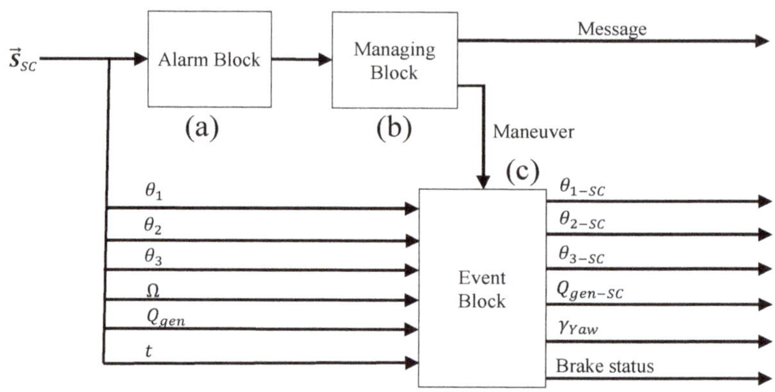

Figure 5. Overview of the supervisory control. It comprises an alarm block, a managing block and an event block. The latter can be used to trigger specific events such as controller faults that are required for a full load calculation.

The third block is the event generator block (Figure 5(c)). This block creates the necessary controller signals to simulate any given event. These signals could be a controller fault required by a specific design load case or the normal stop maneuver requested from the manager block. The input signals for this block are the current pitch angle of each blade θ_i, the rotor speed Ω and torque Q_{gen} and the current time t. The last sensor is needed if an event is to occur at a specific simulation time. The output of the event block are the overridden controller signals. Note that depending on the specific event, not all control signals are necessarily overridden.

3.3.1. Considered Alarms and Reaction Maneuvers

Table 1 lists all the considered alarms of the supervisory control. The first column shows the measured sensor, the second column the limit and the third column the requested maneuver. In the third column, NS stands for normal stop and ES for emergency stop. These are the only two currently implemented maneuvers of the controller and are explained below.

The alarm block considers five sensors: rotor speed Ω, electrical power P, yaw misalignment γ, pitch angles θ and grid status. Both Ω and P have two limits (n4, p4 and nA, pA respectively) that, when surpassed, trigger instantly the appropriate stop maneuver. The same simple trip-logic is implemented for the grid status signal. At the instant that the grid status is measured as offline, a normal stop maneuver is triggered.

For the other sensors, the alarm is triggered if the sensor surpasses the limit for a given time period. The alarm block foresees two limits for γ with two different duration times. The first limit γ_{max-1} considers long-time yaw misalignments. The measured yaw misalignment γ has to be larger than this limit for a time interval greater than $\tau_{\gamma-1}$, which should be fairly large. In contrast, γ_{max-2} considers short-term extreme yaw misalignments. The yaw misalignment signal used to trigger the alarms is filtered with a first order low pass filter (Equation (A1)) to even out variations coming from inflow turbulence.

Table 1. Limits and reaction maneuvers of the supervisory control. Depending on the pitch control strategy, only $\Delta\theta_{ij}$ or $\Delta\theta_{i-CP}$ can be active. NS = normal stop; ES = emergency stop.

Sensor	Limit	Maneuver
Ω	>n4	NS
Ω	>nA	ES
P	>p4	NS
P	>pA	ES
γ	$>\gamma_{max-1}$ for $\tau_{\gamma-1}$ seconds	NS
γ	$>\gamma_{max-2}$ for $\tau_{\gamma-2}$ seconds	NS
$\Delta\theta_{ij}$	$>\Delta\theta_{max-CPC}$ for $\tau_{\Delta\theta-CPC}$ seconds	NS
$\Delta\theta_{i-CP}$	$>\Delta\theta_{max-IPC}$ for $\tau_{\Delta\theta-IPC}$ seconds	NS
$\Delta\theta_{set}$	$>\Delta\theta_{set-max}$ for $\tau_{\Delta\theta-set}$ seconds	NS
Grid	Offline	NS

For the pitch angle group, there are two conditions that can trigger alarms. The first one is the difference between the individual pitch angles ($\Delta\theta_{ij}$ or $\Delta\theta_{i-CP}$). This difference is calculated differently depending on the active pitch control strategy. For the CPC strategy (Section 3.2.1), the alarm block checks the difference between all three angles. If the difference between any two pitch angles is greater than the fixed limit $\Delta\theta_{max-CPC}$ for a given time $\tau_{\Delta\theta-CPC}$, then the alarm is triggered and a normal stop maneuver is requested. For the IPC strategy (Section 3.2.2), the pitch angle differences are measured against the collective pitch angle signal. The limit $\Delta\theta_{max-IPC}$ is not constant but varies depending on the short-term mean difference between the IPC angle and the collective pitch angle (plus a used-defined offset angle). The short-term mean difference of the individual pitch angles is obtained by applying a first-order low pass filter to the instantaneous mean difference. The time constant for this filter is a user-defined parameter. By using this triggering strategy, the supervisory control can detect pitch angle faults even when the difference between two pitch angles lies below the maximum allowed pitch difference for the IPC strategy.

The second condition included in the pitch angle alarm group is the difference between the measured and the demanded pitch angle for each blade $\Delta\theta_{set}$. If the difference between the two signals in any blade is greater than $\Delta\theta_{set-max}$ for a given time $\tau_{\Delta\theta-set}$, then the alarm is triggered and a stop maneuver is requested.

The two stop maneuvers used in this controller are defined by a sequence of actions that are taken (almost) independently of the controller signals. The only exception is the normal stop maneuver that additionally uses the pitch angle differences, as explained below. The parameters for both maneuvers are listed and explained in Table 2. Note that there are two sets of parameters: one for the normal stop and one for the emergency stop procedure.

The implemented stop maneuvers foresee that the collective pitch angle will increase at a rate $\dot{\theta}_S$ up to a maximum value θ_{max-S}. This pitching towards feather can be delayed for a given time using the parameter τ_θ. Parallel to the pitch controller, the torque controller decreases linearly to Q_{min-S} at a rate given by the parameter $\dot{Q}_S$. This procedure can also be delayed for a given time using the parameter τ_Q. In case of a grid loss, a chopper can keep the generator torque constant for a given time $\tau_{Chopper}$ in order to avoid very high rotor speeds. There is also the option to enable the turbine brake during the stop maneuvers. This is controlled with an enable flag and a minimum rotor speed ($\Omega_{\text{Brake}-S}$) at which the brake is activated.

In addition to these fixed parameters, the normal stop procedure checks the pitch angle differences between all blades. If the absolute value of a given difference is above 0.5°, then the pitch rate $\dot{\theta}_S$ is replaced by the minimum value between $\dot{\theta}_{max}$ and $1.5 \cdot \dot{\theta}_S$. This is done for all blades except the one with the highest pitch angle. This way the pitch angle differences are minimized during the stop maneuver avoiding unnecessary oscillations due to aerodynamic imbalances of the rotor. The emergency stop does not have this feature as it is designed to be hard coded in the turbine's safety mechanism.

Table 2. Parameters for the normal and emergency stop maneuvers.

Parameter	Explanation
$\dot{Q}_S$	Torque rate during stop.
Q_{max-S}	Maximum allowed torque during stop.
Q_{min-S}	Minimum torque during stop. Generator torque ramps to this value.
τ_Q	Delay time at which the torque is kept constant before ramping down.
$\tau_{Chopper}$	Delay time at which the torque is kept constant in case of grid loss (only NS).
$\dot{\theta}_S$	Collective pitch rate during stop.
θ_{max-S}	Maximum allowed pitch angle during stop.
θ_{min-S}	Minimum allowed pitch angle during stop.
τ_θ	Delay time at which the pitch angle is kept constant before pitching.
BrakeFlag_S	Flag to enable the brake in the maneuver
$\Omega_{\text{Brake}-S}$	Rotor speed at which the brake is enabled

3.3.2. Considered Events

The event block allows the controller to simulate the events required by the current industry standards for a full load calculation [21]. Table 3 presents and explains all the events that the controller is able to perform. All of the listed events are controlled by a set of parameters passed to the controller.

Table 3. Events included in the event block.

Name	Explanation
Normal stop	Trigger normal stop maneuver.
Emergency stop	Trigger emergency stop maneuver.
Pitch runaway to feather/fine	Runaway of one or multiple blades to the 90°/0° position.
Speed transducer fault	Generator speed signal is erroneously measured.
Grid loss	Sudden loss of the electrical grid.
Yaw runaway	Runaway of the yaw motor.

Both the normal stop and the emergency stop events require two parameters: a flag to indicate if the event will be triggered and the time at which the maneuver is triggered.

The pitch runaway to feather/fine fault simulates an error in one or several pitch actuators. In this event, one or several blades suddenly pitch with a given pitch rate to the respective extreme value. The pitch rate is one of the parameters for this event. The sign of the pitch rate defines if the event is a pitch-to-feather (positive rate) or a pitch-to-fine (negative rate) fault event. There is one activation flag and one activation time parameter for each blade. This gives the flexibility to simulate the single pitch fault events and the collective pitch fault events with a limited set of parameters.

The pitch transducer fault simulates an error in the measurement of the generator speed. Instead of the actual speed, the transducer measures a constant value and passes it to the pitch and torque controller. The parameters for this event are a flag to enable the event, the time at which the event starts and the constant generator speed passed to the controller once the fault occurs.

The grid loss event simulates a sudden loss of the electrical grid. As a consequence, the generator cannot feed the generated power to the grid and the torque drops to 0 Nm. This event also raises the grid loss flag so that the alarm block recognizes the event and triggers the appropriate maneuver. The parameters that control this event are a flag that enables it and the time at which the grid loss event happens.

Finally, the yaw runaway event simulates a malfunction in the yaw system that causes the yaw motor to activate. The parameters for this event are an activation flag, a time to start yawing, a yaw rate and a time to stop yawing.

3.4. Actuator Models

Once the set-point values of the blade pitch angles and generator torque are calculated by the respective controllers, they are passed to the actuator models of the controller (see Figure 2). The actuator models are used to model the additional inertial effects of the physical actuators. These models can be very accurate depending on the individual needs of the calculation.

For the TUB Controller, a simple second-order linear model was chosen to model both the pitch actuators and the generator model. The second-order model has the same mathematical description as a second-order low-pass filter, namely Equation (A2). This model is also used for the yaw actuator in the yaw runaway event. The individual cut-off frequencies and damping ratios for the actuators and the generator model are supplied as parameters by the user. For the brake actuator, a simple first-order linear model as described by Equation (A1) is used. The time constant is supplied by the user.

3.5. Supervisory Control Event Examples

Figure 6 shows four examples of events and stop maneuvers using the DTU 10 MW RWT in turbulent wind simulations. In Figure 6a, a collective pitch-to-fine fault at a simulation time of 125 s leads to the triggering of the first and second over-speed alarms. The resulting maneuvers (normal and emergency stop) can be distinguished by the different pitch rates. Figure 6b shows a single pitch-to-fine event during power production using the IPC strategy. At 120 s of simulation time, θ_1 receives a faulty signal and starts pitching towards 0°. This is caught by the supervisory controller and the normal stop procedure is triggered. The normal stop feature that minimizes aerodynamic imbalances can be clearly seen in this figure, as the difference in individual pitch angles is reduced to 0° at 130 s of simulation time. Figure 6c shows a grid loss scenario and the resulting stop maneuver. Finally, Figure 6d shows a speed transducer fault at 120 s simulation time. As a consequence, the generator torque and pitch angle signals freeze and a normal stop is only triggered because of the first over-power alarm (comparable to the first over-speed threshold).

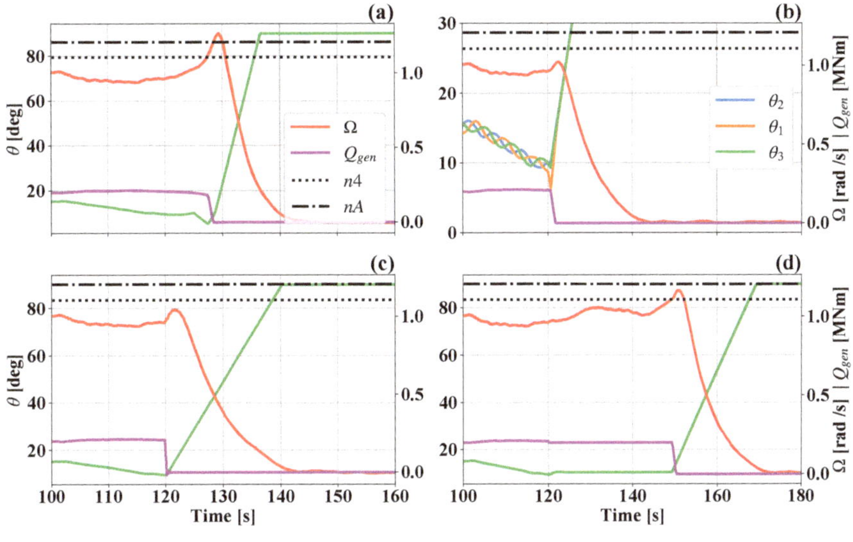

Figure 6. Events triggered by the controller: (**a**) Collective pitch-to-fine; (**b**) Individual pitch-to-fine; (**c**) Grid loss; (**d**) Speed transducer fault. In all cases, the event is captured by the supervisory controller and an appropriate stop maneuver is triggered.

4. Controller Validation

The TUB Controller was tested and validated in a series of load cases to verify its functionality. This was done in steady and turbulent wind conditions, summarized in Table 4. The validation was done by comparing TUBCon to the established Basic DTU Wind Energy Controller [16]. The parameters for both controllers were taken from the LIFE50+ report [38].

Table 4. Main Parameters of the Simulations used for Controller Verification.

Parameter	Steady Calculations	Turbulent Calculations
Mean V_{Hub}	4–24 m/s	4–24 m/s
Wind model	steady	IEC NTM
Wind shear exp.	0	0.2
Upflow angle	0°	8°
Yaw angle	0°	−8°, 0°, 8°
Sim. time	350 s	600 s

4.1. Steady Wind Simulations

Figure 7 shows the behavior of both controllers in the steady wind simulation.

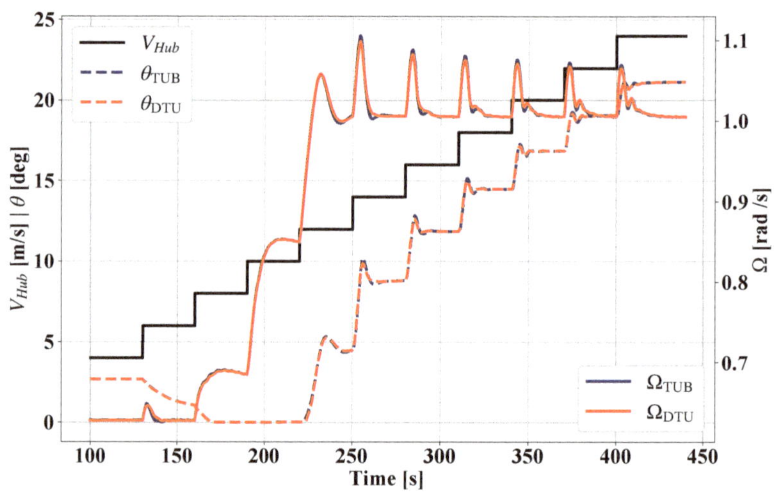

Figure 7. Validation of the controller in steady wind simulations. Pitch angle θ and rotor speed Ω of both controllers are compared for ascending wind steps.

This simulation features a series of wind speed steps from 4 m/s to 24 m/s with a 2 m/s step height. Between two steps, the wind speed is held constant for 30 s. It can be seen that both controllers behave in an almost identical manner. This is due to the fact that the collective pitch and torque controllers from TUBCon were directly adapted from the Basic DTU Controller. With the chosen parameters, TUBCon manages to stabilize the turbine without large overshoots in the pitch and rotor speed signals in spite of the sudden increases in wind speed. This is true for all simulated wind speeds.

4.2. Turbulent Wind Simulations

The turbulent wind simulations were done using the Normal Turbulent wind Model (NTM) from [21]. Because of this, the controller validation has to be done in a statistical manner. The setup is given in Table 4 under the column "Turbulent calculations". For each wind speed bin, we did six simulations with three different yaw angles of the turbine and two turbulence seeds. In total, there were 66 simulations, each 600 s in length for each controller. Because we are using QBlade's LLFVW method, the wake needs to first develop behind the rotor in order to have accurate results. We added 200 s of pre-simulation time to allow the wake to develop. This time was discarded in the analysis. Figure 8 shows an example of an aero-servo-elastic simulation of the DTU 10 MW RWT within QBlade. The incoming turbulent wind speed has an average hub-height value of 14 m/s and is shown on the left. The wake is modeled by vortex elements that are allowed to convect freely downstream of the turbine.

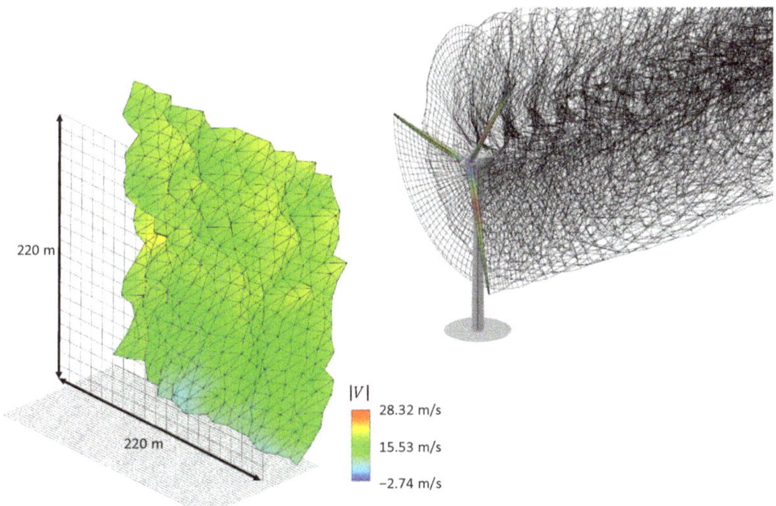

Figure 8. Turbulent wind aero-servo-elastic simulation with QBlade's Lifting Line Free Vortex Wake (LLFVW) aerodynamic model. The DTU 10MW Reference Wind Turbine (RWT) is simulated in a turbulent wind field with an average hub-height wind speed of 14 m/s.

To compare the controllers we chose a selection of turbine sensors that are summarized in Table 5. These sensors give a good estimate of the overall loading of the turbine blades, tower top and tower bottom.

Table 5. Considered sensors for turbulent wind simulations

Sensor Name	Symbol
Blade pitch angle/rotor speed/generator power	$\theta/\Omega/P$
Blade root in-plane/out-of-plane/torsional bending moment	$M_X^{BR}/M_Y^{BR}/M_Z^{BR}$
Yaw bearing roll/tilt/yaw moment	$M_X^{YB}/M_Y^{YB}/M_Z^{YB}$
Tower base side-side/fore-aft/torsional bending moment	$M_X^{TB}/M_Y^{TB}/M_Z^{TB}$

We used two metrics to analyze the performance of the controllers, depending on the nature of the sensor. For the load sensors our metric is the lifetime Damage Equivalent Loads (DELs). These are calculated using the rainflow counting algorithm combined with the Palmgren-Miner linear damage accumulation hypothesis, as described in [39]. The rainflow count was done using the software Crunch [40]. For the controller signals, we use the averaged standard deviation of each signal. The average is taken from the individual standard deviations of all simulations in each of the wind speed bins. To calculate the lifetime DELs we used a wind speed distribution that corresponds to a wind class IA turbine [21], which is the design wind class of the DTU 10MW RWT.

4.2.1. Controller Signals

Figure 9 shows the averaged normalized standard deviation of the controller signals as a function of the average wind speed for the calculations with the DTU and the TUB controllers. The normalization is done with respect to the calculations with the DTU controller. We can see in this figure that the averaged standard deviations of the rotor speed and pitch angle are very similar with both controllers. The main difference is seen for Ω in wind speed bins of 12 m/s and higher. Here, the $\bar{\sigma}(\Omega)$ of the calculations with TUBCon are consistently higher than with the DTU controller. The reason for this can be seen in Figure 9c. Here we can see that $\bar{\sigma}(P)$ is considerably smaller for the TUBCon calculations compared to the calculations with the DTU controller. The torque controller objective was constant power for both controllers. The fact that $\bar{\sigma}(P)$ is smaller in the TUBCon simulations indicates a more aggressive constant power enforcement of this controller. This comes at the expense of larger torque and rotor speed fluctuations when compared to the calculations with the DTU controller. We note that even though the differences in the normalized $\bar{\sigma}(P)$ are large, the absolute differences in the variation of the generator power were small. The large relative differences seen in wind speed bins of 14 m/s and higher arise because the generator power is mostly constant and the values of $\bar{\sigma}(P)$ are small for both controllers.

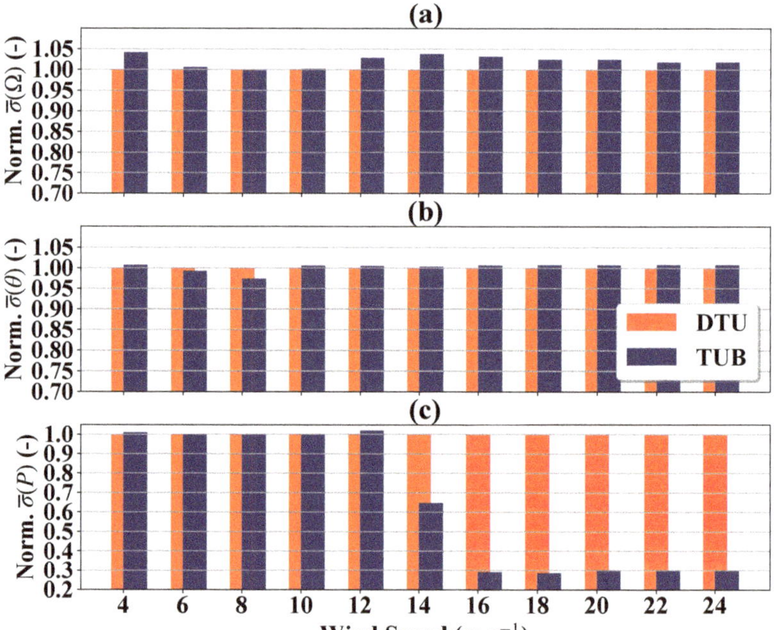

Figure 9. Normalized average standard deviations of controller signals vs. wind speed. (**a**) Rotor speed; (**b**) Pitch angle; (**c**) Generator power. Nomenclature of the signals is found in Table 5.

4.2.2. Fatigue Loads

The effect on the fatigue loads can be seen in Figure 10. The only significant differences occur for M_X^{TB} and M_Y^{TB}, with TUBCon simulations showing a 3% and 2% increase respectively, compared to simulations with the DTU controller. The differences in M_Y^{TB} come from the larger oscillations of Ω in the TUBCon simulations, driven by the more aggressive constant power controller. The increased fatigue loads of M_X^{TB} are only indirectly affected by controller action and it is therefore more difficult to find the source of this difference. It is assumed that the different variations of Ω lead to differences in the induced tower oscillations. For the fore-aft direction, the tower vibrations are damped out quickly due to aerodynamic damping. This is not the case in the side-side direction as the damping is much lower and hence these vibrations cause different amounts of M_X^{TB} fatigue loads. Further investigation is needed to corroborate this assumption.

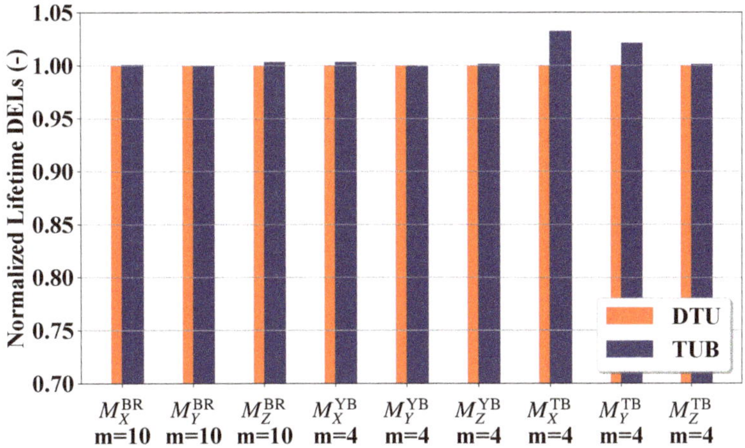

Figure 10. Normalized lifetime Damage Equivalent Loads (DELs) for considered sensors. Nomenclature is found in Table 5.

All in all the differences in controller behavior and turbine loading remain small and the performance of both controllers is very similar.

5. Fatigue Load Reduction through Advanced Control Action

As discussed in [27], many implementations of the BEM aerodynamic model average the axial induction factor across the turbine rotor. This leads to inaccuracies in the local induced velocities on the blades and hence to significant loading differences when compared to the more accurate LLFVW aerodynamic method.

These differences will necessarily affect the performance of advanced control strategies such as IPC. Firstly, the input signals of the IPC controller will be different in LLFVW simulations because of the more accurate estimation of M_Y^{BR} in each blade compared to BEM-based simulations. Secondly, the controller action will affect the turbine loads differently due to the more accurate representation of the non-homogeneous induction field in LLFVW simulations caused by the individual pitch positions.

In this section we explore the load reduction potential of the IPC control strategy (Section 3.2.2) using the LLFVW method from QBlade. This will give us a more accurate picture of the load reduction capabilities of this strategy. The IPC strategy focuses on reducing the once-per-revolution (1P) M_Y^{BR} loads. The parameters were taken from the report [9] and slightly adapted to account for the different coordinate system in which the input is measured. Since we are interested mainly in fatigue load reduction, we considered the same load cases from Table 4 (column "Turbulent calculations"). These load cases correspond to the DLC group 1.2 from [21]. For onshore wind turbines, the DLC group 1.2 is the main contributor of lifetime fatigue loads for most of the turbine components, since the turbine spends most of its operating time in these conditions [41]. Evaluating the fatigue loads from this group will therefore give a close estimate of the real fatigue loads. To keep consistency, we considered the same load sensors and metrics as the ones described in Section 4.2.

5.1. Results

Figure 11 shows a selection of representative time series from a baseline simulation and a simulation with active IPC strategy. Both simulations use the TUB Controller with the same controller parameters. The baseline simulation uses the CPC strategy and the IPC simulation uses the CPC and IPC strategies. In this figure we can see the effect of the IPC strategy on the controller signals and on the M_Y^{BR} loads. From Figures 11b,e it can be seen that the influence of the IPC strategy on Ω and P is small. This is mainly due to the frequency separation between the CPC and IPC strategies. Figure 11c shows the time series of θ_1. We can clearly see the 1P oscillation of the IPC strategy on top of the CPC pitch signal. The load reducing effect of this strategy can be seen in Figure 11d at around 700 s of simulation time. The 1P variation of M_Y^{BR} are clearly reduced due to the additional pitch actuation. The below-rated power limitation of the IPC strategy can be seen in Figure 11c around 650 s of simulation time. When the mean pitch angle decreases to 0°, the amplitude of the IPC signal is also reduced 0° to maximize power capture.

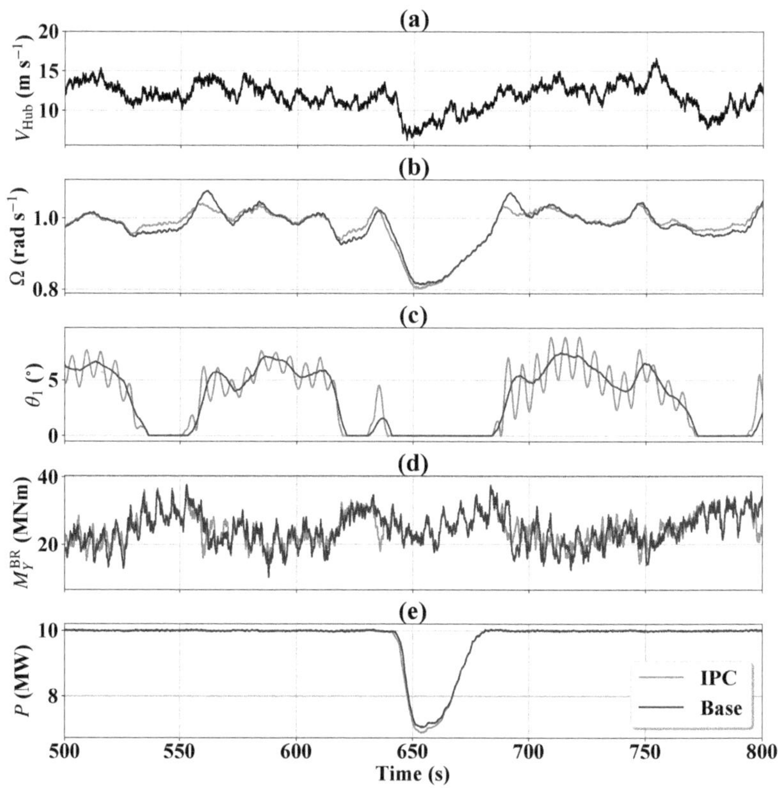

Figure 11. Time series selection of simulations with and without Individual Pitch Control (IPC) strategy from the 12 m/s wind speed bin. (**a**) Wind speed; (**b**) Rotor speed; (**c**) Blade 1 pitch angle; (**d**) Out-of-plane BRBM blade 1; (**e**) Generator power. Base = Collective Pitch Control (CPC) strategy.

5.1.1. Controller Signals

The effect of the IPC strategy on the controller signals can be seen in Figure 12. It shows the normalized averaged standard variations of Ω, θ and P respectively. Figure 12a shows that the normalized $\bar{\sigma}(\Omega)$ of the IPC strategy is practically 1 for all wind speed bins except the bins corresponding to 12 and 14 m/s average speed. There we see that Ω from the IPC simulations oscillates less than Ω from the CPC simulations. This can be understood if we look at Figure 11b. The additional 1P fluctuation of θ does influence the value of Ω by reducing slightly the sensitivity of the rotor to changes in the wind speed. This is particularly marked around rated wind, where the pitch angle is close to 0° and there are high angles of attack in the outer span of the blades. The latter lead to large values of lift forces and changes in the lift force due to pitching have an increased effect on rotor thrust and torque.

Figure 12b shows the normalized $\bar{\sigma}(\theta)$ vs. the wind speed bin. This is the controller signal that shows the largest differences. From 10 m/s onward, the value of $\bar{\sigma}(\theta)$ increases up to values larger than 1.2. It is this region where the IPC strategy starts to function. The increase in pitch activity from the IPC strategy is directly linked to the fact that the 1P fluctuation of M_Y^{BR} increases with increasing wind speed [36].

Finally, Figure 12c shows the normalized values of $\bar{\sigma}(P)$. Here, there is almost no difference between the two strategies. The lower value in the IPC simulations seen in the wind speed bin of 14 m/s comes from the relatively small reference value of $\bar{\sigma}(P)$ for the CPC calculations. Because P is fairly constant in above-rated wind simulations, differences in the drop of P for temporarily low wind speeds (as seen in Figure 11e) will have a large effect on the normalized $\bar{\sigma}(P)$.

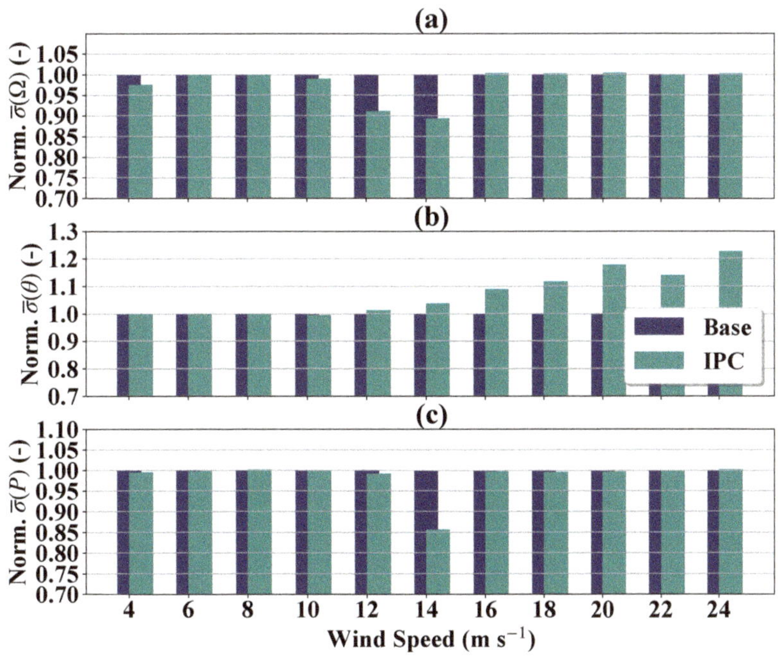

Figure 12. Averaged standard deviations of controller signals vs. wind speed. (**a**) Rotor speed; (**b**) Pitch angle; (**c**) Generator power. Nomenclature of the signals is found in Table 5. Base = CPC strategy.

5.1.2. Fatigue Loads

Figure 13 shows the normalized lifetime DELs of the simulations with the CPC and the IPC strategies. We can see that the main effect of the IPC strategy is to reduce the lifetime DELs of M_Y^{BR} and M_Z^{BR} by 14% and 9.4% respectively. There are also smaller effects on other load sensors. The lifetime DELs of M_X^{BR} are 2.4% lower in the simulations with the IPC strategy compared to the CPC strategy. The load differences for these three sensors come directly from the IPC strategy. Since the strategy is minimizing the yaw and tilt moments of the rotor, it indirectly minimizes the load fluctuations of the individual blade root moments.

These fluctuations arise from oscillations of the effective lift force on the individual blades. So, indirectly, the IPC strategy is also minimizing the fluctuation of the effective lift force on the individual blades. Although the out-of-plane component dominates in the lift force composition, there is also a smaller in-plane component. The former affects M_Y^{BR} and the latter M_X^{BR}. Reducing the lift fluctuations also reduces the pitching moment oscillations and oscillations in the out-of-plane deflection of the blades. These two quantities directly affect M_Z^{BR}. So by reducing the lift force fluctuation, all three load sensors are affected.

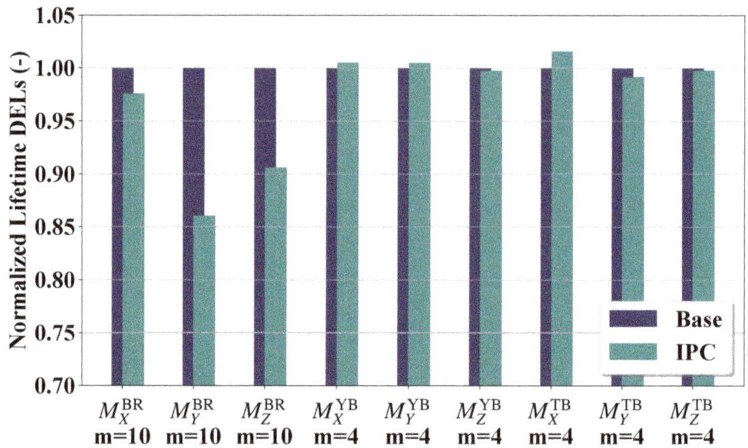

Figure 13. Normalized lifetime Damage Equivalent Loads (DELs) for considered sensors. Nomenclature is found in Table 5. Base = CPC strategy.

The yaw bearing sensors are practically unaffected by the IPC strategy. This is due to the nature of our metric. The 1P based IPC strategy focuses on reducing the steady or slow varying yaw and tilt moments of the rotor. Our calculation of the DELs does not consider the means but only the ranges of the load cycles. A reduction of the load averages is therefore not captured by our metric. An IPC strategy that additionally targets the 2P load fluctuations would also reduce the yaw bearing DELs, ([10,20], (pp. 501–503)). The tower base lifetime DELs show that the IPC strategy also has a small impact on the tower fatigue loads. The IPC strategy increases the M_X^{TB} DELs by 1.5% and decreases the M_Y^{TB} DELs by 0.9% compared to the CPC strategy.

6. Conclusions

This paper presented and detailed TUBCon, an advanced open-source wind turbine controller that includes pitch, torque and supervisory controllers. The controller can therefore be used to perform a complete load calculation according to industry standards. It is compatible with the common aeroelastic simulation codes including QBlade, which features the higher order LLFVW aerodynamic method in addition to a standard unsteady BEM aerodynamic model.

TUBCon was validated against an established turbine controller from the literature by comparing aeroelastic simulations of the DTU 10 MW RWT in steady and turbulent wind conditions. For steady wind conditions, the controllers show practically the same performance. For turbulent wind conditions, TUBCon shows a more aggressive constant power regulation. This reduces the mean standard deviation of the generator power for the above-rated wind bins by up to 70%, but increases the standard deviation of the rotor speed by up to 3.7%. This difference also affects the lifetime DELs of the tower base bending moments. The lifetime DELs of M_X^{TB} and M_Y^{TB} are increased in the TUBCon simulations by 3% and 2% respectively.

We also investigated the advanced load reduction capabilities of TUBCon by comparing the performances of the IPC strategy against a baseline CPC strategy. This was done using QBlade's LLFVW aerodynamic method, which is able to calculate the aerodynamic effects of a non-homogeneous induction field due to individual pitching more accurately compared to most BEM-based aerodynamic methods. The results show that the IPC strategy is able to reduce the lifetime DELs of M_Y^{BR} and M_Z^{BR} by 14% and 9.4% respectively when compared to the baseline simulations. This comes at the cost of increased pitch activity from the IPC controller. The normalized $\bar{\sigma}(\theta)$ of the IPC strategy increase to values up to 22.6% higher than the values of the CPC strategy for wind speed bins above rated wind speed.

Future work will include further development of TUBCon to add new features such as tower vibration damping and rotor speed exclusion. It is also planned to add a fully featured trailing edge flap controller and the necessary features so that TUBCon can be used as a controller in offshore floating wind turbine simulations. Furthermore, the controller will be extended to include power curtailment and wake steering capabilities so that it can be used in conjunction with a wind farm controller. In addition, the effect of QBlade's LLFVW aerodynamic model on advanced controller performance will be further analyzed by comparing the results from LLFVW-based simulations to more common BEM-based simulations and considering more DLC groups.

Author Contributions: Conceptualization, S.P.-B.; methodology, S.P.-B. and D.M.; software, S.P.-B. and D.M.; validation, S.P.-B. and D.M.; formal analysis, S.P.-B.; investigation, S.P.-B. and D.M.; resources, C.N.N. and C.O.P.; data curation, S.P.-B. and D.M.; writing–original draft preparation, S.P.-B.; writing—review and editing, D.M., C.N.N. and C.O.P.; visualization, S.P.-B.; supervision, C.N.N. and C.O.P.; project administration, C.N.N. and C.O.P.; All authors have read and agreed to the published version of the manuscript.

Funding: This research received no external funding.

Data Availability Statement: TUBCon and QBlade are open-source codes available online. The latest version of QBlade is available at https://www.qblade.org. The latest version of TUBCon is available at https://github.com/s-perez-becker/TUBCon. The time series for the load calculations are stored in the HAWC2 binary format. They can be made available upon request.

Acknowledgments: SPB wishes to thank WINDnovation Engineering Solutions GmbH for supporting his research.

Conflicts of Interest: The authors declare no conflict of interest.

Appendix A. Equations of Used Filters and Switches

This section includes the continuous functions of the used filters and switches. All filters are given in the Laplace domain. The equivalent form for discrete time can be found in [16].

The first order low-pass filtered value of a general variable is given by:

$$x_{LP}(s) = \frac{1}{\tau \cdot s + 1} \cdot x(s), \tag{A1}$$

where τ is the filter time constant.

The second order low-pass filter of a general signal takes the form

$$x_{LP}(s) = \frac{\omega^2}{s^2 + 2\xi\omega \cdot s + \omega^2} \cdot x(s). \tag{A2}$$

Here x and x_{LP} denote the original and low-passed signal respectively. ω is the filter frequency and ξ the damping factor.

A band pass filter of a general variable is implemented as:

$$x_{BPF}(s) = \frac{2\xi\omega \cdot (s + \tau \cdot s^2)}{s^2 + 2\xi\omega \cdot s + \omega^2} \cdot x(s), \tag{A3}$$

where ω is the center frequency, ξ the damping ratio and τ a time constant.

The notch filter of a general signal takes the form

$$x_{NF}(s) = \frac{s^2 + 2\xi_2\omega \cdot s + \omega^2}{s^2 + 2\xi_1\omega \cdot s + \omega^2} \cdot x(s), \quad (A4)$$

where ξ_1 and ξ_2 are the damping ratios and ω the filter frequency.

The general form of the switching function used by the controller is given by the equation

$$\sigma(x_0, x_1, x) = \begin{cases} 0 & \text{for } x < x_0 \\ a_3x^3 + a_2x^2 + a_1x + a_0 & \text{for } x_0 \leq x < x_1 \\ 1 & \text{for } x \geq x_1, \end{cases} \quad (A5)$$

where the coefficients take following values:

$$\begin{aligned} a_3 &= \frac{2}{(x_0 - x_1)^3} \\ a_2 &= \frac{-3(x_0 + x_1)}{(x_0 - x_1)^3} \\ a_1 &= \frac{6x_0x_1}{(x_0 - x_1)^3} \\ a_0 &= \frac{(x_0 - 3x_1)x_0^2}{(x_0 - x_1)^3}. \end{aligned} \quad (A6)$$

Appendix B. Abbreviations

Table A1 lists the abbreviations used in this work.

Table A1. Abbreviations used in this work.

Abbreviation	Explanation
BEM	Blade element momentum
BRBM	Blade root bending moment
CPC	Collective pitch control
DEL	Damage equivalent load
DLC	Design load case
DTU	Technical University of Denmark
IPC	Individual pitch control
LLFVW	Lifting line free vortex wake
NREL	National renewable energies laboratory
NTM	Normal turbulence model
RWT	Reference wind turbine
TUB	Technical University of Berlin
TUBCon	TUB Controller

References

1. Burger, B. *Power Generation in Germany—Assesment of 2017*; Technical Report; Fraunhofer Institute for Solar Energy Systems ISE: Freiburg, Germany, 2018.
2. Jamieson, P. *Innovation in Wind Turbine Design*, 2nd ed.; John Wiley & Sons Ltd.: West Sussex, UK, 2018; p. 416.
3. Bossanyi, E.A. Individual Blade Pitch Control for Load Reduction. *Wind Energy* **2003**, *6*, 119–128. doi:10.1002/we.76.
4. Larsen, T.J.; Madsen, H.A.; Thomsen, K. Active Load Reduction using individual Pitch, based on local Blade Flow Measurements. *Wind Energy* **2005**, *8*, 67–80. doi:10.1002/we.141.
5. Jones, B.L.; Lio, W.H.; Rossiter, J.A. Overcoming fundamental limitations of wind turbine individual blade pitch control with inflow sensors. *Wind Energy* **2018**, *21*, 922–936. doi:10.1002/we.2205.

6. Wilson, D.; Berg, D.; Resor, B.; Barone, M.; Berg, J. Combined Individual Pitch Control and Active Aerodynamic Load Controller Investigation for the 5MW Upwind Turbine. In Proceedings of the AWEA Wind Power Conference & Exhibition, Chicago, IL, USA, 4–7 May 2009; pp. 1–12.
7. Lackner, M.; van Kuik, G.A.M. A Comparison of Smart Rotor Control Approaches using Trailing Edge Flaps and individual Pitch Control. *Wind Energy* **2010**, *13*, 117–134. doi:10.1002/we.353.
8. Plumley, C.; Graham, M.; Leithead, W.; Bossanyi, E.A.; Jamieson, P. Supplementing Wind Turbine Pitch Control with a Trailing Edge Flap Smart Rotor. In Proceedings of the 3rd Renewable Power Generation Conference (RPG 2014), Naples, Italy, 24–25 September 2014; pp. 1–6. doi:10.1049/cp.2014.0919.
9. Jost, E.; Barlas, T.; Riziotis, V.; Navalkar, S.T. INNWIND Deliverable 2.32: Validation of New Control Concepts by Advanced Fluid-Structure Interaction Tools. Technical Report. INNWIND.eu. 2015. Available online: http://www.innwind.eu/-/media/Sites/innwind/Publications/Deliverables/2015-09-02_INNWIND_EU_D2_3_2_final2.ashx?la=da&hash=B357FF565D1B203869C1016F06D9CCE191CAD434 (accessed on 5 June 2020).
10. Chen, Z.; Stol, K.; Mace, B. System Identification and Controller Design for individual Pitch and Trailing Edge Flap Control on upscaled Wind Turbines. *Wind Energy* **2016**, *19*, 1073–1088. doi:10.1002/we.1885.
11. Navalkar, S.T.; Van Wingerden, J.W.; Van Solingen, E.; Oomen, T.; Pasterkamp, E.; Van Kuik, G.A.M. Subspace Predictive Repetitive Control to Mitigate Periodic Loads on Large Scale Wind Turbines. *Mechatronics* **2014**, *24*, 916–925. doi:10.1016/j.mechatronics.2014.01.005.
12. Asgharnia, A.; Jamali, A.; Shahnazi, R.; Maheri, A. Load mitigation of a class of 5-MW wind turbine with RBF neural network based fractional-order PID controller. *ISA Trans.* **2020**, *96*, 272–286. doi:10.1016/j.isatra.2019.07.006.
13. Jonkman, J.; Butterfield, S.; Musial, W.; Scott, G. *Definition of a 5-MW Reference Wind Turbine for Offshore System Development*; Technical Report TP-500-38060; NREL: Golden, CO, USA, 2009.
14. Bossanyi, E.A. The Design of closed loop controllers for wind turbines. *Wind Energy* **2000**, *3*, 149–163. doi:10.1002/we.34.
15. Hansen, M.H.; Hansen, A.; Larsen, T.J.; Oye, S.; Sorensen, P.; Fuglsang, P. *Control Design for a Pitch-Regulated, Variable Speed Wind Turbine*; Technical Report R1500; Risoe National Laboratory: Roskilde, Denmark, 2005.
16. Hansen, M.H.; Henriksen, L.C.; Hartvig, M.; Christian, L. *Basic DTU Wind Energy Controller*; Technical Report E-0028; DTU Wind Energy: Roskilde, Denmark, 2013.
17. Mulders, S.P.; Van Wingerden, J.W. Delft Research Controller: An open-source and community-driven wind turbine baseline controller. *J. Phys. Conf. Ser.* **2018**, *1037* 032009. doi:10.1088/1742-6596/1037/3/032009.
18. Abbas, N.J.; Wright, A.; Pao, L. An Update to the National Renewable Energy Laboratory Baseline Wind Turbine Controller. *J. Phys. Conf. Ser.* **2020**, *1452* 012002. doi:10.1088/1742-6596/1452/1/012002.
19. Meng, F.; Lio, W.H.; Barlas, T. DTUWEC: an open-source DTU Wind Energy Controller with advanced industrial features. *J. Phys. Conf. Ser.* **2020**, *1618*, 022009. doi:10.1088/1742-6596/1618/2/022009.
20. Burton, T.; Jenkins, N.; Sharpe, D.; Bossanyi, E.A. *Wind Energy Handbook*, 2nd ed.; John Wiley & Sons Ltd.: West Sussex, UK, 2011; pp. 1–742.
21. *IEC 61400-1: Wind Turbines—Part 1: Design Requirements*; International Electrotechnical Commission: Geneva, Switzerland, 2005.
22. Madsen, H.A.; Riziotis, V.; Zahle, F.; Hansen, M.O.L.; Snel, H.; Larsen, T.J.; Politis, E.; Rasmussen, F. Blade Element Momentum Modeling of Inflow with Shear in Comparison with Advanced Model Results. *Wind Energy* **2012**, *15*, 63–81, doi:10.1002/we.493.
23. Hauptmann, S.; Bülk, M.; Schön, L.; Erbslöh, S.; Boorsma, K.; Grasso, F.; Kühn, M.; Cheng, P.W. Comparison of the Lifting-Line Free Vortex Wake Method and the Blade-Element-Momentum Theory regarding the simulated Loads of Multi-MW Wind Turbines. *J. Phys. Conf. Ser.* **2014**, *555*, 012050. doi:10.1088/1742-6596/555/1/012050.
24. Boorsma, K.; Hartvelt, M.; Orsi, L. Application of the Lifting Line Vortex Wake Method to Dynamic Load Case Simulations. *J. Phys. Conf. Ser.* **2016**, *753*, 022030. doi:10.1088/1742-6596/753/2/022030.
25. Madsen, H.A.; Larsen, T.J.; Pirrung, G.R.; Li, A.; Zahle, F. Implementation of the Blade Element Momentum Model on a Polar Grid and its Aeroelastic Load Impact. *Wind Energy Sci.* **2020**, *5*, 1–27. doi:10.5194/wes-5-1-2020.
26. Perez-Becker, S.; Saverin, J.; Marten, D.; Alber, J.; Pechlivanoglou, G.; Paschereit, C.O. Investigations on the Fatigue Load Reduction Potential of Advanced Control Strategies for Multi-MW Wind Turbines using a Free Vortex Wake Model. In Proceedings of the ASME Turbo Expo 2018, Oslo, Norway, 11–15 June 2018; p. V009T48A008. doi:10.1115/GT2018-76078.
27. Perez-Becker, S.; Papi, F.; Saverin, J.; Marten, D.; Bianchini, A.; Paschereit, C.O. Is the Blade Element Momentum theory overestimating wind turbine loads?—An aeroelastic comparison between OpenFAST's AeroDyn and QBlade's Lifting-Line Free Vortex Wake method. *Wind Energy Sci.* **2020**, *5*, 721–743. doi:10.5194/wes-5-721-2020.
28. Marten, D.; Lennie, M.; Pechlivanoglou, G.; Nayeri, C.N.; Paschereit, C.O. Implementation, Optimization and Validation of a Nonlinear Lifting Line-Free Vortex Wake Module within the Wind Turbine Simulation Code QBlade. *ASME J. Eng. Gas Turbines Power* **2015**, *138*, 072601. doi:10.1115/GT2015-43265.
29. Bak, C.; Zahle, F.; Bitsche, R.; Kim, T.; Yde, A.; Henriksen, L.C.; Andersen, P.B.; Natarajan, A.; Hansen, M.H. *Design and Performance of a 10 MW Wind Turbine*; Technical Report I-0092; DTU Wind Energy: Roskilde, Denmark, 2013.
30. Bergami, L.; Gaunaa, M. *ATEFlap Aerodynamic Model, a Dynamic Stall Model Including the Effects of Trailing Edge Flap Deflection*; Technical Report Risø-R-1792; DTU Wind Energy: Roskilde, Denmark, 2012.

31. Wendler, J.; Marten, D.; Pechlivanoglou, G.; Nayeri, C.N.; Paschereit, C.O. An Unsteady Aerodynamics Model for Lifting Line Free Vortex Wake Simulations of HAWT and VAWT in QBlade. In Proceedings of the ASME Turbo Expo: Turbine Technical Conference and Exposition GT2016, Orlando, FL, USA, 13–17 June 2016; p. V009T46A011. doi:10.1115/GT2016-57184.
32. Tasora, A.; Serban, R.; Mazhar, H.; Pazouki, A.; Melanz, D.; Fleischmann, J.; Taylor, M.; Sugiyama, H.; Negrut, D. Chrono: An Open Source Multi-Physics Dynamics Engine. In Proceedings of the International Conference on High Performance Computing in Science and Engineering, Salt Lake City, UT, USA, 13 November 2016; pp. 19–49. doi:10.1007/978-3-319-40361-8_2.
33. Bossanyi, E.A. Further Load Reductions with Individual Pitch Control. *Wind Energy* **2005**, *8*, 481–485. doi:10.1002/we.166.
34. Plumley, C.; Leithead, W.; Jamieson, P.; Bossanyi, E.A.; Graham, M. Comparison of individual Pitch and Smart Rotor Control Strategies for Load Reduction. *J. Phys. Conf. Ser.* **2014**, *524*, 012054. doi:10.1088/1742-6596/524/1/012054.
35. Plumley, C. The Smart Rotor Wind Turbine. Ph.D. Thesis, University of Strathclyde: Glasgow, UK, 2015.
36. Bergami, L.; Gaunaa, M. Analysis of Aeroelastic Loads and their Contributions to Fatigue Damage. *J. Phys. Conf. Ser.* **2014**, *555*, 012007. doi:10.1088/1742-6596/555/1/012007.
37. Iribas, M.; Hansen, M.H.; Mahmood, M.; Tibaldi, C.; Natarajan, A.; Bossanyi, E.; Stock, A.; Jamieson, P.; Leithead, W.; Schlipf, D. INNWIND Deliverable 1.42: Methodology for Feed-Forward Control Strategies Using Nacelle or Blade Based Sensors and Distributed Control. Technical Report. INNWIND.eu. 2015. Available online: http://www.innwind.eu/publications/deliverable-reports (access on 5 June 2020).
38. Borg, M.; Mirzaei, M.; Bredmose, H. *LIFES50+ Deliverable D1.2: Wind Turbine Models for the Design*; Technical Report E-101; DTU Wind Energy: Roskilde, Denmark, 2015.
39. Hayman, G.J. *MLife Theory Manual for Version 1.00*; Technical Report; NREL: Golden, CO, USA, 2012.
40. Buhl, M.L.J. Crunch Software. Available online: https://nwtc.nrel.gov/Crunch (accessed on 27 January 2018).
41. Bergami, L. Adaptive Trailing Edge Flaps for Load Reduction. In Proceedings of the 7th PhD Seminar on Wind Energy in Europe, Delft, The Netherlands, 27–28 October 2011.

energies

MDPI

Article

An Acoustic Source Model for Applications in Low Mach Number Turbulent Flows, Such as a Large-Scale Wind Turbine Blade

Hui Tang [1,2], Yulong Lei [1,2,*] and Xingzhong Li [1,2]

1 College of Automotive Engineering, Jilin University, Renmin Street No. 5988, Changchun 130012, China; tanghui15@mails.jlu.edu.cn (H.T.); whdxjx123@126.com (X.L.)

2 State Key Laboratory of Automotive Simulation and Control, Jilin University, Renmin Street No. 5988, Changchun 130012, China

* Correspondence: leiyl@jlu.edu.cn

Received: 2 November 2019; Accepted: 29 November 2019; Published: 3 December 2019

check for updates

Abstract: Aerodynamic noise from wind turbine blades is one of the major hindrances for the widespread use of large-scale wind turbines generating green energy. In order to more accurately guide wind turbine blade manufacturers to optimize the blade geometry for aerodynamic noise reduction, an acoustic model that not only understands the relation between the behavior of the sound source and the sound generation, but also accounts for the compressibility effect, was derived by rearranging the continuity and Navier–Stokes equations as a wave equation with a lump of source terms, including the material derivative and square of the velocity divergence. Our acoustic model was applied to low Mach number, weakly compressible turbulent flows around NACA0012 airfoil. For the computation of flow fields, a large-eddy simulation (LES) with the dynamic Smagorinsky subgrid scale (SGS) model and the cubic interpolated pseudo particle (CIP)-combined unified numerical procedure method were conducted. The reproduced turbulent flow around NACA0012 airfoil was in good agreement with the experimental data. For the estimation of acoustic fields, our acoustic model and classical sound source models, such as Lighthill and Powell, were performed using our LES database. The investigation suggested that the derived material derivative of the velocity divergence plays a dominant role as sound source. The distribution of the sources in our acoustic model was consistent with that of the classical sound source models. The sound pressure level (SPL) predicted based on the above-mentioned LES and our newly derived acoustic model was in reasonable agreement with the experimental data. The influence of the increase of Mach number on the acoustic field was investigated. Our acoustic source model was verified to be capable of treating the influence of Mach numbers on the acoustic field.

Keywords: large-scale wind turbine balde; computational aeroacoustics; sound source detection; low Mach number turbulent flows; NACA0012 airfoil

1. Introduction

At the present time, wind energy is a renewable, sustainable source of power, and one of the most rapidly developing electricity production fields worldwide [1]. Based on the European Union's (EU's) report of the gross electricity consumption from wind power, a more than threefold increase between 2004 and 2014 took place, and to fulfill EU climate goals for 2030 it can be expected that this trend will continue in the future [2]. Wind energy increase will mean that many more wind turbines will be installed, inevitably closer to more people and their residences. Wind turbine noise is one of the major hindrances for the widespread use of wind energy. Surveys [3] show that noise from a wind turbine is annoying to people and that is perceived to be more annoying than other forms of industrial noise at the same level. To accommodate the expected increase in the number of installed wind farms and to reduce public disquiet, there is need to reduce wind turbines' noise.

In Western Europe alone, an estimated 1.0–1.6 million healthy life years are lost each year because of environmental noise based on the report of the World Health Organization [4]. The wind turbine noise is playing a more and more important role in environmental noise as wind turbines are increasingly installed worldwide [5]. The sleep disturbance of the wind turbine noise is the greatest influence to long-term health [2]. Wind turbine noise has aerodynamic and mechanical origins. For a modern, large-scale wind turbine, aerodynamic noise from the blades is generally considered to be the dominant noise source, in which the turbulent flow around an airfoil that induces aerodynamic noise typically has a high Reynolds number flow at a low Mach number [6]. Empirical or semi-empirical models have been developed to predict the overall noise emitted by a wind turbine. However current empirical or semi-empirical models do not contain an accurate description of the wind turbine blade geometry and its relations to emitted noise. Furthermore, wind turbine blade manufacturers are interested in small modifications of given blade geometries and their exact influences on the aerodynamic noise. It is, therefore, necessary to develop techniques that take the correct blade geometry into account to predict the aerodynamic noise. As a result, computational fluid dynamics (CFD) and computational aeroacoustics (CAA) have become useful tools to numerically simulate the complex flow and aerodynamic noise for engineering applications. Various numerical investigations of aerodynamic wind turbine noise using CFD and CAA have been conducted [7–10]. Since the pioneering paper of Lighthill [11,12], computational techniques to deal with flow-induced noise have been classified into two categories: direct methods [13] and indirect methods [14–17]. In a direct method, sound sources and sound propagations are obtained as a result of the numerical simulation based on the compressible Navier–Stokes equations. The direct method can to reproduce the sound generation mechanism exactly and is suitable when a strong interaction between the flow and acoustic fields exists. This means that the order of magnitude of the pressure fluctuation of the flow field closes to the sound pressure; namely, the case of high Mach number flows. However, for the objective of the present research, a large-scale wind turbine blade, use of the direct method is inappropriate due to the low Mach number. Furthermore, it is also very difficult for practical applications to apply the direct method in the industry due to the high computational cost.

On the other hand, Lighthill–Curle acoustic analogy [18] has been widely used for predicting a far-field sound in engineering practice. In this sort of indirect method, unsteady flows are simulated by the incompressible scheme, usually with Reynolds averaged numerical simulation (RANS), large eddy simulation (LES), or LES/RANS hybrid method. Then, the acoustic field is predicted based on the theoretically estimated sound source; e.g., Lighthill–Curle acoustic analogy [18] and Powell [19]. Thus, there is no mutual interaction between the flow field and the sound field; that is, it is assumed that the noise is determined by the information of the flow field, and the sound generated does not influence the flow field. This assumption is valid in low Mach number flow since the sound pressure is small compared to the pressure fluctuation of the flow field. Actually, in the case that the effect of feedback from the sound field to the flow field is not important, this method has been widely used in industrial applications, since the far-field sound is reproduced successfully.

For estimation of the acoustic field around objects in fluid flows, the Lighthill–Curle acoustic analogy [18] is most widely used; see Amiet [20] and Wang [21]. In this method, a pressure fluctuation of the object surface obtained from the incompressible flow field is used as a sound source, and then a far-field sound is estimated separately. This method, however, makes it difficult to understand the relationship between the behavior of the sound source in the flows field and the radiated sound. As a result, this method can not guide engineers and manufacturers to optimize the large-scale wind turbine blade for the reduction of the aerodynamic noise. On the other hand, the theory of vortex sound proposed by Powell [19] and then extended by Howe [22] is also widely used: take for example, Mohring [23], Takaishi et al. [24], and Ewert et al. [25]. In this method, the sound source is estimated from the behavior of vortices, and the far-field sound pressure is computed by using the compact green's function [22]. Although the prediction accuracy of the far-field sound is affected by the range of volume integral regions, this method is able to treat the sound source that is distributed to the space, qualitatively. The behavior of vortex might be a true sound source so that this method is probably suitable for predicting the aerodynamic noise generated from large-scale wind turbine blades and then guiding manufacturers in optimizing the blade geometry. However, there is an important assumption for this method; i.e., the sound source is derived under the assumption of incompressible flows. The compressibility effect that appears even in low Mach number, weakly compressible turbulent flows is, therefore, not taken into account. However, even a small fluctuation of density affects the flow filed and the flow-induced sound field around the object, such as that from an airfoil. Hutcheson et al. [26,27] showed that the peak frequency in the profile of sound pressure level is different from the change of Mach number even in the low Mach number range. Since this characteristic is not reproduced by the classical indirect method mentioned above, this sort of indirect method becomes less accurate in low Mach number turbulent flows, especially with increasing Mach numbers. In order to predict the acoustic field accurately for applications in low Mach number turbulent flow, such as for large-scale wind turbine, it is necessary to improve the acoustic model of considering the compressibility effect even in low Mach number flows.

The purpose of this study was to seek for a more accurate acoustic model for large-scale wind turbine blade manufacturers to optimize the blade geometry for aerodynamic noise reduction. To attain that end, an new acoustic model was required, one that not only understood what kind of fluctuations of the flow field cause the aerodynamic noise but also accounted for the small fluctuation of density in the noise source (namely, compressibility effect). In the derivation of the new acoustic theory, we rearranged the continuity and Navier–Stokes equations as a wave equation with a lump of source terms including the material derivative and square of the velocity divergence. These source terms are used for sound source detection and the estimation of the far-field sound.

In this study, our acoustic model was applied to low Mach number, weakly compressible turbulent flows around NACA0012 airfoil. For the computation of flow fields and considering the weak compressibility in flow fields, an LES with the dynamic Smagorinsky model [28,29] and the cubic interpolated pseudo particle (CIP)-combined unified numerical procedure method [30] were conducted. Our LES technique was verified by comparison its results with the experimental results performed by Miyazawa et al. [31]. The reproduced turbulent flow around NACA0012 airfoil was in good agreement with the experimental data. For the estimation of acoustic fields, different acoustic models were performed using our LES database. The distribution of the sources obtained by our acoustic model was compared with classical sound source models, such as Lighthill [11] and Powell [19], in the case of very low fluctuation of density. Then, the sound pressure level (SPL) predicted based on the above-mentioned LES and our newly derived acoustic model was compared with the SPLs obtained by the Lighthill–Curle's equation [18] using our LES database and the experimental data by Miyazawa et al. [31]. Finally, our acoustic source model was verified to treat the influence of Mach numbers on the acoustic field, and the influence of the increase of Mach number on the acoustic field was investigated.

2. LES of Low Mach Number Turbulent Flows around NACA0012 Airfoil

2.1. Basic Equations and Smagorinsky Subgrid Scale (SGS) Model

The Favre-averaged filter and spatial filter are expressed as $\bar{()}$ and $\hat{()}$ respectively, and are used for the continuity and Navier–Stokes equations, which include the continuity equation, momentum equations, and equation of state. In Equation (2), the eddy viscosity assumption is used for SGS turbulence stress. A non-dimensionalization is applied to all variables by means of the streamwise velocity U_0 and chord length C. Since the boundary-fitted-grid is employed in our computation, general curvilinear coordinates have to be applied and is represented as ξ, η, and ς. Thus,

$$\frac{\partial \hat{\rho}}{\partial t} + \frac{1}{J}\frac{\partial}{\partial \xi^k}\left(J\hat{\rho}\bar{U}^k\right) = 0, \tag{1}$$

$$\frac{\partial(\hat{\rho}\bar{u}_i)}{\partial t}+\frac{1}{J}\frac{\partial}{\partial\xi^k}\left(J\hat{\rho}\bar{u}_i\bar{U}^k\right)=-\frac{1}{J}\frac{\partial}{\partial\xi^k}\left[J\frac{\partial\xi^k}{\partial x_i}\left(\hat{p}+\frac{2}{3}\hat{\rho}k_{sgs}\right)\right]+\frac{1}{J}\frac{\partial}{\partial\xi^k}\left(J\frac{\partial\xi^k}{\partial x_i}\sigma_{ij}\right), \tag{2}$$

$$\hat{p}=\hat{\rho}RT, \tag{3}$$

where

$$\sigma_{ij}=2\left(\frac{1}{Re}+\hat{\rho}\nu_{sgs}\right)\left(\bar{S}_{ij}-\frac{1}{3}\delta_{ij}\bar{S}_{kk}\right), \tag{4}$$

where J represents the Jacobian of the coordinate transformation, ρ denotes the density, $\hat{U}^k$ means the contravariant velocity, k_{sgs} is the SGS kinetic energy, δ_{ij} is the Kronecker symbol, T is the absolute temperature, R is the ideal gas constant, Re means the Reynolds number, as in $\rho_0 U_0/\nu$, $\bar{S}_{ij}$ denotes the strain rate tensor

$$\bar{S}_{ij}=\frac{1}{2}\left(\frac{\partial\xi^k}{\partial x_i}\frac{\partial\bar{u}_j}{\partial\xi^k}+\frac{\partial\xi^k}{\partial x_j}\frac{\partial\bar{u}_i}{\partial\xi^k}\right), \tag{5}$$

and ν_{sgs} means the eddy viscosity since the effect of SGS turbulence on the large-scale motion is expected to be estimated by an SGS model.

Using the local equilibrium assumption, i.e., that the dissipation rate of SGS energy is in balance with the production rate, the dynamic Smagorinsky model can be obtained for LES:

$$\nu_{sgs}=C\hat{\Delta}^2|\overline{S}|, \tag{6}$$

in which $|\overline{S}|$ denotes the norm of the strain rate tensor and is defined as $\sqrt{2\overline{S}_{ij}\overline{S}_{ij}}$; C is a variable to be dynamically determined from the grid-scale velocity field $\bar{u}$.

Applying the test filter on the grid-filtered N-S equations, the Germano identity can be defined as

$$L_{ij}=T_{ij}-\widetilde{\tau}_{ij}=\widetilde{\overline{u}_i\overline{u}_j}-\widetilde{\overline{u}}_i\widetilde{\overline{u}}_j, \tag{7}$$

where L_{ij} can be calculated based on the resolved scales, and $T_{ij}=\widetilde{\overline{u_i u_j}}-\widetilde{\overline{u}}_i\widetilde{\overline{u}}_j$ represents the residual turbulent stress at a test-filter scale $\widetilde{\Delta}$ and can be given as

$$T_{ij}-\frac{1}{3}\delta_{ij}T_{kk}=-2C\widetilde{\Delta}^2|\widetilde{\overline{S}}|\widetilde{\overline{S}}_{ij}. \tag{8}$$

Upon substituting Equations (4) and (8) into Equation (7) and assuming $\hat{\Delta}$ and C are constant inside the test filter, an equation for determining C was obtained:

$$L_{ij}-\frac{1}{3}\delta_{ij}L_{kk}=-2C\hat{\Delta}^2M_{ij}, \tag{9}$$

where

$$M_{ij}=\frac{\widetilde{\Delta}^2}{\hat{\Delta}^2}|\widetilde{\overline{S}}|\widetilde{\overline{S}}_{ij}-\widetilde{|\overline{S}|\overline{S}_{ij}}. \tag{10}$$

Minimization of the error of Equation (9) over all independent tensor components [29], and over some averaging region of statistical homogeneity, leads to

$$C=-\frac{1}{2\hat{\Delta}^2}\frac{\langle L_{ij}M_{ij}\rangle}{\langle M_{ij}M_{ij}\rangle}. \tag{11}$$

Finally, substituting the dynamically determined variable C of Equation (11) into Equation (6), the eddy viscosity ν_{sgs} can be computed.

Through using the CIP-combined unified numerical procedure method [30], the fractional step method is applied for time-marching of Equation (2):

$$(\hat{\rho}\bar{u})^F = \frac{\Delta t}{2}\left(\frac{2}{\Delta t}(\hat{\rho}\bar{u})^n + 3\nabla\cdot[-(\hat{\rho}\bar{u}\bar{u}) + \tau]^n - \nabla\cdot[-(\hat{\rho}\bar{u}\bar{u}) + \tau]^{n-1}\right), \quad (12)$$

$$(\hat{\rho}\bar{u})^{n+1} = (\hat{\rho}\bar{u})^F - \Delta t\nabla\hat{p}^{n+1}, \quad (13)$$

in which τ represents the viscous stress, Δt the time increment, and n the time step count. The advancement method of time for Equation (1) can be written as:

$$\hat{\rho}^{n+1} = \hat{\rho}^n - \Delta t\nabla\cdot(\hat{\rho}\bar{u})^{n+1}. \quad (14)$$

Applying the divergence for Equation (13) and then substitution of this divergence into Equation (14) leads to the pressure elliptic equation for $\hat{p}^{n+1}$.

$$-\hat{\rho}^{n+1} + \hat{\rho}^n - \Delta t\nabla\cdot(\hat{\rho}\bar{u})^F = -\Delta t^2\nabla^2\hat{p}^{n+1}. \quad (15)$$

In order to take the compressibility into account for the pressure equation, Equation (3) can be written as the following for weakly compressible flows:

$$\hat{p}^{n+1} - \hat{p}^n = (\hat{\rho}^{n+1} - \hat{\rho}^n)RT. \quad (16)$$

Substituting Equation (16) into Equation (15) results in a pressure equation that accounts for the compressibility; thus,

$$(\Delta t)^2RT\nabla^2\hat{p}^{n+1} - \hat{p}^{n+1} = (\Delta t)RT\nabla\cdot(\hat{\rho}\bar{u})^F - \hat{p}^n. \quad (17)$$

Therefore, the second term in two sides of Equation (17) accounts for the compressibility effect. In a word, using Equation (17), $\hat{p}^{n+1}$ is solved; and then, substituting $\hat{p}^{n+1}$ into Equation (12), $(\hat{\rho}\bar{u})^{n+1}$ is calculated; and finally, applying (1), $\hat{\rho}^{n+1}$ is obtained. Note that he applicability of this procedure is limited with weakly compressible flows due to the fact that $\Delta\hat{p}$ and $\Delta\hat{\rho}$ change very slightly in one time step for the low Mach number turbulent flow.

2.2. Validation of LES for Low Mach Number Turbulent Flows around NACA0012 Airfoil

The calculation target of this study is a three-dimensional flow around a two-dimensional wing of the NACA0012 airfoil, which is a typical streamlined object. NACA0012 airfoil is defined by the following equation and its geometry is given in Figure 1.

$$\pm y/C = 0.6\times\left[0.2969\sqrt{x/C} - 0.1260(x/C) - 0.3516(x/C)^2 + 0.2843(x/C)^3 - 0.01015(x/C)^4\right]. \quad (18)$$

Figure 1. The profile of NACA0012 airfoil.

In order to validate our method, we conducted a LES of the compressible flow around NACA0012 airfoil with the computational setup matching the experimental setup of Miyazawa et al. [31], in which the Mach number was 8.75×10^{-2}, the angel of attack was 9°, and the Reynolds number was 2×10^5, which was based on the uniform velocity in the streamwise direction and the chord length. Figure 2 shows the computational domain and computational boundary conditions in the present study. For the Cartesian coordinate system, x, z, and y were defined in the streamwise direction, spanwise direction, and vertical direction, respectively. We used a C-type boundary-fitted grid in the x–y plane. Thus, a general curvilinear coordinate system had to be applied and represented as ξ, η, ζ for all computations, in which the direction following the mainstream surface of the airfoil is denoted as ξ, the direction away from the surface of airfoil is defined as η, and the spanwise direction is represented as ζ. The computational domain size is defined as the following: the length of the wake, the semidiameter of the C-type grid, and the spanwise extent were 11C, 11C, and 0.5%C, respectively. Here, C is chord length. For the mesh, there were 1600 grid points in the ξ direction, in which 800 was on the airfoil surface, 160 grid points were in the η direction, and 60 grid points were in the ζ direction. As shown in Figure 2, the non-slip boundary condition was applied on the airfoil surface. The inflow was the uniform stream without disturbance. The convective outflow was applied for the outflow boundary condition. The spanwise direction boundary condition is defined as the periodic. For the variables in the η direction, zero gradient was employed at the op and bottom boundary. Particular attention was paid for the non-reflective boundary condition by [32] at the far boundary.

A second-order central finite-difference discretization scheme was applied in the equation of motion for the diffusion terms. In order to improve the numerical stability in high Reynolds number flow, a QUICK method was applied for the convective terms in the general curvilinear coordinate system. For coupling the pressure field and the continuity equation, the fractional method was applied. The second-order Adams–Bashforth method is used to the convective term and the viscous term for the advancement method of time in the equation of motion. For the dynamic procedure, the test filter was used in the streamwise direction and spanwise direction with second-order accuracy and the test-to-grid filter ratio $\tilde{\Delta}/\hat{\Delta} = 2$. The present numerical method and computer program have been tested extensively in several turbulent flows [33–37].

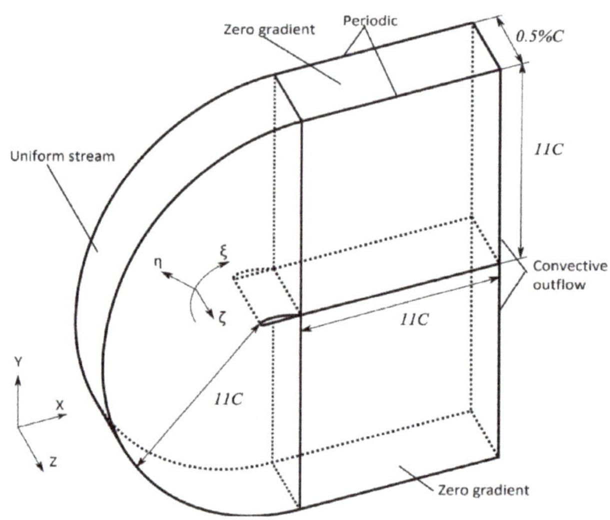

Figure 2. Computational domain and boundary conditions.

All simulations were computed on an NEC SX-8R supercomputer of Cybermedia Center, Osaka University with the time step $dt = 0.0495H/U_c$. The greater part of the total effort toward calculations was spent on solving the pressure equation through the residual cutting method [38]. All simulations were run until the flow fields that were fully developed and the first-order, and second-order statistics exhibited adequate convergence. The time-averaging, and, the spatial averaging in the spanwise direction were used for collecting the data. Figures 3 and 4 show the distribution of the mean pressure coefficient C_p on the airfoil surface and the average pressure coefficient fluctuation $C_{p\,rms}$ on the suction side of the airfoil, in which C_p is defined by the freestream pressure p_0.

$$C_p = \frac{p - p_0}{\frac{1}{2}\rho U_0^2}. \tag{19}$$

Then, the root-mean-square of C_p is defined as $C_{p\,rms}$. In order to demonstrate the validity of our model, the results of our LES are compared with the measurement by [31]. In Figures 3 and 4, our computational results are in good agreement with the experimental data.

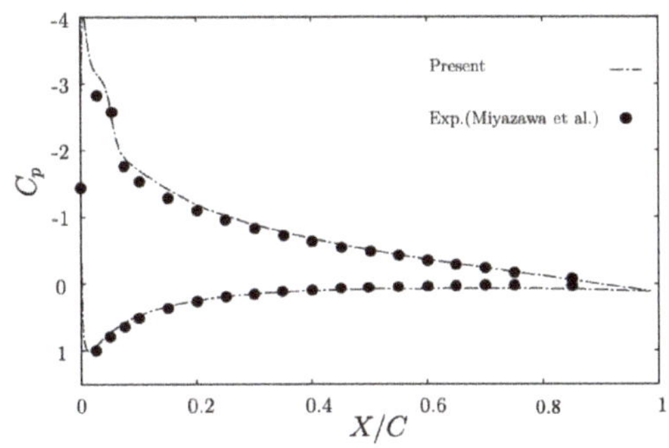

Figure 3. Mean pressure coefficient C_p.

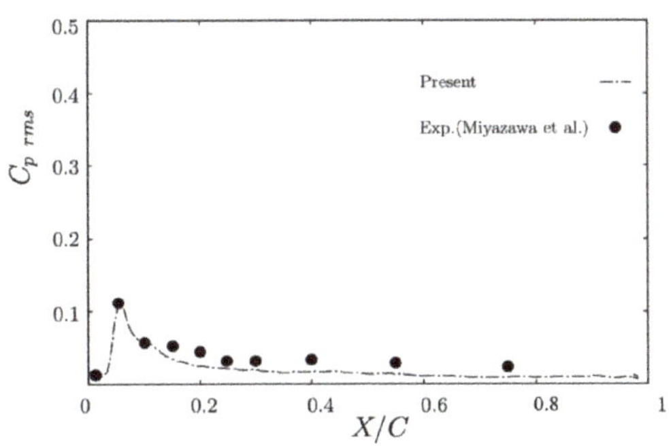

Figure 4. Mean fluctuation of C_p.

3. Derivation of a New Acoustic Model

A new acoustic model was derived by using the compressible continuity and Navier–Stokes equations of the flow field; i.e.,

$$\frac{\partial \rho}{\partial t} + \frac{1}{J}\frac{\partial}{\partial \xi^k}\left(J\frac{\partial \xi^k}{\partial x_i}\rho u_j\right) = 0, \tag{20}$$

$$\frac{\partial u_i}{\partial t} + \frac{1}{J}\frac{\partial}{\partial \xi^k}\left(J\frac{\partial \xi^k}{\partial x_j}\rho u_j u_i\right) - \frac{\partial \xi^k}{\partial x_i}\frac{\partial p}{\partial \xi^k} = \frac{\partial \xi^k}{\partial x_j}\frac{\partial \sigma_{ij}}{\partial \xi^k}, \tag{21}$$

$$\sigma_{ij} = 2\mu\left(S_{ij} - \frac{1}{3}\delta_{ij}S_{kk}\right), \quad S_{ij} = \frac{1}{2}\left(\frac{\partial \xi^k}{\partial x_i}\frac{\partial u_j}{\partial \xi^k} + \frac{\partial \xi^k}{\partial x_j}\frac{\partial u_i}{\partial \xi^k}\right). \tag{22}$$

Variables ρ and p are decomposed as

$$\rho = \rho_0 + \rho', \quad p = p_0 + p', \tag{23}$$

where (0) indicates the mean component of variables; ($'$) means the perturbation component of variables, due to in low Mach number turbulent flow

$$\rho_0 \gg \rho', \quad p_0 \gg p'. \tag{24}$$

Rewriting Equations (20) and (21) by using Equation (23) leads to

$$\frac{\partial \rho'}{\partial t} + \frac{1}{J}\frac{\partial}{\partial \xi^k}\left(J\frac{\partial \xi^k}{\partial x_i}\rho' u_j\right) = -\rho_0 \frac{\partial \xi^k}{\partial x_j}\frac{\partial u_j}{\partial \xi^k}, \tag{25}$$

$$\frac{\partial u_i}{\partial t} + \frac{1}{J}\frac{\partial}{\partial \xi^k}\left(J\frac{\partial \xi^k}{\partial x_j}u_j u_i\right) - \frac{1}{\rho}\frac{\partial \xi^k}{\partial x_i}\frac{\partial p'}{\partial \xi^k} = \frac{1}{\rho}\frac{\partial \xi^k}{\partial x_j}\frac{\partial \sigma_{ij}}{\partial \xi^k}. \tag{26}$$

After multiplying u_i to Equation (25) and ρ' to Equation (26), Equations (25) and (26) are added. Namely,

$$\frac{\partial \rho' u_i}{\partial t} + \frac{1}{J}\frac{\partial}{\partial \xi^k}\left(J\frac{\partial \xi^k}{\partial x_i}\rho' u_i u_j\right) + \frac{\rho'}{\rho}\frac{\partial \xi^k}{\partial x_i}\frac{\partial p'}{\partial \xi^k} = \frac{\rho'}{\rho}\frac{\partial \xi^k}{\partial x_j}\frac{\partial \sigma_{ij}}{\partial \xi^k} - \rho_0 u_i \frac{\partial \xi^k}{\partial x_j}\frac{\partial u_j}{\partial \xi^k}. \tag{27}$$

Taking the divergence of Equation (27), subtracting $\frac{\partial}{\partial t}$ of Equation (25), and then adding $-c_0^2 \frac{1}{J}\frac{\partial}{\partial \xi^k}\left(\gamma^{kl}\frac{\partial \rho'}{\partial \xi^l}\right)$ concerning two sides of the equation obtained, the wave equation with source terms is finally obtained as the following

$$\begin{aligned}\frac{\partial^2 \rho'}{\partial t^2} - c_0^2\frac{1}{J}\frac{\partial}{\partial \xi^k}\left(\gamma^{kl}\frac{\partial \rho'}{\partial \xi^l}\right) &= \frac{1}{J}\frac{\partial}{\partial \xi^k}\left(\gamma^{kl}\frac{\partial}{\partial \xi^l}\rho' u_i u_j\right) - c_0^2\frac{1}{J}\frac{\partial}{\partial \xi^k}\left(\gamma^{kl}\frac{\partial \rho'}{\partial \xi^l}\right)\\
+ & \frac{\rho'}{\rho}\frac{1}{J}\frac{\partial}{\partial \xi^k}\left(\gamma^{kl}\frac{\partial \rho'}{\partial \xi^l}\right) + \frac{\partial \xi^l}{\partial x_i}\frac{\partial p'}{\partial \xi^l}\frac{\partial \xi^k}{\partial x_i}\frac{\partial}{\partial \xi^k}\frac{\rho'}{\rho} - \rho_0\frac{\partial}{\partial t}\left(\frac{\xi^k}{\partial x_i}\frac{\partial u_i}{\partial \xi^k}\right) - \frac{\rho'}{\rho}\frac{1}{J}\frac{\partial}{\partial \xi^k}\left(\gamma^{kl}\frac{\partial}{\partial \xi^l}\sigma_{ij}\right)\\
- & \frac{\partial \xi^k}{\partial x_i}\frac{\partial \xi^l}{\partial x_j}\frac{\partial \sigma_{ij}}{\partial \xi^k}\frac{\partial}{\partial \xi^l}\left(\frac{\rho'}{\rho}\right) + \rho_0\frac{\partial \xi^l}{\partial x_i}\frac{\partial \xi^k}{\partial x_j}\frac{\partial u_i}{\partial \xi^l}\frac{\partial u_j}{\partial \xi^k} + \rho_0 u_i\frac{1}{J}\frac{\partial}{\partial \xi^k}\left(\gamma^{kl}\frac{\partial u_j}{\partial \xi^l}\right).\end{aligned} \tag{28}$$

The first term and the last term of the right-hand side of Equation (29) can be rewritten as

$$\frac{1}{J}\frac{\partial}{\partial \xi^k}\left(\gamma^{kl}\frac{\partial}{\partial \xi^l}\rho' u_i u_j\right) + \rho_0 u_i\frac{1}{J}\frac{\partial}{\partial \xi^k}\left(\gamma^{kl}\frac{\partial u_j}{\partial \xi^l}\right) = \frac{1}{J}\frac{\partial}{\partial \xi^k}\left(\gamma^{kl}\frac{\partial}{\partial \xi^l}\rho u_i u_j\right) + \rho_0 u_j\frac{\partial \xi^k}{\partial x_j}\frac{\partial}{\partial \xi^k}\left(\frac{\partial \xi^l}{\partial x_i}\frac{\partial u_i}{\partial \xi^l}\right). \tag{29}$$

Upon substituting Equation (29) into Equation (29), the wave equation with a lump of source terms is obtained as follows

$$\begin{aligned}\frac{\partial^2 \rho'}{\partial t^2} - c_0^2\frac{1}{J}\frac{\partial}{\partial \xi^k}\left(\gamma^{kl}\frac{\partial \rho'}{\partial \xi^l}\right) &= \frac{1}{J}\frac{\partial}{\partial \xi^k}\left(\gamma^{kl}\frac{\partial}{\partial \xi^l}\rho u_i u_j\right) - c_0^2\frac{1}{J}\frac{\partial}{\partial \xi^k}\left(\gamma^{kl}\frac{\partial \rho'}{\partial \xi^l}\right) + \frac{\rho'}{\rho}\frac{1}{J}\frac{\partial}{\partial \xi^k}\left(\gamma^{kl}\frac{\partial \rho'}{\partial \xi^l}\right)\\
&+ \frac{\partial \xi^l}{\partial x_i}\frac{\partial p'}{\partial \xi^l}\frac{\partial \xi^k}{\partial x_i}\frac{\partial}{\partial \xi^k}\frac{\rho'}{\rho} - \frac{\rho'}{\rho}\frac{1}{J}\frac{\partial}{\partial \xi^k}\left(\gamma^{kl}\frac{\partial}{\partial \xi^l}\sigma_{ij}\right)\\
&- \frac{\partial \xi^k}{\partial x_i}\frac{\partial \xi^l}{\partial x_j}\frac{\partial \sigma_{ij}}{\partial \xi^k}\frac{\partial}{\partial \xi^l}\left(\frac{\rho'}{\rho}\right) - \rho_0\frac{D}{Dt}(\nabla\cdot\mathbf{u}) + \rho_0(\nabla\cdot\mathbf{u})^2,\end{aligned} \tag{30}$$

where $\frac{D}{Dt} = \frac{\partial}{\partial t} + u_j\frac{\partial}{\partial x_j}$ means the material derivative. Here, assuming a low Mach number flow ($\rho_0 \gg \rho'$) and a high Reynolds number flow ($\delta_{ij} \ll 1$), a wave equation with source terms can be finally obtained as

$$\frac{\partial^2 \rho'}{\partial t^2} - c_0^2\frac{1}{J}\frac{\partial}{\partial \xi^k}\left(\gamma^{kl}\frac{\partial \rho'}{\partial \xi^l}\right) = \frac{1}{J}\frac{\partial}{\partial \xi^k}\left(\gamma^{kl}\frac{\partial}{\partial \xi^l}\rho u_i u_j\right) - c_0^2\frac{1}{J}\frac{\partial}{\partial \xi^k}\left(\gamma^{kl}\frac{\partial \rho'}{\partial \xi^l}\right) - \rho_0\frac{D}{Dt}(\nabla\cdot\mathbf{u}) + \rho_0(\nabla\cdot\mathbf{u})^2. \tag{31}$$

The form of our wave Equation (31) is similar to the form of the following Lighthill's equation [11]. Lighthill's equation is expressed as

$$\frac{\partial^2 \rho'}{\partial t^2} - c_0^2\frac{1}{J}\frac{\partial}{\partial \xi^k}\left(\gamma^{kl}\frac{\partial \rho'}{\partial \xi^l}\right) = \frac{1}{J}\frac{\partial}{\partial \xi^k}\left(\gamma^{kl}\frac{\partial T_{ij}}{\partial \xi^l}\right), \tag{32}$$

where

$$T_{ij} = \rho u_i u_j + \delta_{ij}\left[(p - p_0) - c_0^2(\rho - \rho_0)\right] - \sigma_{ij}. \tag{33}$$

Equation (31) and Lighthill's Equation (32) both seem to be wave equations having the source terms on the right hand side. But there are unknowns in T_{ij} for Lighthill's Equation (32). Especially in case of low Mach number flows, the simplification of $T_{ij} \approx \rho_0 u_i u_j$ works well. Hence the indirect method, which solves the acoustic field using T_{ij} given by the incompressible solver, becomes reasonable. On the other hand, from our wave Equation (31), the utilization of $\nabla \cdot \mathbf{u}$ for the sound source is possible. Actually, the divergence of the velocity vector plays an important role in the flow field even in the low Mach number [34]. Therefore, we rearranged the acoustic wave equation by exposing $\nabla \cdot \mathbf{u}$ intentionally and then obtained Equation (31), so that the indirect method could be extended from the zero Mach number region to low or moderate Mach number region.

For easily applying to engineering practices in the industry, such as large-scale wind turbine blades and underlining the key role of $\nabla \cdot \mathbf{u}$ in the flow field, our acoustic model Equation (31) can be further approximated as

$$\frac{\partial^2 \rho'}{\partial t^2} - c_0^2 \frac{1}{J}\frac{\partial}{\partial \xi^k}\left(\gamma^{kl}\frac{\partial \rho'}{\partial \xi^l}\right) \approx -\rho_0 \frac{D}{Dt}(\nabla \cdot \mathbf{u}) + \rho_0 (\nabla \cdot \mathbf{u})^2. \tag{34}$$

The Lighthill's Equation (32) is exact, but some kind of turbulence modeling for computing the sound source is required when the Reynolds number is high. Especially in case of low Mach number flows, the simplification of $T_{ij} \approx \rho_0 u_i u_j$ is utilized so that the solver becomes incompressible. Powell's sound source model [19] is derived under the assumption of incompressible flows. On the other hand, although some approximation is used in the derivation of the acoustic equation, our sound source model (34) does not need any modeling process. Furthermore, it can consider the influence of the compressibility effect. Thus, the behavior of sound source due to the variation of Mach number is expected to be reproduced appropriately.

Applying our acoustic theory and the compact green function [22], the sound pressure in the far-field is solved from the wave Equation (34). That is, assuming that the observation point x is sufficiently far from the sound source area of the objet y, the object keep stationary and the velocity of the surface S of the object is zero, the wave Equation (34) is able to be solved applying the compact green function:

$$p_a(\boldsymbol{x}, t) = \rho_0 \iint \frac{D}{Dt}(\nabla \cdot \mathbf{u}) \cdot \nabla G(\boldsymbol{x}, \boldsymbol{y}, t - \boldsymbol{\tau}) d^3\boldsymbol{y} d\boldsymbol{\tau}, \tag{35}$$

where p_a denotes the pressure sound, G means the compact green function and is represented as

$$G(\boldsymbol{x}, \boldsymbol{y}, t - \boldsymbol{\tau}) = \frac{1}{4\pi|\boldsymbol{x}|}\delta\left(t - \boldsymbol{\tau}\frac{|\boldsymbol{x}|}{c_0}\right) + \frac{\boldsymbol{x} \cdot \boldsymbol{Y}}{4\pi c_0 |\boldsymbol{x}|^2}\frac{\partial}{\partial t}\delta\left(t - \boldsymbol{\tau} - \frac{|\boldsymbol{x}|}{c_0}\right). \tag{36}$$

Here, $\boldsymbol{Y}$ means Kirchhoff vector and is calculated through $\nabla^2 \boldsymbol{Y}_i = 0$. Then substitution of Equation (36) into Equation (35) results in

$$p'_a(\boldsymbol{x}, t) = \frac{\rho_0 x_i}{4\pi c_0 |\boldsymbol{x}|^2}\frac{\partial}{\partial t}\int \frac{D}{Dt}(\nabla \cdot \mathbf{u})\left(\boldsymbol{y}, t - \frac{|\boldsymbol{x}|}{c_0}\right)\nabla \boldsymbol{Y}_i(\boldsymbol{y}) d^3\boldsymbol{y}. \tag{37}$$

From the Equation (37), the sound pressure p'_a can be obtained from conducting the Fourier transform. Finally the sound pressure level (SPL) can be calculated from the following equation through applying the solved p'_a:

$$SPL(dB) = 10 \times log\left(\frac{p'^2_a}{P_b^2}\right), \tag{38}$$

where P_b means the reference sound pressure. In general, the magnitude of P_b is $P_b = 2 \times 10^{-5} Pa$.

4. Results of the Acoustic Field around NACA0012 Airfoil

4.1. Comparison of Different Sound Source Detection

In this section, using LES databases obtained in Section 2 for the weakly compressible flow field around NACA0012 airfoil under the conditions of $Re = 2 \times 10^5$, $M = 0.0875$, and $\alpha = 9^\circ$, results of the sound field due to the flow field are discussed. The relationship between our proposed sound source model (34) and the classical sound source models by Lighthill [11] and Powell [19] is discussed through the comparison of the distribution of sound source terms around the airfoil using our LES database, as shown in Figure 5. Figure 5a,b show the instantaneous and cross-sectional profiles of the well-know classical sound source models of $\nabla \cdot (\nabla \cdot \mathbf{T})$ by Lighthill [11] and $\nabla \cdot (\boldsymbol{\omega} \times \boldsymbol{u})$ by Powell [19], respectively; Figure 5c,d show the instantaneous and cross-sectional profiles of our derived sound source of Equation (34) for $-\rho_0 \frac{D}{Dt}(\nabla \cdot \mathbf{u})$ and $\rho_0(\nabla \cdot \mathbf{u})^2$, respectively. Because the experimentally estimated sound source which was caused by the separation bubble was confirmed near the leading-edge, we focused on that region. In Figure 5a–c, the distributions of $\nabla \cdot (\nabla \cdot \mathbf{T})$, $\nabla \cdot (\boldsymbol{\omega} \times \boldsymbol{u})$, and $-\rho_0 \frac{D}{Dt}(\nabla \cdot \mathbf{u})$ are locally similar, near the leading edge. Pairs of positive and negative patterns are observed near the leading edge in the suction side, and intense regions correspond to the experimentally estimated sound source region. But, the distribution of $\rho_0(\nabla \cdot \mathbf{u})^2$ differs from the other terms and has a relatively very small value. Therefore, from Figure 5, it is considered that $\rho_0(\nabla \cdot \mathbf{u})^2$ does not nearly contribute to the sound field, while $\rho_\infty \frac{D}{Dt}(\nabla \cdot \mathbf{u})$ plays an important role in generation of sound.

According to the theory of vortex sound by Powell [19] and Howe [22], the main sound source of the aerodynamic noise is related to the behavior of vortex. Thus, the comparison of the behavior of the spanwise vorticity ω_z and the main sound source $-\rho_0 \frac{D}{Dt}(\nabla \cdot \mathbf{u})$ was conducted. Figure 6 shows time evolution of instantaneous and cross-sectional profiles of $-\rho_0 \frac{D}{Dt}(\nabla \cdot \mathbf{u})$ and ω_z near the leading edge. The significant distribution of ω_z exists in the suction side, and its region is similar to the distribution region of $-\rho_0 \frac{D}{Dt}(\nabla \cdot \mathbf{u})$, apart from the area of strong compressibility. Moreover, the period of moving of $-\rho_0 \frac{D}{Dt}(\nabla \cdot \mathbf{u})$ corresponds to that of ω_z: its value is 2.3×10^{-4}. From results that time evolution of $-\rho_0 \frac{D}{Dt}(\nabla \cdot \mathbf{u})$ is similar to the time evolution of the unsteady vortex ω_z near the leading edge, and the distribution of $-\rho_0 \frac{D}{Dt}(\nabla \cdot \mathbf{u})$ is similar to the distribution of $\rho_0 \nabla \cdot (\boldsymbol{\omega} \times \boldsymbol{u})$, it is considered that our sound source model might reproduce the vortex sound appropriately.

We compared sound pressure levels (SPL) for different acoustic models using our LES database and the published results from computation and measurement in the far-field; see Figure 7. The sound pressure in the far-field for our acoustic model was obtained by Equation (37) and converted to SPL by Equation (38). It is necessary to perform volume integration when determining the sound pressure in our sound source model. And the range of volume integration is determined by reference to the distribution of $\rho_0 \nabla \cdot [(\boldsymbol{u} \cdot \nabla)\boldsymbol{u}]$ in Figure 5 and in consideration of the calculation costs. In Figure 7, the SPL profile calculated from the Lighthill–Curle's equation [18] employing our LES database was computed through following equation

$$p'_a(\boldsymbol{x}, t) = \frac{\rho_0 x_i}{4\pi c_0 |\boldsymbol{x}|^2} \frac{\partial}{\partial t} \int n_j p' \delta_{ij} \left(\boldsymbol{y}, t - \frac{|\boldsymbol{x}|}{c_0} \right) \nabla \mathbf{Y}_i(\boldsymbol{y}) d^3\boldsymbol{y}, \tag{39}$$

where p' is the sound pressure, $\mathbf{x}$ the observation point, $\mathbf{y}$ the sound source point, n_j the component of the outward pointing unit normal of the surface, and c_0 the speed of sound. Finally, the SPL is converted by Equation (38). In order to correspond to the experimental conditions of Miyazawa et al. [31] in the prediction of sound pressure, for the computed SPL, the representative velocity in the main flow direction is set to $U_0 = 30 \times m/s$, the chord length is $C = 0.1m$, and the observation point is $1m$ away from the leading edge in the direction normal to the streamwise direction.

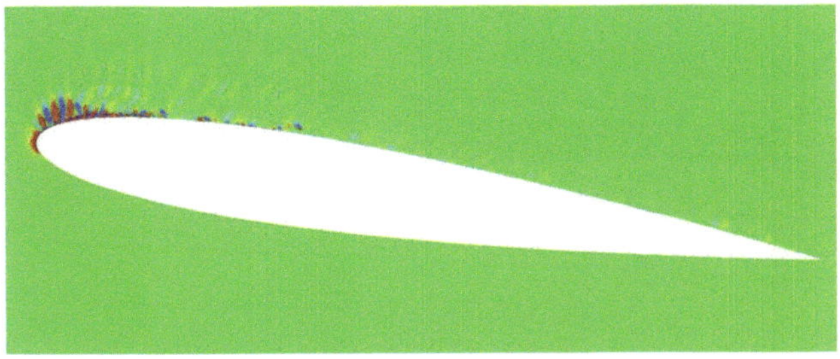

(a)

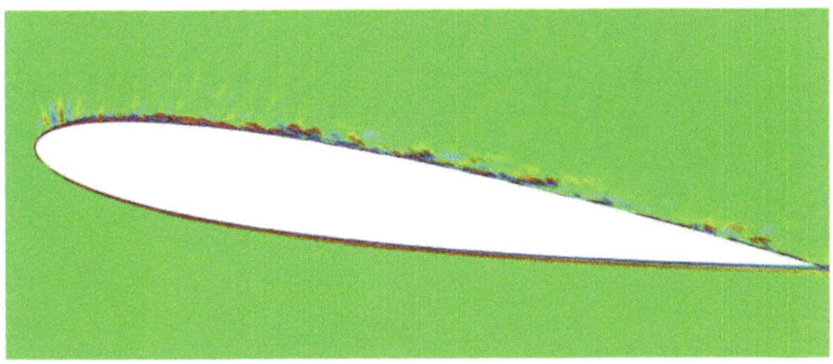

(b)

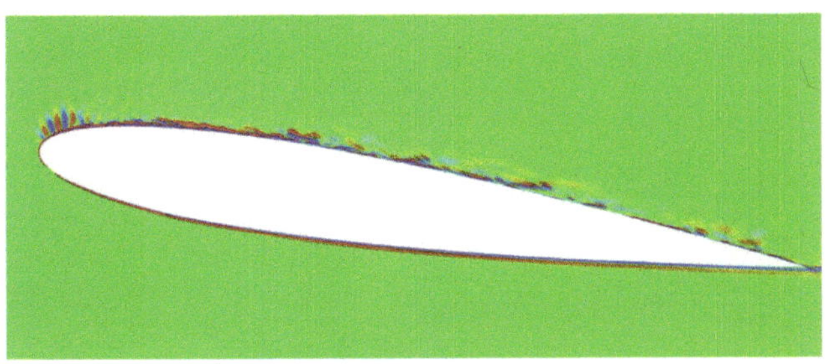

(c)

(d)

Figure 5. Instantaneous, cross-sectional profile of the sound source terms around NACA0012 airfoil at $M = 0.0875$: (**a**) $\nabla \cdot (\nabla \cdot \mathbf{T})$; (**b**) $\nabla \cdot (\boldsymbol{\omega} \times \boldsymbol{u})$; (**c**) $-\rho_0 \frac{D}{Dt}(\nabla \cdot \mathbf{u})$; (**d**) $\rho_0(\nabla \cdot \mathbf{u})^2$.

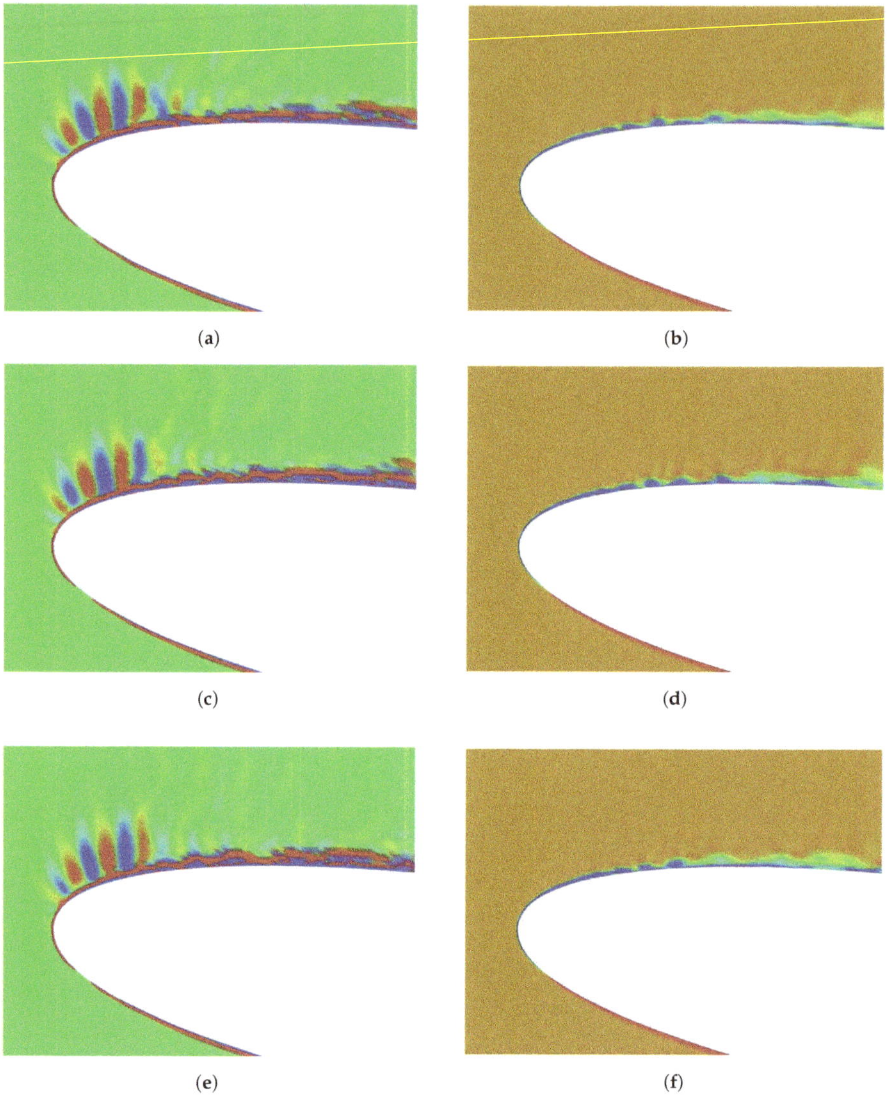

Figure 6. Time evolution of instantaneous and cross-sectional profiles of $-\rho_0 \frac{D}{Dt}(\nabla \cdot \mathbf{u})$ and ω_z near the leading edge at $M = 0.0875$: (**a**,**c**,**e**) denote $-\rho_0 \frac{D}{Dt}(\nabla \cdot \mathbf{u})$ at different moments, where the time interval is 3.5×10^{-2} s; (**b**,**d**,**f**) denote ω_z at different moments corresponding to the times of (**a**,**c**,**e**), respectively.

Overall, the SPL profiles by $-\rho_0 \frac{D}{Dt}(\nabla \cdot \mathbf{u})$ and the Lighthill–Curle's method obtained by our LES do not agree well with the experimental data. However, the SPL profile obtained by the calculation of Miyazawa et al. [31] does not agree with their experimental data either. The reason for the discrepancies between the experimental data and the results obtained by the numerical calculations may be lack of resolution in the simulation of the flow field, or the compact assumption that the sound source area is regarded as a point source in sound pressure prediction. Thus to verify our sound source model, we focused on comparing the SPL profile calculated from the Lighthill–Curle's method which is widely used for sound pressure prediction, and the SPL profile obtained by our sound source model $-\rho_0 \frac{D}{Dt}(\nabla \cdot \mathbf{u})$. Both SPL profiles were in reasonable agreement, especially in high-frequency regions. On the other hand, in both cases, the peak value existed at about 4300 Hz. This frequency is in good agreement with the period of the moving of positive and negative patterns near the leading edge in the suction side confirmed in Figure 5c. The computational cost to use $-\rho_0 \frac{D}{Dt}(\nabla \cdot \mathbf{u})$ as a sound source increases approximately 40% against the Lighthill–Curle analogy, due to the volume integral required in our method, while the Lighthill–Curle formulation needs only surface integral. But our method enables the understanding of the relationship between the behavior of the sound source and the sound generation. Moreover, our method might be applicable to higher Mach numbers.

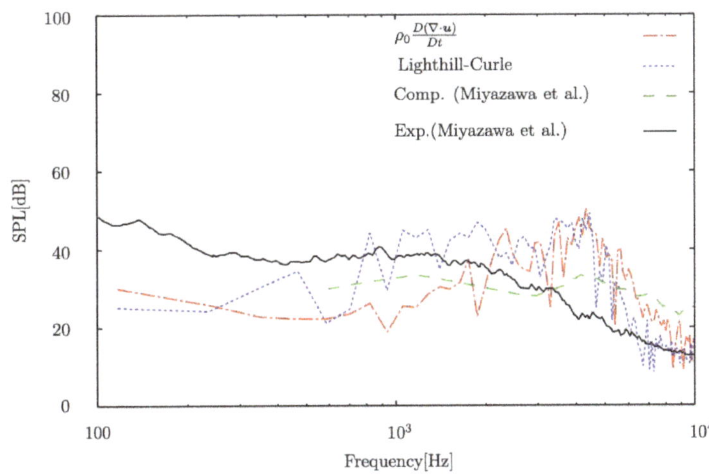

Figure 7. Sound pressure level at $M = 0.0875$.

4.2. Comparison of Different Mach Numbers

In this subsection, using the same computational parameters such as Reynolds number $Re = 2 \times 10^5$ and the angle of attack $\alpha = 9°$ in Section 2, we conducted a LES of the turbulent flow field around the NACA0012 airfoil at Mach number $M = 0.15$. In order to verify if our acoustic method could reproduce the behavior of sound source due to the variation of Mach number, and, investigate the influence of the increase of Mach number on the acoustic field, the results of the sound field in case of $M = 0.15$ are compared with that in case of $M = 0.0875$. Figure 8 shows the instantaneous distribution of $\nabla \cdot \boldsymbol{u}$ around NACA0012 airfoil colored by $|\nabla \cdot \boldsymbol{u}| \leq 0.1$ at Mach numbers 0.0875 and 0.15. In the case of $M = 0.15$, the remarkable distribution of $\nabla \cdot \boldsymbol{u}$ was observed near the leading edge in the suction side, while a relatively small distribution was shown in case of $M = 0.0875$. From this observation, the compressibility effect due to the increase of Mach number was represented by $\nabla \cdot \boldsymbol{u}$, and thus the change of the behavior of sound source was reproduced.

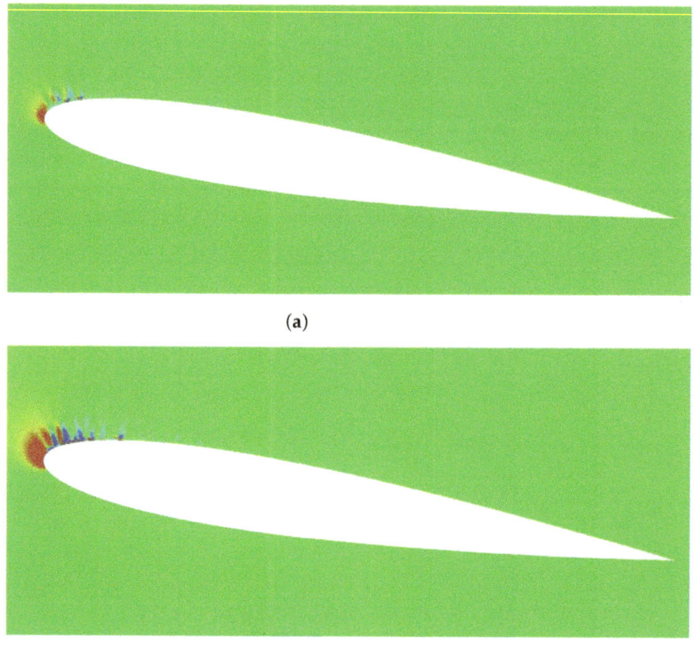

Figure 8. Instantaneous, cross-sectional profiles of $\nabla \cdot \boldsymbol{u}$ around NACA0012 airfoil at different Mach numbers: (a) $M = 0.0875$; (b) $M = 0.15$.

Figure 9 shows the instantaneous distribution of $-\rho_0 \frac{D}{Dt}(\nabla \cdot \mathbf{u})$ near the leading edge at Mach numbers $M = 0.0875$ and 0.15. The noticeable difference of the distribution for $-\rho_0 \frac{D}{Dt}(\nabla \cdot \mathbf{u})$ between Figure 9a,b was not observed in the immediate vicinity of the leading edge. However, in comparison with the case of $M = 0.0875$, the clear patterns of $-\rho_0 \frac{D}{Dt}(\nabla \cdot \mathbf{u})$ were observed in the circle region in case of $M = 0.15$. From this observation, our acoustic theory was proven to be able to reproduce the influence of Mach number on the sound field.

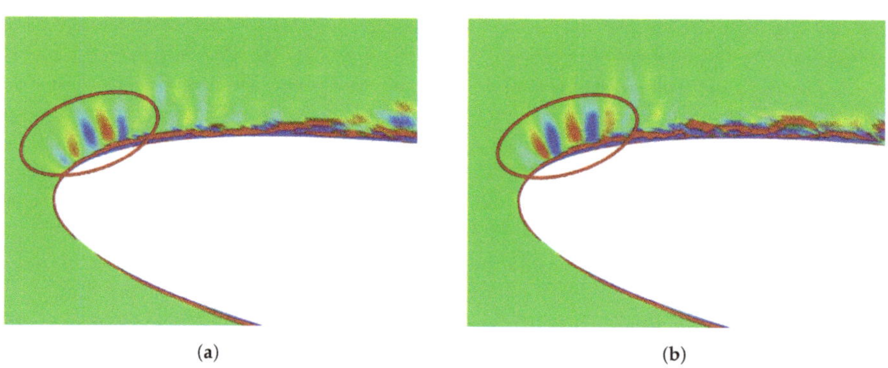

Figure 9. Instantaneous, cross-sectional profiles of $-\rho_0 \frac{D}{Dt}(\nabla \cdot \mathbf{u})$ near the leading edge: (a) $M = 0.0875$; (b) $M = 0.15$.

Figure 10 compares the SPL at Mach number $M = 0.15$ and 0.0875 measured at point 1 m from the leading edge in the direction normal to the streamwise direction. Hutcheson et al. [26] measured the aerodynamic noise generated from the flow field around the NACA0015 airfoil under the condition of the angle of attack $\alpha = 10°$, and the Mach number at $M = 0.09$, 0.11, and 0.127. Their investigation shows that the position of the peak frequency of SPL tends to move from the high-frequency region to the low-frequency region as the Mach number increases. In Figure 10, the maximum peak value obtained by $M = 0.15$ is observed in a low frequency region, as it was against the profile of SPL obtained by $M = 0.0875$. In other words, the tendency of SPL due to the increase of Mach number is reproduced by our sound source model.

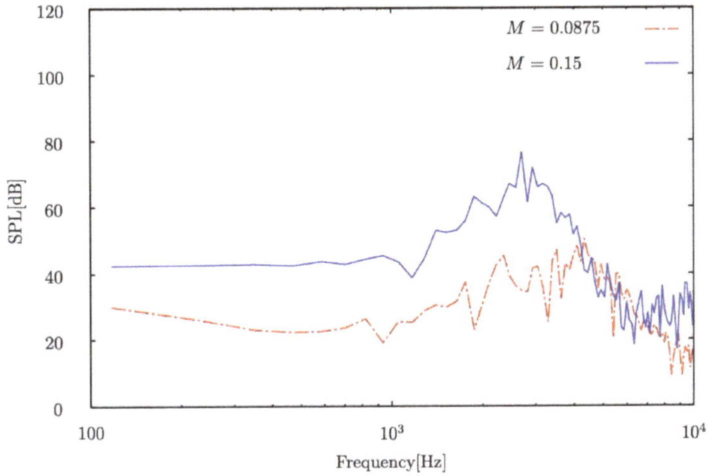

Figure 10. Sound pressure level obtained by using $\rho_0\nabla \cdot [(\boldsymbol{u} \cdot \nabla)\boldsymbol{u}]$ at $M = 0.0875$ and 0.15.

5. Conclusions

Aerodynamic noise from wind turbine blades is one of the major hindrances for the widespread use of large-scale wind turbines to generated green energy. Generally, the weakly compressible turbulent flows around large-scale wind turbine blades that induce the aerodynamic noise, are typical, high Reynolds number flows at low Mach numbers. In order to accurately guide wind turbine blade manufacturers to optimize the blade geometry for aerodynamic noise reduction, an acoustic model that not only understands the relationship between the behavior of sound source and the sound generation but accounts for the compressibility effect, is required. To that end, in this study, a new acoustic theory was proposed, in which we rearranged the continuity and Navier–Stokes equations as a wave equation with a lump of source terms, including the material derivative and square of the velocity divergence. These source terms were used for sound source detection and the estimation of the far-field sound. Our acoustic model was applied to low Mach number, weakly compressible turbulent flows around NACA0012 airfoil. For the computation of flow fields and considering the weak compressibility in flow fields, a LES with the dynamic Smagorinsky subgrid-scale model [28,29] and the CIP-combined unified numerical procedure method [30] was conducted. The reproduced turbulent flow around NACA0012 airfoil was in good agreement with the experimental data. For the estimation of acoustic fields, different acoustic models were performed using our LES database. The distribution of the sources obtained from our acoustic model was compared with the classical sound source models, such as Lighthill [11] and Powell [19] in the case of very low fluctuation of density. The investigation suggested that the derived material derivative of the velocity divergence plays a dominant role as a sound source, and the distribution of our derived source model is consistent with that of the classical sound models. The sound pressure level predicted based on the above-mentioned LES and our newly derived acoustic model was compared with the SPLs calculated from the Lighthill–Curle's equation [18] employing our LES database and the experimental data by Miyazawa et al. [31]. The results showed that the SPL from our acoustic model was in reasonable agreement with the experimental data. The influence of the increase of Mach number on the acoustic field was investigated. From the observation, our acoustic source model was verified to be capable of treating the influence of Mach numbers on the acoustic field. As a result, noise prediction of the large-scale wind turbine blade using our acoustic source model is more accurate at a low Mach number, and further, can more accurately guide wind turbine blade manufactures to optimize the blade geometry for aerodynamic noise reduction. At this stage, further validation of sound rated by comparisons with experimental data is necessary. But we believe our proposal contributes to the development of computational aeroacoustics for applications in low Mach number turbulent flows, such as large-scale wind turbine blades.

Author Contributions: H.T. and Y.L. conceived the original ideal; H.T. wrote and edited the manuscript; and Y.L. and X.L. supervised the study.

Funding: This research was funded by a project of the Chinese National Natural Science Foundation (grant number: 51575220), a project of the Key Scientific and Technological Project of Jilin Province (grant numbers: 20160519008JH and 20170204073GX), and a project of National Key R&D Program of China (grant number: 2016YFB0101402).

Acknowledgments: The authors thank Takeo Kajishima for his help with the numerical method proposed and simulations conducted.

Conflicts of Interest: The authors declare no conflict of interest.

Abbreviations

The following abbreviations are used in this manuscript:

CFD	computational Fluid Dynamics
CIP	Cubic interpolated pseudo particle
DNS	Direct numerical simulation
DSM	Dynamic Smagorinsky model
EU	European Union
LES	Large eddy simulation
RANS	Reynolds averaged numerical simulation
SGS	Subgrid-scales

References

1. Bai, W.; Lee, D.; Lee, K. Stochastic dynamic AC optimal power flow based on a multivariate short-term wind power scenario forecasting model. *Energies* **2017**, *10*, 2138. [CrossRef]
2. Ageborg Morsing, J.; Smith, M.; Ogren, M.; Thorsson, P.; Pedersen, E.; Forssen, J.; Persson Waye, K. Wind turbine noise and sleep: Pilot studies on the influence of noise characteristics. *Int. J. Environ. Res. Public Health* **2018**, *15*, 2573. [CrossRef] [PubMed]
3. Zhu, W.J.; Shen, W.Z.; Barlas, E.; Bertagnolio, F.; Sorensen, J.N. Wind turbine noise generation and propagation modeling at DTU Wind Energy: A review. *Renew. Sustain. Energy Rev.* **2018**, *88*, 133–150. [CrossRef]
4. World Health Organization. *Burden of Disease from Environmental Noise: Quantification of Healthy Life Years Lost in Europe*; World Health Organization, WHO Regional Office for Europe: Copenhagen, Denmark, 2017.
5. Onakpoya, I.J.; OSullivan, J.; Thompson, M.J.; Heneghan, C.J. The effect of wind turbine noise on sleep and quality of life: A systematic review and meta-analysis of observational studies. *Environ. Int.* **2015**, *82*, 1–9. [CrossRef]
6. Wagner, S.; Bareiss, R.; Guidati, G. *Wind Turbine Noise*; Springer Science & Business Media: New York, NY, USA, 1996.
7. Ghasemian, M.; Ashrafi, Z.N.; Sedaghat, A. A review on computational fluid dynamic simulation techniques for Darrieus vertical axis wind turbines. *Energy Convers. Manag.* **2017**, *149*, 87–100. [CrossRef]
8. Li, Q.A.; Maeda, T.; Kamada, Y.; Murata, J.; Kawabata, T.; Shimizu, K.; Ogasawara, T.; Nakai, A.; Kasuya, T. Wind tunnel and numerical study of a straight-bladed Vertical Axis Wind Turbine in three-dimensional analysis (Part II: For predicting flow field and performance). *Energy* **2016**, *104*, 295–307. [CrossRef]

9. Chen, Y.; Lian, Y. Numerical investigation of vortex dynamics in an H-rotor vertical axis wind turbine. *Eng. Appl. Comput. Fluid Mech.* **2015**, *9*, 21–32. [CrossRef]
10. Oerlemans, S. Reduction of wind turbine noise using blade trailing edge devices. In Proceedings of the 22nd AIAA/CEAS Aeroacoustics Conference, Lyon, France, 30 May–1 June 2016.
11. Lighthill, M.J. On sound generated aerodynamically I. General theory. *Proc. R. Soc. Lond. Ser. A* **1952**, *211*, 564–587.
12. Lighthill, M.J. On sound generated aerodynamically II. Turbulence as a source of sound. *Proc. R. Soc. Lond. Ser. A* **1954**, *222*, 1–32.
13. Sandberg, R.D.; Jones, L.E. Direct numerical simulations of low Reynolds number flow over airfoils with trailing-edge serrations. *J. Sound Vib.* **2011**, *330*, 3818–3831. [CrossRef]
14. Wolf, W.R.; Lele, S.K. Trailing-edge noise predictions using compressible large-eddy simulation and acoustic analogy. *AIAA J.* **2012**, *50*, 2423–2434. [CrossRef]
15. Sandberg, R.D.; Sandham, N.D. Direct numerical simulation of turbulent flow past a trailing edge and the associated noise generation. *J. Fluid Mech.* **2008**, *596*, 353–385. [CrossRef]
16. Bogey, C.; Bailly, C.; Juve, D. Computation of flow noise using source terms in Linearized Euler's equations. *AIAA J.* **2002**, *40*, 235–243. [CrossRef]
17. Jawahar, H.K.; Lin, Y.; Savill, M. Large eddy simulation of airfoil self-noise using OpenFOAM. *Aircraft Eng. Aerosp. Technol.* **2018**, *90*, 126–133.
18. Curle, N. The influence of solid boundaries upon aerodynamic sound. *Proc. R. Soc. Lond. Ser. A* **1955**, *231*, 505–514.
19. Powell, A. Theory of vortex sound. *J. Acoust. Soc. Am.* **1964**, *36*, 177–195 [CrossRef]
20. Amiet, R.K. Noise due to turbulent flow past a trailing edge. *J. Sound Vib.* **1976**, *47*, 387–393. [CrossRef]
21. Wang, M.; Freund, J.B.; Lele, S.K. Computational prediction of flow-generated sound. *Annu. Rev. Fluid Mech.* **2006**, *38*, 483–512. [CrossRef]
22. Howe, M.S. *Theory of Vortex Sound*; Cambridge University Press: Cambridge, UK, 2003; Volume 33.
23. Möhring, W. On vortex sound at low Mach number. *J. Fluid Mech.* **1978**, *85*, 685–691. [CrossRef]
24. Takaishi, T.; Ikeda, M.; Kato, C. Method of evaluating dipole sound source in a finite computational domain. *J. Acoust. Soc. Am.* **2004**, *116*, 1427–1435. [CrossRef]
25. Ewert, R.; Schroder, W. On the simulation of trailing edge noise with a hybrid LES/APE method. *J. Sound Vib.* **2004**, *270*, 509–524. [CrossRef]
26. Hutcheson, F.V.; Brooks, T.F.; Stead, D.J. Measurement of the noise resulting from the interaction of turbulence with a lifting surface. *Int. J. Aeroacoust.* **2012**, *11*, 675–700. [CrossRef]
27. Hutcheson, F.V.; Brooks, T.F. Noise radiation from single and multiple rod configurations. *Int. J. Aeroacoust.* **2012**, *11*, 291–333. [CrossRef]
28. Germano, M.; Piomelli, U.; Moin, P.; Cabot, W.H. A dynamic subgrid-scale eddy viscosity model. *Phys. Fluids A* **1991**, *3*, 1760–1765. [CrossRef]
29. Lilly, D.K. A proposed modification of the Germano subgrid-scale closure method. *Phys. Fluids A* **1992**, *4*, 633–635. [CrossRef]
30. Yabe, T.; Wang, P.Y. Unified numerical procedure for compressible and incompressible fluid. *J. Phys. Soc. Jpn.* **1991**, *60*, 2105–2108. [CrossRef]
31. Miyazawa, M. Large eddy simulation of flow around an isolated aerofoil and noise prediction. In Proceedings of the 5th JSME-KSME Joint Fluids Engineering Conference, Nagoya, Japan, 14–18 July 2002; pp. 546–551.
32. Okita, K.; Kajishima, T. Numerical simulation of unsteady cavitating flows around a hydrofoil. *Trans. Jpn. Soc. Mech. Eng. Ser. B* **2002**, *68*, 637–644. [CrossRef]
33. Kajishima, T.; Ohta, T.; Okazaki, K.; Miyake, Y. High-order finite-difference method for incompressible flows using collocated grid system. *JSME Int. J. Ser. B* **1998**, *41*, 830–839. [CrossRef]
34. Han, C.; Kajishima, T. Large eddy simulation of weakly compressible turbulent flows around an airfoil. *J. Fluid Sci. Technol.* **2014**, *9*, JFST0063. [CrossRef]
35. Tang, H.; Lei, Y.; Li, X.; Fu, Y. Large-Eddy Simulation of an Asymmetric Plane Diffuser: Comparison of Different Subgrid Scale Models. *Symmetry* **2019**, *11*, 1337. [CrossRef]
36. Tang, H.; Lei, Y.; Fu, Y. Noise Reduction Mechanisms of an Airfoil with Trailing Edge Serrations at Low Mach Number. *Appl. Sci.* **2019**, *9*, 3784. [CrossRef]

37. Tang, H.; Lei, Y.; Li, X.; Fu, Y. Numerical Investigation of the Aerodynamic Characteristics and Attitude Stability of a Bio-Inspired Corrugated Airfoil for MAV or UAV Applications. *Energies* **2019**, *12*, 4021. [CrossRef]
38. Tamura, A.; Kikuchi, K.; Takahashi, T. Residual cutting method for elliptic boundary value problems. *J. Comput. Phys.* **1997**, *137*, 247–264. [CrossRef]

energies

Article

Aerodynamic Investigation of a Horizontal Axis Wind Turbine with Split Winglet Using Computational Fluid Dynamics

Miguel Sumait Sy [1], Binoe Eugenio Abuan [2] and Louis Angelo Macapili Danao [2,*]

1 Energy Engineering Graduate Program, University of the Philippines, Diliman 1101, Philippines; emikesy@gmail.com
2 Department of Mechanical Engineering, University of the Philippines, Diliman 1101, Philippines; beabuan@up.edu.ph
* Correspondence: louisdanao@up.edu.ph

Received: 19 August 2020; Accepted: 17 September 2020; Published: 22 September 2020

Abstract: Wind energy is one of the fastest growing renewable energy sources, and the most developed energy extraction device that harnesses this energy is the Horizontal Axis Wind Turbine (HAWT). Increasing the efficiency of HAWTs is one important topic in current research with multiple aspects to look at such as blade design and rotor array optimization. This study looked at the effect of wingtip devices, a split winglet, in particular, to reduce the drag induced by the wind vortices at the blade tip, hence increasing performance. Split winglet implementation was done using computational fluid dynamics (CFD) on the National Renewable Energy Lab (NREL) Phase VI sequence H. In total, there are four (4) blade configurations that are simulated, the base NREL Phase VI sequence H blade, an extended version of the previous blade to equalize length of the blades, the base blade with a winglet and the base blade with split winglet. Results at wind speeds of 7 m/s to 15 m/s show that adding a winglet increased the power generation, on an average, by 1.23%, whereas adding a split winglet increased it by 2.53% in comparison to the extended blade. The study also shows that the increase is achieved by reducing the drag at the blade tip and because of the fact that the winglet and split winglet generating lift themselves. This, however, comes at a cost, i.e., an increase in thrust of 0.83% and 2.05% for the blades with winglet and split winglet, respectively, in comparison to the extended blade.

Keywords: winglet; computational fluid dynamics (CFD), wind energy; renewable energy; rotor blade; tip vortices

1. Introduction

Wind energy is one of the fastest growing renewable energy sources. Data from 2009–2019 indicates that global installed capacity increased by 409%, i.e., to a total of 651 GW in installed capacity as of 2019. Average yearly growth for the same time period is around 16.84% [1]. Data for the Philippines from 2005–2018 shows that as of 2018 the Philippines had 426.9 MW of installed capacity for wind energy. Majority of the additions for the Philippines wind energy capacity came from 2014–2015 with the National Renewable Energy Plan foreseeing a total value of 2378 MW by 2030 [2,3]. Wind energy resource assessments showed that there are around 11,055 km^2 of land area that are rated good to excellent for wind energy use. Using conservative assumptions of about 7 MW/km^2, these areas could support more than 76,000 MW of potential installed capacity, delivering more than 195 billion kWh per year [4].

Wind energy can be extracted from the air through the use of wind turbines. Wind turbines work by converting the kinetic energy in the wind as it flows into mechanical energy through the rotor, which is connected to the generator to produce electricity. A lot of effort has been put to improve the performance of wind turbines through modifying the turbine blade profile to increase the turbine's efficiency. Instead of changing the blade profile, this study will modify the tip of the turbine blade by adding a structure called the split winglet, in an attempt of increasing the efficiency by improving the aerodynamics of the blade tip.

Winglets are structures attached to the tip of the wing that counter the drag created by wingtip vortices. These vortices form when the high-pressure air from the pressure side of the wing flows spanwise around the wingtip, moving toward the suction side of the wing. This flow creates a vortex rotating upwards and inwards toward the root of the wing as seen on Figure 1. These vortices increase the drag by changing the wings' lift backward, converting the usable lift force into drag. Winglets are small wing-shaped structures that generate lift perpendicular to the relative wind. When used in airplanes and similar applications, winglets reduce the induced drag at the wing tip. Winglets modify the direction of the relative wind at the wingtip by adding a component of wind that flows toward the root of the wing, resulting in a lift vector that has a component pointing forward. This forward lift counters the drag produced by the vortices at the tip of the winglet [5].

Figure 1. Wingtip vortices with and without winglets. Image source: Olivier Cleynen, Wikimedia Foundation Inc.—Wingtip Device, 2020 [6].

Split winglets, as can be seen in Figure 2, are simply a combination of a blended winglet and a wingtip fence. The lower winglet works in the same way as the upper winglet. In this study, the blade tip modification will be tested with both the single winglet and the split winglet and a comparison of the rotor performance will be presented.

Figure 2. The advance technology (AT) split winglet used for the development of the Boeing 737 MAX wings. Image source: Oleg V. Belyakov, Wikimedia Foundation Inc.—Wingtip Device, 2020 [6].

Whitcomb, considered to be the father of winglet design, started his experiments on winglets in 1974 and later published his report in 1976. He found that the strength of wing tip vortices is reduced when a near vertical wing-like surface is attached to the wingtip [7]. The application of such a wingtip device was still limited to aircrafts and has not been adopted in wind turbines until 1996 when Hasegawa et al. performed numerical simulations on wind turbines with winglets, using a vortex lattice method with a free wake model [8]. Their calculations have shown a higher rotor power coefficient and a lower flap bending moment compared to a longer wing case. Johansen conducted three related studies regarding the application of winglets on wind turbines. The first one looked at the effects of winglets using computational fluid dynamics bending in upstream and downstream directions and of varying twist angle. (The upstream direction means that the winglet is facing against the flow of the wind.) Results showed that winglets increased power by around 1.3% for winds greater than 6 m/s but at the same time also increased thrust by around 1.6%. It also showed that upwind facing winglets performed better than the downwind facing counterpart [9]. The second study looked at the effects of winglets by changing four parameters: winglet height, curvature radius, sweep and twist. The results showed that twist had a very little effect, a larger curvature radius decreased power gains, a sweep angle of 30° also decreased power gains and that winglet height influenced the performance of winglets the most wherein taller was better, although the author pointed out the challenge in the viability of such design due to the loads on the winglets [10]. The last study looked at both the theoretical considerations and computational results on the nature of using winglets on wind turbines. The results showed that the increase in power that can be obtained with winglets was due to a reduction in tip-effects, and was not due to the shift in downwind vorticity due to downwind winglets. The numerical results, on the other hand, showed that downwind facing winglets performed better than upwind facing ones and that the power increase from winglets were smaller than simply extending the rotor blades, although shorter winglets <2% of the rotor radius came close to it [11]. Elfarra et al. found that the k-ε Launder–Sharma turbulence model predicted power more accurately compared to Shear Stress Transport (SST) k-ω. Optimizations done showed that a cant angle of 84° and a twist of 2° for the winglets is the best shape for maximum torque. Results showed that the optimized winglets provided a boost of 9% in power and only a 1.3% increase in axial thrust [12].

A similar study by Vasjaliya et al. used CFD for qualitative comparisons between turbine with and without winglets and wind tunnel testing on a scaled model for quantitative results. The qualitative comparison showed the vortices at the wing tips was split into two smaller vortices by the winglet, one at the wingtip and the other slightly away, in comparison to one large vortex of at the wing tip by an ordinary rotor blade. Quantitative results showed that on average the model with winglet generated 1.57% more power than the one without [13].

In more recent years, though numerous studies have been carried out on the effects of winglets on wind turbine performance, very few tackle the problem of split winglets. In 2016, Nedyalkov et al. studied the effects of plain wing tip, generic winglet, and split winglet in tidal turbine blades [14]. They conducted experimental and numerical tests on a model turbine and found that the split winglets have notably decreased the vortices formed at the tip. However, no significant increase in performance was observed. Pratilastiaro et al. carried out experimental tests on a wind tunnel scale wind turbine with and without winglets [15]. Their study contradicts the current literature on the performance improvement effects of winglets. Their results showed a significant decrease in power coefficient when winglets were used. This effect was further studied by the same group when they conducted a numerical study on the same turbine and found that the blade with the split winglet produced a torque lower than that produced by the plain blade [16]. They concluded that, in their tests, the device did not give a better performance.

It can be seen from the literature that in some studies the addition of the single winglet increases the power extracted in the wind turbines, but very few studies presented the effects of split winglets. In theory, the lower winglet works the same way as a single winglet, which can lead one to conclude that it can also affect the performance of the turbine in a positive way. This is the effect that this study wants to investigate—to compare the effect of a split winglet in a HAWT to that of a HAWT with no winglet and a HAWT with a single winglet through CFD. However, there is no consensus on the effects of split winglet in the performance of wind turbines. This is a strong justification that further studies on split winglets as applied to wind turbines must be carried out to provide significant evidence of its effects, positive or negative, on the performance of these energy conversion devices.

2. Methods

To conduct a CFD simulation to investigate the performance of a HAWT, a numerical mesh must be made first, with the wind turbine as the focus of the model. The rotor blade used in this study is based on the National Renewable Energy Laboratory Unsteady Aerodynamics Experiment Phase VI. NREL Phase VI involves large scale experiments conducted at the NASA Ames wind tunnel facilities. Among the series of tests and sequences, the blade from sequence H was selected as the baseline of this study [17]. The experimental data was acquired from NREL through Mr. Lee Fingersh (personal communication, 19 September 2018) and used for validation and comparison purposes. The wind turbine is an upwind, 2-bladed HAWT and the blades are tapered and twisted. The wind turbine utilizes the NREL S809 airfoil section. Specifications can be seen in Table 1.

Table 1. NREL Phase VI Sequence H blade description. Data source: NREL blade design, 2020 [12].

Number of Blades	2
Rotor diameter	10.06 m
RPM	71.63
Cone angle	0 degrees
Rotor location	Upwind
Power regulation	Stall regulated
Blade tip pitch angle	3 degrees (down)
Blade profile	S809
Blade chord length	0.358–0.728 m (linearly tapered)
Twist angle	Non-linear twist along the span
Blade thickness	t/c = 21% throughout the span

In this study, the blade was modeled using a computer aided design software by creating each airfoil section, and then using the loft command to create the body of the 3D model. Four different blade models were made in total—the base blade, an extended blade, the blade with winglet and blade with split winglet as presented in Figure 3. Since the winglets add an additional 1.5% to the length of the blade, an extended blade was also modeled to account for the length difference. The blade models were then imported to Ansys Workbench (ANSYS Inc., Canonsburg, PA, USA) to produce the mesh that will be used for simulation. The imported blade model is then used as a reference to create the rotating domain, which was then further enclosed by a stationary domain in ANSYS Design Modeller (ANSYS Inc., Canonsburg, PA, USA). The rotating domain has a radius of 2 times and a depth of 2.5 times the blade length, while the stationary domain has a radius of 5 times and a depth 9 times the blade length.

In order to simplify the CFD analysis and to save computational resources, the domain was sliced in half and the core where the rotor hub is located was then hollowed out to create a 180° wedge model, using periodic boundary conditions on both ends of the domain. The model was then meshed using ANYS Workbench Meshing. The model was meshed using the curvature size function with relevance center set to fine. Minimum size was set to 2 mm while the curvature normal angle was set to 2.5 degrees. Growth rate was then set to 1.4. An inflation layer was also used. It contains five layers with a growth rate of 1.1 and a first layer thickness of 2.5 mm. The first layer thickness was calculated to have a y+ that stays within the 30–300 range. In addition to this setting, a few body-of-influence size functions were used to ensure that areas in the leading and trailing edge of the blade have a fine enough mesh. Match control was then used on the periodic surfaces. The resulting outer mesh is shown on Figure 4a.

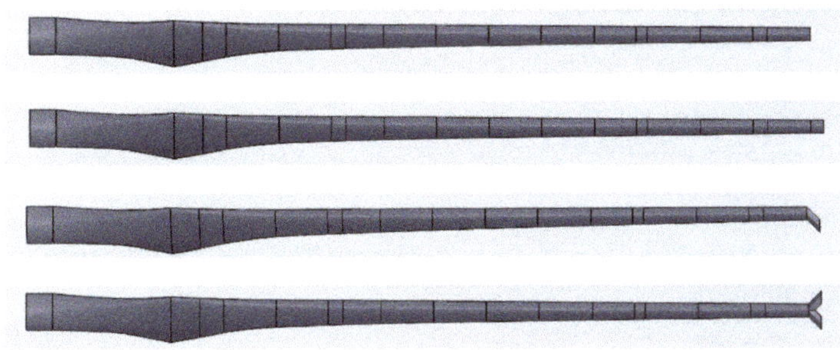

Figure 3. Top to bottom: NREL Phase VI Seq. H blade, NREL Phase VI Seq. H blade with 1.5% extension, extension with 45° offset to suction side to create a winglet and extension with 45° offset to both the suction and pressure side to create a split winglet.

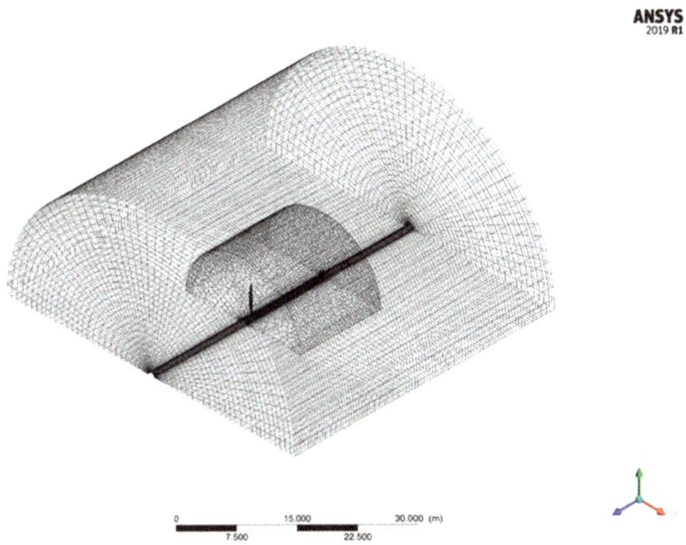

(a)

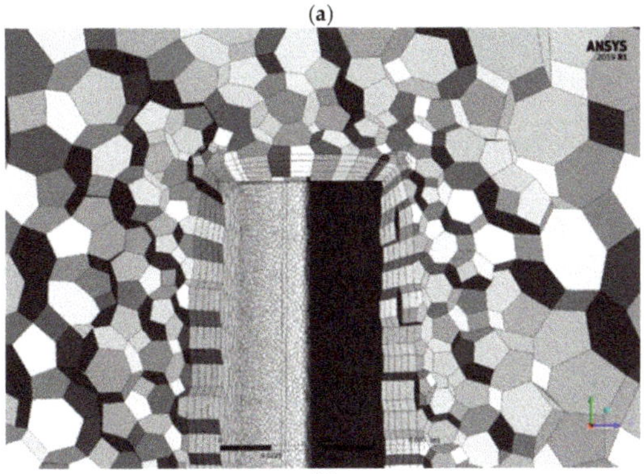

(b)

Figure 4. Mesh generated for the computational domain. (**a**) The entire domain and (**b**) a close up of the mesh near the blade tip after converting to polyhedra on Ansys Fluent (ANSYS Inc., Canonsburg, PA, USA).

The converted mesh has 1.35 million cells with 6.14 million faces and 3.81 million nodes with a minimum orthogonal quality of 0.166 and a maximum aspect ratio of 39.82. The simulation uses a pressure-based steady-state Reynolds-averaged Navier-Stokes (RANS) solver. A steady-state solver was used to reduce computation time as early tests made using a transient solver did not appreciably differ from the steady-state solver. Both the stationary and rotational domain were set as fluid domains with standard air as fluid material. Turbulence model was set to Standard K-epsilon using an enhanced wall function after sample runs against realizable K-epsilon, Standard K-omega and K-omega SST have shown that Standard K-epsilon predicted power coefficient (CP) are the closest to the experimental data. Boundary conditions include non-slip walls for the surface of the rotor blade, slip walls for the inner and outer walls, periodic for the bottom surfaces, velocity inlet for the tunnel entrance and pressure outlet for the tunnel exit. Turbulence parameters was set to intensity and length scale with a value of 0.5% and 0.358 m. Mesh interface between the two domains is then set to periodic repeat. In the solution section, coupling scheme is set to "coupled" while discretizations were all set to second order to improve the accuracy of the solution. Solution controls are kept on their default values except for higher wind speeds where the courant was set 50 to aid convergence. A report definition to monitor the moment on the rotor blade along the axis of rotation was created. Simulations are deemed to be converged when residual values fell below 10^{-4}.

3. Results and Discussions

The following are the results from the NREL phase VI sequence H experiment and from the simulation using Ansys Fluent of the four types of rotor blade modeled: base rotor blade, extended rotor blade, rotor blade with winglet and rotor blade with split winglet. Each rotor blade was run on seven different wind velocities, 7 m/s, 8 m/s, 9 m/s, 10 m/s, 11 m/s, 13 m/s and 15 m/s, totaling to 35 data points, 28 of which are simulated and 7 are data from the NREL phase VI experiment.

Wind power through a rotor disk (P) can be computed using the air density (ρ), wind velocity (V) and area of the rotor disk (A) as seen in Equation (1). The power coefficient (C_P), a non-dimensional number that represents the fraction of the power in the wind that is extracted by the rotor can then be computed by dividing the output rotor power (P_{out}) to P. Cp, however, has a theoretical maximum called the Betz limit, which is $C_{p,max} = 16/27$ due to the fact that not all the wind flowing through the rotor disk loses its kinetic energy upon contact. Similar to the power, the thrust on a wind turbine can be characterized by a non-dimensional thrust coefficient as seen in Equation (3) [18]. Output rotor power for the simulated blades are obtained by converting moment obtained along the axis of rotation (M), which is equal to torque (τ) that is equal to power divided by the rotors' angular velocity (ω) as seen in Equation (4).

$$P = \frac{1}{2}\rho A V^3 \tag{1}$$

$$C_P = \frac{P_{out}}{\frac{1}{2}\rho A V^3} = \frac{\text{Output rotor power}}{\text{Power in the wind}} \tag{2}$$

$$C_T = \frac{T}{\frac{1}{2}\rho V^2 A} = \frac{\text{Thrust force}}{\text{Dynamic force}} \tag{3}$$

$$M = \tau = \frac{P}{\omega} \tag{4}$$

Figure 5 shows how each of the simulated rotor blades can be compared to each other and how closely they follow the experimental data in terms of power generation. The simulated results closely follow the results from the experimental runs done by NREL on the phase VI sequence H experiment. Data from the simulated results of the base rotor blade show that P_{out} is within ±9% from the experimental results. The experiment values show a flattening of the power output from 9 m/s onwards. This behavior can be attributed to the turbine being a stall-regulated rotor with the performance limited to 10 kW.

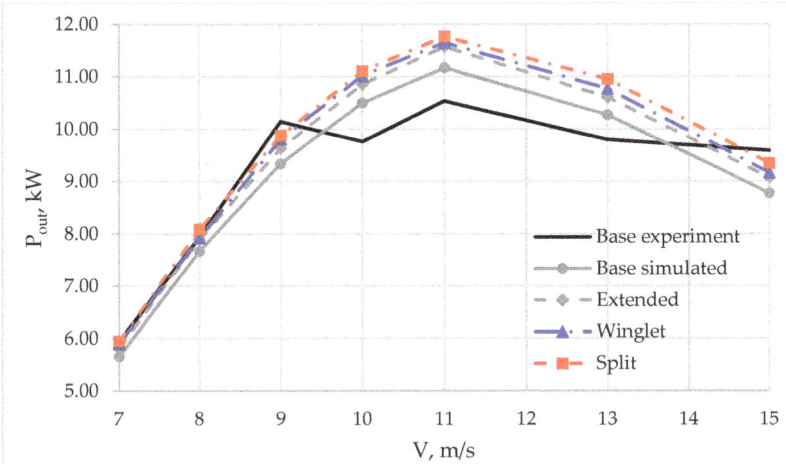

Figure 5. Power generation results from the experimental runs done by NREL and from the simulated runs.

The extended blade was added to eliminate the advantage gained by the blades with winglet and split winglet due to the increase in rotor disk area. Comparisons between the base and extended blades are made to ensure that the results are in line with the base blade. P_{out} increased on an average, across the seven wind speeds that were simulated, by 1.23% with the winglet and by 2.53% with the split winglet when compared with the extended rotor blade. The results show that the split winglet doubled the improvement in power generation brought by the winglet.

Calculating the C_P for the wind turbine shows that the peak occurs on or before the 7 m/s wind velocity mark. It also shows the simulated results following the experimental results closely. Results from the base rotor blade compared to those of the extended one are generally within striking distance of each other with the average being only 0.21% higher and the largest gap being only 0.56% higher. Using the extended rotor blade as the basis for the other comparisons, on an average, the power coefficient of the rotor blade with winglet is 1.22% higher while the one with split winglet is 2.51% higher. The results show that the winglets increased the power production of the wind turbine by increasing its efficiency.

Additional runs were made to create a tip speed ratio sweep at V = 11 m/s. Data shows that the winglet and split winglet decreased the wind turbine's performance at low tip speed ratios, tip speed ratio (tsr) less than 4.4. The impact of this decrease is minimal, however, on the total performance of the wind turbine, since at these tip speed ratios, the blade is generally at stall condition due to friction at the drive train or producing very little power. At higher tip speed ratios, tsr greater than or equal to 4.4, the winglet and split winglet started producing additional power as seen in Figure 6.

Using a plane just behind the trailing edge of the tip of the rotor blade, the vortices formed at the tip can be seen by using line vectors representing the tangential wind velocity of the plane as shown in Figure 7. Comparing the base and extended rotor blade, the results are generally the same for the two with the wind forming a swirling motion at the suction side of the rotor blade. Maximum tangential wind velocity is also practically the same with the only difference within 1–3% of each other. Looking at the winglet, the swirling motion of air, while still present, is not nearly as pronounced as that of the previous two rotor blades nor is it as close to the main body of the rotor blade. The maximum tangential wind velocity for the winglet is also around 10–20% lower than the previous two blades. Lastly, the results from the rotor blade with the split winglet shows similar results to the one with winglet wherein the swirling motion of the air is not as pronounced as the first two blades. The maximum tangential wind velocity for the split winglet is also similar to the one with winglet up to wind velocity of 11m/s after which it starts increasing more. The results show that the winglet and split winglet had quite a dramatic effect on the tangential wind velocity, lowering it by up to around 20%, and visually reducing the swirling motion at the suction side.

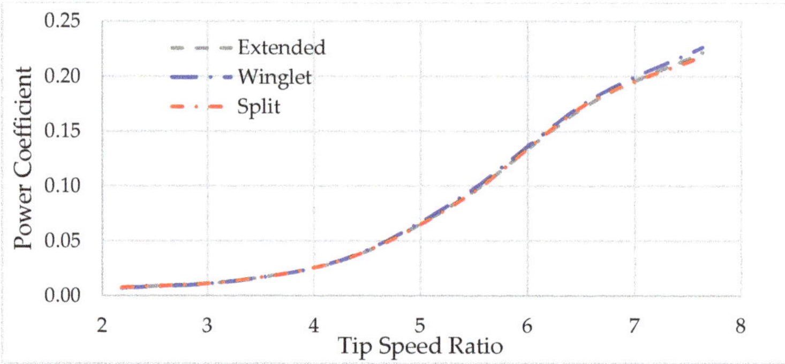

Figure 6. Tip speed ratio sweep at V = 11 m/s.

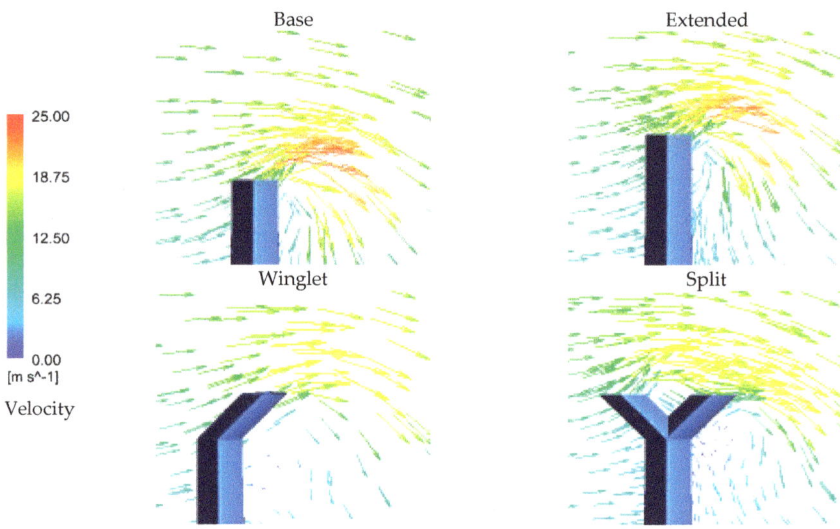

Figure 7. Tangential wind velocity at a plane just behind the trailing edge of the tip of the rotor blade and V = 11 m/s.

To help validate that the winglet and split winglet have indeed reduced the vortices at the blade tip, snapshots of the vortex core region using the swirling strength detection method using a level value of 0.01 was taken for comparison, which can be seen in Figure 8. The effect of the winglet and split winglet to the vortices formed at the blade tip are immediately seen. The vortices formed at the tips of both the base rotor blade and extended rotor blade are remarkably larger at higher wind velocities than the other two rotor blades.

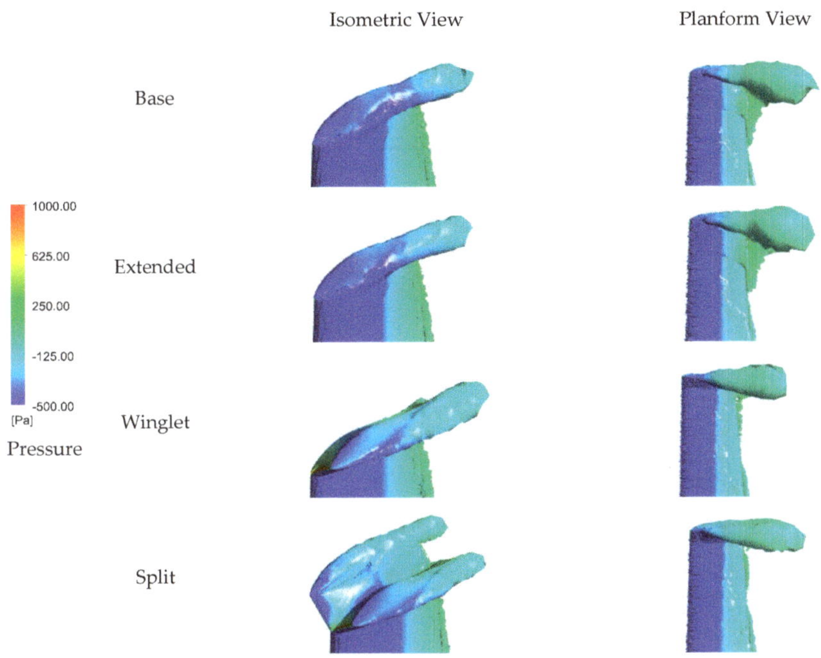

Figure 8. Vortices at the rotor blade tips at V = 11 m/s.

Looking at the pressure contour at the rotor blade tips, Figure 9, the results for the base and extended blades are practically the same. For the blades with winglet and split winglet the low-pressure area at the suction side near the trailing edge is smaller in magnitude and intensity compared to the first two blades. Looking more at the split winglet, for the part that bends toward the pressure side, the high-pressure area at the leading edge shows greater intensity and magnitude compared to the other blades. Further, the low-pressure area in the pressure side is almost non-existent, and as for the part that bends toward the suction side, the pressure contour looks very similar to the blade with winglet with the exception of much smaller low-pressure area in the trailing edge. The results show that the winglet and split winglet significantly reduce the magnitude of the low-pressure area, caused by the wingtip vortices, at the trailing edge of the rotor blade tips that causes additional drag. In addition, for the blade with split winglet, the pressure difference on the left winglet (winglet closer to the pressure side) appears to be relatively bigger, suggesting good lift generation. The winglet near the suction side showed the same pressure contour as the blade with the single winglet configuration. Surface pressure contours on the rotor blades shows that pressure distribution across the blades are largely similar among the four blades simulated. There are small differences like the high-pressure area in the pressure side varying slightly in magnitude and intensity across the blades but nothing that immediately stands out. Most of the notable differences happened near the tip of the rotor blade.

To see the effect of the winglets, the results near the wingtip for the extended, winglet and split winglet blades were scrutinized. Looking at Figure 10, at the leading edge, all four blade models followed the same pattern—big pressure difference between the pressure and suction side at the leading edge that tapers down. The change, however, occurs at $x/c \leq 0.5$, where the pressure difference increases after that tapers off slightly again creating a hump instead of continuing to taper off. This shows the effect of the vortices at the blade tip as corroborated by Figures 7 and 8. The formation of the vortices creates a low-pressure area, which further reduces the pressure at the suction side resulting to the low-pressure hump at $x/c \leq 0.5$.

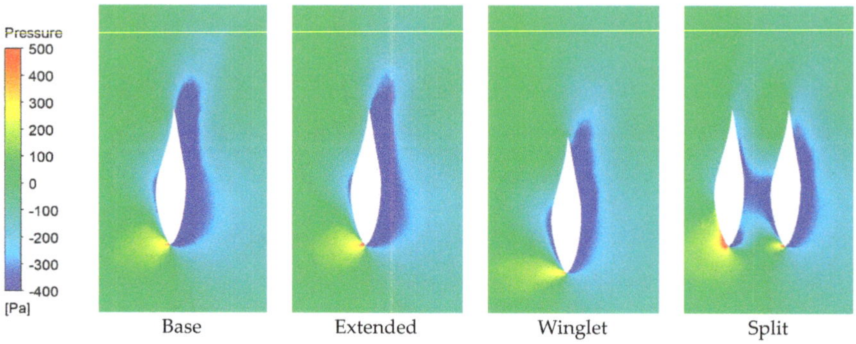

Figure 9. Pressure contour at the rotor blade tip at V = 11 m/s.

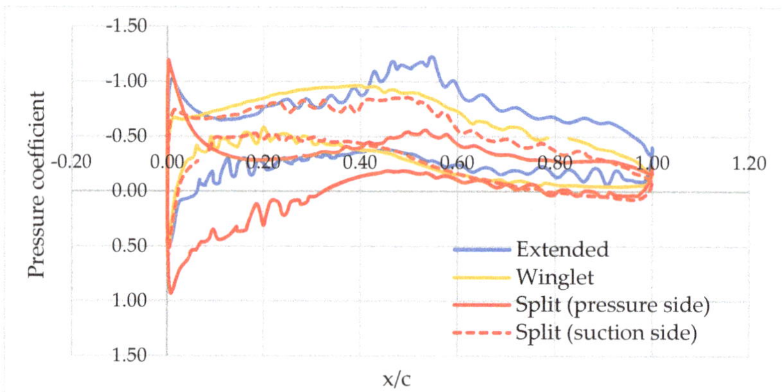

Figure 10. Surface Pressure Coefficient at y = 5.1 m with V = 11 m/s.

Comparing the extended and the blade with winglets shows two main differences. First, the pressure difference at the leading edge for the winglet is smaller than that of the extended. Second, the pressure at the suction side at $x/c \leq 0.5$ for the winglet is not as low as that of the extended blade. This suggests that the winglet does not create additional lift at the wingtip but less drag is experienced at the trailing edge. Compared with the split winglet, the pressure side winglet's pressure profile is quite similar to the results at 95% of the blade span of the base blade, suggesting good lift generation. In addition, the pressure difference at the trailing edge is smaller than even the one with winglet, suggesting reduced drag due to blade tip vortices. For the suction side winglet, the results follow the winglet pressure profile closely with the main difference that the low-pressure area does not go as low as that of the winglet, suggesting less lift generation compared to the one with winglet but also less drag due to blade tip vortices.

While the previous results have been largely positive, it can be seen that one of the side effects of having a winglet was the increase in thrust (T) at the wind turbine blades as presented in Table 2. Using the extended rotor blade for comparison shows that the winglet increased thrust on average by 0.83% while the split winglet increased it on average by 2.05%. This increase in thrust therefore means that the rotor blade should have a more robust structural framework capable of handling the additional loads if a winglet or split winglet is to be used.

Table 2. Relative thrust of the simulated wind turbines.

Wind Speed	Winglet vs. Extended	Split vs. Extended
7	101.24%	102.40%
8	100.71%	102.38%
9	101.33%	102.36%
10	101.16%	102.18%
11	100.91%	101.82%
13	100.62%	101.90%
15	99.88%	101.31%

4. Conclusions

To extract as much potential as possible, there is a need to increase the wind turbines efficiency. One way of increasing the efficiency of wind turbines is by reducing the sources of drag and for wind turbines one such source are the blade tip vortices.

The study aimed to find out if the application of split winglets at the tip of HAWT blades would improve its aerodynamic performance, and the results showed that adding a split winglet to the NREL phase VI sequence H design can increase its power generation. Accounting for the difference in length due to adding a split winglet, power generation on an average increased by 2.53%. In comparison, a winglet that bends toward the suction side improves power generation on an average by 1.23%. Results also show that while winglets reduce performance at low tip speed ratios, the performance benefits at higher tip speed ratios where the majority of the power is generated more than make up for it.

The study also shows that the increase is achieved through two factors. First is that the winglet and split winglet drastically reduced the low-pressure area at the trailing edge of the blade tip that pulls back the blade, creating drag. This reduction in vortex core region was a result of the large drop in the tangential wind velocity going from the pressure to the suction side of the blade tip due to the winglets. Second is that the winglets themselves generate lift as can be seen on the pressure contour plot at the blade tip. The difference in pressure at the blade tip between the pressure and suction side responsible for creating lift is still present on the winglet and in case of the split winglet improved.

The results, however, are not all positive. One negative is that the winglets increased the thrust experienced by the rotor blade. Accounting for the difference in blade span of the base blade and ones with winglet, thrust increased by 0.83% and 2.05% for the blade with winglet and split winglet, respectively. This means that additional structural support is required to help support the additional load brought by the addition of winglet to the blade.

In the case of this study, attaching a split winglet to a wind turbine rotor blade increased power generation by a small but not insignificant amount. The study also showed that the split winglet is better than a winglet in increasing power generation. The opposite is true, on the other hand, in terms of the additional thrust experienced by the blade.

Author Contributions: Investigation, M.S.S.; methodology, M.S.S. and B.E.A.; supervision, L.A.M.D.; writing—original draft, M.S.S.; writing—review & editing, B.E.A. and L.A.M.D. All authors have read and agreed to the published version of the manuscript.

Funding: There is no funding that was availed in the conduct of this study.

Acknowledgments: The authors would like to thank Lee Fingersh of NREL for the experimental data provided in this study.

Conflicts of Interest: The authors declare no conflict of interest.

References

1. Renewable Energy Policy Network for the 21st Century (REN21). *Renewables 2020 Global Status Report*; REN21 Secretariat: Paris, France, 2020.
2. The International Renewable Energy Agency. Trends in Renewable Energy. Available online: https://public.tableau.com/views/IRENARETimeSeries/Charts?:embed=y&:showVizHome=no&publish=yes&:toolbar=no (accessed on 10 September 2020).
3. Federal Ministry for Economic Affairs and Energy (BMWi). *Wind Energy in the Philippines: Potential, Opportunities and Challenges*; Federal Ministry for Economic Affairs and Energy (BMWi): Berlin, Germany, 2015.
4. The International Renewable Energy Agency. *Renewables Readiness Assessment: The Philippines*; The International Renewable Energy Agency: Abu Dhabi, UAE, 2017.
5. Nasa Drysden Research Center Fact Sheets—Winglets. Available online: https://www.nasa.gov/centers/dryden/about/Organizations/Technology/Facts/TF-2004-15-DFRC.html (accessed on 10 September 2020).
6. Wikimedia Foundation Inc. Wingtip Device. Available online: https://en.wikipedia.org/wiki/Wingtip_device (accessed on 22 September 2020).
7. Whitcomb, R. *A Design Approach and Selected Wind-Tunnel Results at High Subsonic Speeds for Wing-Tip Mounted Winglets*; NASA: Washington, DC, USA, 1976.
8. Hasegawa, Y.; Kikuyama, K.; Imamura, H. Numerical analysis of a horizontal axis wind turbine rotor with winglets. *Trans. Jpn. Soc. Mech. Eng. Ser. B* **1996**, *62*, 3088–3094. [CrossRef]
9. Johansen, J.; Sørensen, N.N. *Aerodynamic investigation of Winglets on Wind Turbine Blades Using CFD*; Risø National Laboratory: Rosklide, Denmark, 2006.
10. Johansen, J.; Sørensen, N.N. *Numerical Analysis of Winglets on Wind Turbine Blades Using CFD*; Risø National Laboratory: Rosklide, Denmark, 2007.
11. Johansen, J.; Gaunaa, M. Determination of the maximum aerodynamic efficiency of wind turbine rotors with winglets. *J. Phys. Conf. Ser.* **2007**, *75*. [CrossRef]
12. Elfarra, M.A. Horizontal Axis Wind Turbine Rotor Blade: Winglet and Twist Aerodynamic Design and Optimization Using CFD. Ph.D. Thesis, Middle East Technical University, Ankara, Turkey, 2011.
13. Vasjaliya, N.G. *Fluid-Structure Interaction and Multidisciplinary Design Analysis Optimization of Composite Wind Turbine Blade*; Embry-Riddle Aeronautical University: Daytona Beach, FL, USA, 2013.
14. Nedyalkov, I.; Gagnon, J.; Shull, J.; Brindley, J.; Wosnik, M. Wingtip Devices for Tidal Turbines: Performance Improvement and Cavitation Mitigation. In Proceedings of the ASME 2016 Fluids Engineering Division Summer Meeting collocated with the ASME 2016 Heat Transfer Summer Conference and the ASME 2016 14th International Conference on Nanochannels, Microchannels, and Minichannels, Washington, DC, USA, 10–14 July 2016; Volume 2.
15. Pratilastiarso, J.; Nugroho, S.; Tridianto, E.; Syifa, I.R. Experimental study on horizontal axis wind turbine with splitted winglets. *IOP Conf. Series Earth Environ. Sci.* **2018**, *105*, 12102. [CrossRef]
16. Syifa, R.I.; Nugroho, S. Vortex formation on horizontal axis wind turbine with splitted winglets. *J. Physics Conf. Ser.* **2018**, *1090*, 012051. [CrossRef]
17. Hand, M.M.; Simms, D.; Fingersh, L.; Jager, D.; Cotrell, J.; Schreck, S.; Larwood, S. *Unsteady Aerodynamics Experiment Phase VI: Wind Tunnel Test Configurations and Available Data Campaigns*; National Renewable Energy Laboratory: Golden, CO, USA, 2001.
18. Manwell, J.F.; McGowan, J.G.; Rogers, A.L. *Wing Energy Explained*, 2nd ed.; John Wiley & Sons Ltd.: Hoboken, NJ, USA, 2009.

energies

MDPI

Article

Numerical Analysis on the Effectiveness of Gurney Flaps as Power Augmentation Devices for Airfoils Subject to a Continuous Variation of the Angle of Attack by Use of Full and Surrogate Models

Piotr Wiśniewski [1], Francesco Balduzzi [2], Zbigniew Buliński [1] and Alessandro Bianchini [2,*]

1 Institute of Thermal Technology, Silesian University of Technology, Konarskiego 22, 44-100 Gliwice, Poland; piotr.wisniewski@polsl.pl (P.W.); zbigniew.bulinski@polsl.pl (Z.B.)

2 Department of Industrial Engineering (DIEF), Università degli Studi di Firenze, via di Santa Marta 3, 50139 Firenze, Italy; francesco.balduzzi@unifi.it

* Correspondence: alessandro.bianchini@unifi.it; Tel.: +39-055-275-8773; Fax: +39-055-275-8755

Received: 11 March 2020; Accepted: 8 April 2020; Published: 12 April 2020

check for updates

Abstract: The disclosing of new diffusion frontiers for wind energy, like deep-water offshore applications or installations in urban environments, is putting new focus on Darrieus vertical-axis wind turbines (VAWTs). To partially fill the efficiency gap of these turbines, aerodynamic developments are still needed. This work in particular focuses on the development of a mathematical model that allows predicting the possible performance improvements enabled in a VAWT by application of the Gurney flaps (GFs) as a function of the blade thickness, the rotor solidity and geometry of the Gurney flap itself. The performance of airfoil with GFs was evaluated by means of detailed simulations making use of computational fluid dynamics (CFD). The accuracy of the CFD model was assessed against the results of a dedicated experimental study. In the simulations, a dedicated method to simulate cycles of variation of the angle of attack similar to those taking place in a cycloidal motion (rather than purely sinusoidal ones) was also developed. Based on the results from CFD, a multidimensional interpolation based on the radial basis functions was conducted in order to find the GF design solution that provides the highest efficiency for a given turbine in terms of airfoil and solidity. The results showed that, for the selected study cases based on symmetric airfoils, the GF positioned facing outwards from the turbine, which provides the upwind part of the revolution, can lead to power increments ranging from approximately 30% for the lower-solidity turbine up to 90% for the higher-solidity turbine. It was also shown that the introduction of a GF should be coupled with a re-optimization of the airfoil thickness to maximize the performance.

Keywords: VAWT; wind turbine; gurney flap; CFD; RBF; power augmentation

1. Introduction

1.1. Background

The interest in cheap and environment friendly electrical energy generation, lately driven also by the need for meeting stricter standards of clean energy production, has resulted in a wide range of scientific research on the subject of renewable energy sources. One of the leading technologies is wind energy, which is reaching a cost of energy competitive (in the case of large rotors) with other conventional sources. Although the majority of installed wind energy power today comes from wind farms made of several large horizontal axis wind turbines (HAWTs), the disclosing of new diffusion frontiers like deep-water offshore applications or installations in densely inhabited environments are putting new focus on different turbine architectures, like Darrieus vertical axis wind turbines (VAWTs) [1]. This technology has some undisputed advantages (e.g., the insensitivity to wind direction, the possibility of putting the generation system on the ground, the lower susceptibility to highly turbulent flows [2,3]), but their efficiency is lower compared to that of HAWTs. This is not only due to intrinsically more complex aerodynamics with a continuous variation of the angle of attack (AoA), often leading to dynamic stall [4], but also due to the lack of systematic scientific research from their conception in the 1920s up to the 1990s [2]. If this efficiency gap is somehow filled, many scientists forecast a significantly more important role of VAWTs in the near future [5].

Among different approaches to reach this scope, lately increasing attention is given to the possibility of applying passive flow control devices to Darrieus blades [6], in order to delay the onset of stalls and improve the lift-to-drag ratio, especially at medium-low Reynolds numbers. Gurney flaps (GFs) are one of the technologies in the spotlight. In the early 1970s, the American racing driver Dan Gurney came out with an idea to fix a short metal bar at the trailing edge on his racing car rear wing. After conducting few tests, he found out that this simple modification allowed approaching turns with higher velocity and also increasing the car speed on straight lines. The simple construction of this device has encouraged researchers to investigate its application in different areas [7–10], and especially in wind turbines, where they do represent one of the most interesting solutions [11,12]. It has been found that the effect of the lift coefficient enhancement of the GF is connected with the change of the flow structure at the trailing edge, as it is shown in Figure 1, which reports the vorticity contours near the trailing edge of the airfoil. The two large separation bubbles around the sharp trailing edge are replaced by two thinner vortices inducing a lower drag. The Gurney flap also delays the flow separation to a higher angle of attack. The gain on the lift coefficient is burdened with increments of the drag coefficient. Thus, it is a particularly good solution in case of applications where the drag force is of minor importance, like in the case of Darrieus VAWTs.

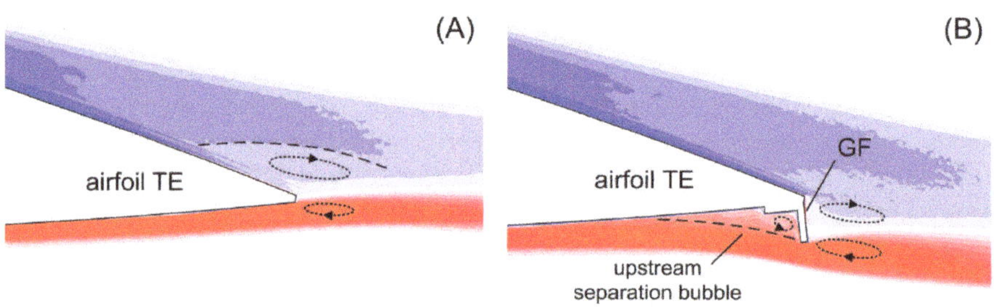

Figure 1. (**A**) Flow field around smooth airfoil; (**B**) flow field around an airfoil with a Gurney flap (field data of vorticity from numerical calculation of the authors).

1.2. Objectives

The aim of the present study is to assess the possible benefits provided by GFs if used on airfoils subject to continuous variations of the angle of attack, as in the blades of Darrieus wind turbines. More specifically, focus is given to the symmetric NACA 4-digits airfoils, which have been shown to be particularly effective in VAWTs [13]. The airfoil thickness represents the first investigated parameter; values of 12% chord (NACA0012), 15% chord (NACA0015), 18% chord (NACA0018), and 21% chord (NACA0021) are considered. Then, the impact of different heights and shapes of the GFs on the performance of these airfoils is evaluated in static conditions, but also in dynamic pitching movements. It is often erroneously thought that the variation of the angle of attack in the Darrieus-type cycloidal motion can be modeled as a pure pitching motion. However, different energy extractions take place upwind and downwind, which in turn impose a notable variation of the AoA in those zones [2].

Moreover, the change of sign of the AoA in proximity of the azimuthal positions of 0° and 180° leads to very sudden variation rates, which are also responsible for the onset of dynamic stall [14].

To meet the objectives described above, the use of computational fluid dynamics (CFD) is mandatory. Due to the complex flow structures taking place behind the GF, the continuous variation of the AoA, and the existence of large separated regions when the stall appears, the simpler modeling methods (e.g., a panel method) are insufficient for this scope [6,15]. In order to limit the computational cost, the unsteady Reynolds-averaged formulation of the Navier–Stokes equations (URANS approach) is used for the presented analyses as the best compromise. Due to the wide range of spatial and temporal scales that need to be captured in the flow features in the presented problem, more accurate methods addressing the turbulent flows such as direct numerical simulation (DNS) or large eddy simulation (LES) would be in fact unusable. In addition, recent examples in the literature showed that the proposed approach is able to properly capture all the effects connected to the use of GFs [6]. A key original model developed for the study presented in this paper is represented by the definition of AoA variations that match exactly the functioning conditions in a broad range of Darrieus VAWTs. These were defined upon combination of detailed full-CFD simulations carried out by the authors and computation of airfoil in pitching motion. Finally, the obtained results were extended using radial basis functions (RBFs) interpolation to provide a comprehensive overview of the effects of GFs installation on the performance of selected airfoils.

1.3. Organization of the Study

The study is organized as follows. Section 2 presents the study cases that have been used both for the calibration and validation of the numerical approach and for the sensitivity analyses. Section 3 presents the methods that were used for the analysis. A description of the CFD settings, including their validation, is first presented; then, the development of the AoA variation trends that mimic the Darrieus functioning is discussed. Section 4 is the main body of the study, including results obtained for static airfoils as well as results obtained for dynamic airfoils in Darrieus-like motion. In this section the multivariate sensitivity analysis based on the radial basis functions (RBFs) interpolation is presented. Section 5 summarizes the main outcomes of the study.

2. Study Cases

2.1. Experimental Validation Benchmark

In order to assess the effectiveness of the numerical techniques prior to proceed with the extended sensitivity analysis on the GF effects on the airfoil in Darrieus-like motion, an experimental benchmark was identified. In particular, the test case presented by researchers from the Technical University (TU) of Berlin in [16] was used. Dedicated experimental studies were conducted inside the laminar wind tunnel of the Hermann Föttinger Institute. The tested airfoil was NACA0021 with the Gurney flap on the pressure side. As discussed, this airfoil was also used in one of the study cases of the sensitivity analysis; therefore the experimental case was then fully representative for the scope of this study.

In [17] the authors presented a variety of tests with different Reynold numbers (Re = 140 k and Re = 180 k) and GF size and mounting configuration. For the sake of brevity, the CFD validation was here reported only for the configuration with the GF conventionally mounted on the pressure side, depicted in Figure 2, which shows a sketch of the Gurney flap geometry used in the experiments. Table 1 reports the chosen test conditions for the validation and the Gurney flap height is given by a percentage value of chord length. For any additional details on experimental measurements (which are not the original content of the present work), please refer to [17].

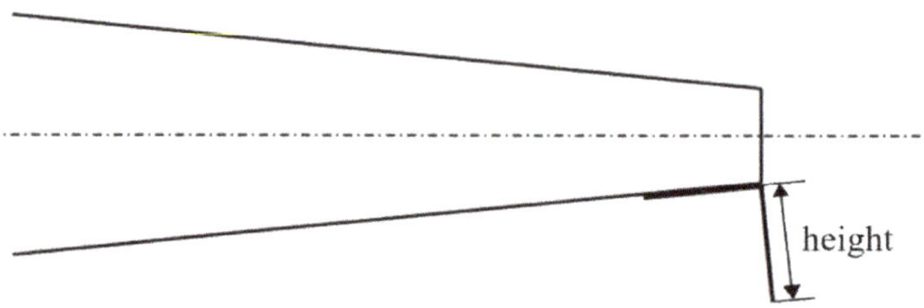

Figure 2. Experimental Gurney flap (GF) geometry.

Table 1. Details of the experimental setup.

Profile	NACA0021
Chord length	0.14 m
Reynolds number	180,000
Gurney flap height	2.5%

2.2. Gurney Flaps

Two different types of Gurney flap mounting were investigated (represented in Figure 3). In further detail, the conventional one-side mounting (A) is the most common method, generally including the GF mounted towards the pressure side of the airfoil. In case of functioning onboard a Darrieus turbine, however, each side of the airfoil acts alternatively as the pressure or suction side depending on the fact that the blade is moving in the upwind or downwind half of the trajectory [18]. On this basis, both the configuration with the GF facing out and the one facing in with respect to the revolution centre were tested. In addition to these configurations, the one presented in Figure 3B, called "fish tail" in the following, was also tested. This configuration was thought to somehow reply to the contrasting requests discussed before, i.e., it is able to provide the power augmentation both for positive and negative incidence angles, partially limiting the additional drag coming from the half working in the suction side by inclining it with respect to the chord. In this sense, it can be considered as an evolution of the "both-side" configuration tested in [6].

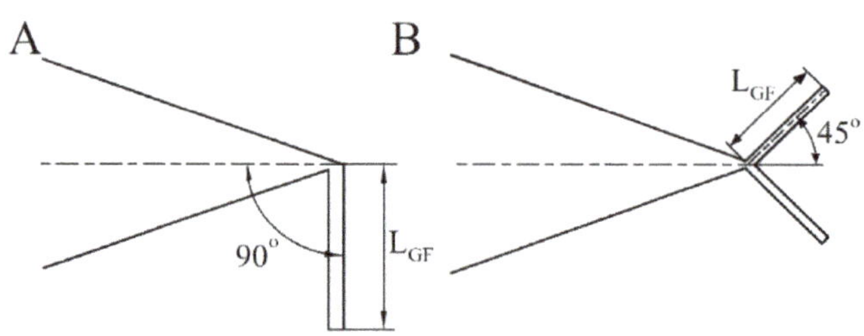

Figure 3. GF configurations: (**A**) one-side GF; (**B**) fish tail GF.

For the scopes of the present work, the two configurations were tested with GFs having a length varying in the range of 0% to 5% of the chord length.

2.3. Test Plan

As discussed, the scope of the study was to evaluate the effectiveness of GFs as power augmentation devices when operating on board Darrieus turbine. To this end, three configurations of an airfoil in cycloidal motion were considered. The idea was to reproduce realistic functioning conditions in terms of Reynolds number, AoA variation trend, and inflow. On this basis, relevant study cases were selected from the literature, with particular attention to those already tested by some of the authors and for which a relevant body of data was available. The configurations are summarized in Table 2. Upon examination of the table, one can notice that one important parameter that has been taken into account is the equivalent turbine solidity, calculated as in Equation (1).

$$\sigma = \frac{Nc}{D} \quad (1)$$

Table 2. Summary of the operational conditions considered for the study cases.

Geometrical Parameters			
Diameter	3.50 m	1.60 m	1.03 m
Number of blades	1	1	3
Chord	0.200 m	0.200 m	0.086 m
Solidity	0.057	0.125	0.250
Original wing profile	NACA0018	NACA0018	NACA0021
Working Conditions			
Free stream velocity	8.235 m/s	8.0 m/s	13.0 m/s
Air density	1.20 kg/m^3	1.20 kg/m^3	1.20 kg/m^3
Air viscosity	1.789×10^{-5} Pa·s	1.789×10^{-5} Pa·s	1.789×10^{-5} Pa·s
TSR	4.45	3.14	3.30
Blade Reynolds Number	482,000	386,000	250,000

The solidity of the rotor is in fact an index of how much the turbine is "permeable" to the flow, thus of how much the energy extraction is unbalanced between the upwind and the downwind portion of the revolution. In more detail, the higher the solidity, the more kinetic energy from the wind is harvested by the upwind part and the less energy can be harvested by the downwind part. The velocity used to calculate the blade Reynolds number, presented in Table 2, is an average value of the relative wind velocity during the revolution. The turbine tip-speed ratio (TSR) is conventionally defined as the ratio between the peripheral speed of the airfoil and the undisturbed wind velocity.

The study cases presented in Table 2 were used in particular to extract realistic trends of variation of the angle of attack and the relative air speed on the airfoils. These curves, presented in Figure 4, were obtained with the procedure described in [19] and slightly smoothed to purge them by unphysical discontinuities arising during the calculation of the induced velocity in areas of macro-vorticity, as discussed in the reference. The trends of Figure 4 were then used as an input for the sensitivity analysis on the GF effects. To this end, they were applied to four different uncambered airfoils of different thickness-to-chord ratios, namely the NACA0012, NACA0015, NACA0018, and NACA0021 (shown in Figure 5).

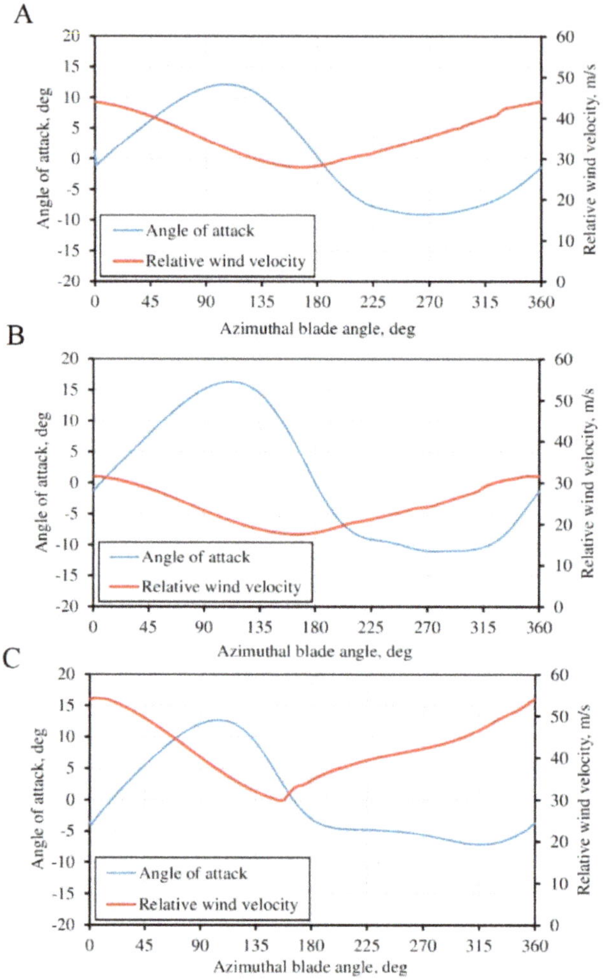

Figure 4. Variation trends for the angle of attack and the relative wind velocity for the selected study cases.

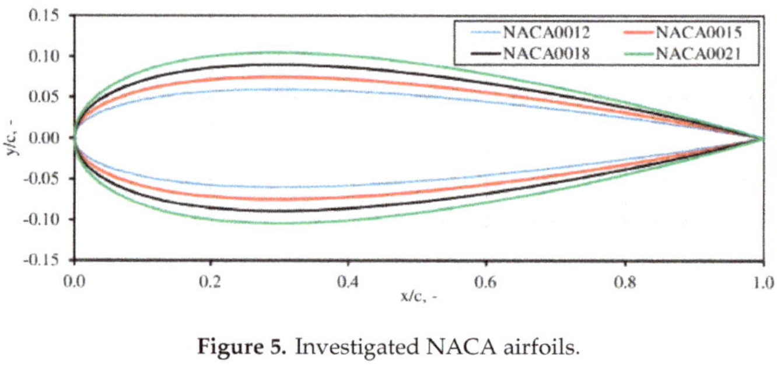

Figure 5. Investigated NACA airfoils.

3. Methods

3.1. Numerical CFD Simulations

The computational domain adopted for the CFD simulations is depicted in Figure 6A. The domain was made in the conventional bullet shape, having a distance of 20 and 40 chords upstream and downstream of the airfoil, respectively. The dimensions of the bullet were in agreement with the most conservative suggestions that can be found in the literature. The choice of the bullet-shaped domain was due to the possibility of imposing only an inlet (at which values of the velocity vector, turbulence intensity, and turbulent viscosity ratio were assumed) and an outlet boundary (at which the value of the gauge pressure was assumed, and for all other quantities zero normal derivative was assumed), with benefits in terms of calculation stability. The same approach was followed successfully in [16]. The works by Balduzzi et al. [20] were taken as the main references in order to select the most suitable numerical settings for the solver. For the sake of completeness, an overview on the main settings of the simulation models is given in the following section.

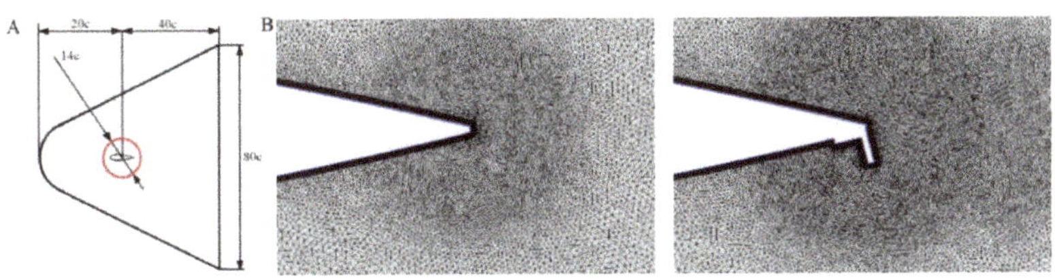

Figure 6. (**A**) Computational domain; (**B**) detailed view of the mesh at the trailing edge of the smooth airfoil (**left**) and near the GF (**right**).

The commercial code ANSYS® FLUENT® was used in the two-dimensional form to solve the time-dependent unsteady Reynolds-averaged Navier–Stokes (URANS) equations in a pressure-based formulation. The fluid was air, modeled as an ideal compressible gas. Based on previous experience, the validation against experiments made use of the four equations Transition SST model. This choice was due to the very low Reynolds number achieved in experiments (max 180 k), which provoked a massive impact of transition phenomena. Conversely, in the sensitivity analysis based on the realistic conditions of Table 2, the authors decided to achieve the turbulence closure by means of Menter's SST k-ω model [21], which is a blend of k-ε and k-ω two-equation formulations. This was due to pretty higher Reynolds numbers, which made the transitional effects less relevant. Moreover, the same study cases were originally simulated with this turbulence model, which proved to be particularly effective and robust. The Coupled algorithm (non-segregated) was employed, where the Navier–Stokes equations set is directly solved through an implicit discretization of pressure in the momentum equations, with benefits in terms of robustness and convergence, as shown in [20]. The second order upwind scheme was used for the spatial discretization of the whole set of URANS and turbulence equations, as well as the bounded second order for time differencing to obtain a good resolution. To allow for the pitching movement of the airfoil, the domain was split into rotating and stationary parts. The interface was circular, having a diameter of 14C. To handle the coupling of the two domains, a general grid interface (GGI) was used. The computational mesh generated for the two domains was of the unstructured type, made with the native mesher of the ANSYS® package. Triangular elements were used throughout the domain, with a massive element refinement within the rotating region and an additional local refinement area around the GF, as shown in Figure 6B. In order to properly capture the flow behavior within the boundary layer, a 30-layer O-grid with prismatic elements was instead created around the airfoil and the GF. The first element height was always chosen so as to guarantee a dimensionless wall distance (y^+) at the grid nodes of the first layer above the blade wall constantly lower than 1. According to the prescriptions of [22], the expansion ratio for the growth of elements starting from the surface was kept below 1.05 to achieve good mesh quality. A mesh dependency study was carried out to ensure that the results were not affected by mesh density and quality. Regarding the grid created for the validation of experiments, please refer to [17] for any additional details. For the spatial discretization of the airfoils with GF three different meshes were created. The y^+ was lower than 1, thus the meshes differed by the number of elements along the airfoil surface and their size in the free stream area. Created meshes had around 160,000, 280,000, and 400,000 elements, respectively. The mesh sensitivity analysis was conducted for the NACA0021 airfoil with Gurney flap lengths equal to 1%, 2%, and 3% of the chord length, respectively (these values are indicated in Figure 7A–C as *GF1*, *GF2*, *GF3*). Different AoAs were tested, even though Figure 7 only reports the case at AoA = 0° at Re = 250 K for brevity.

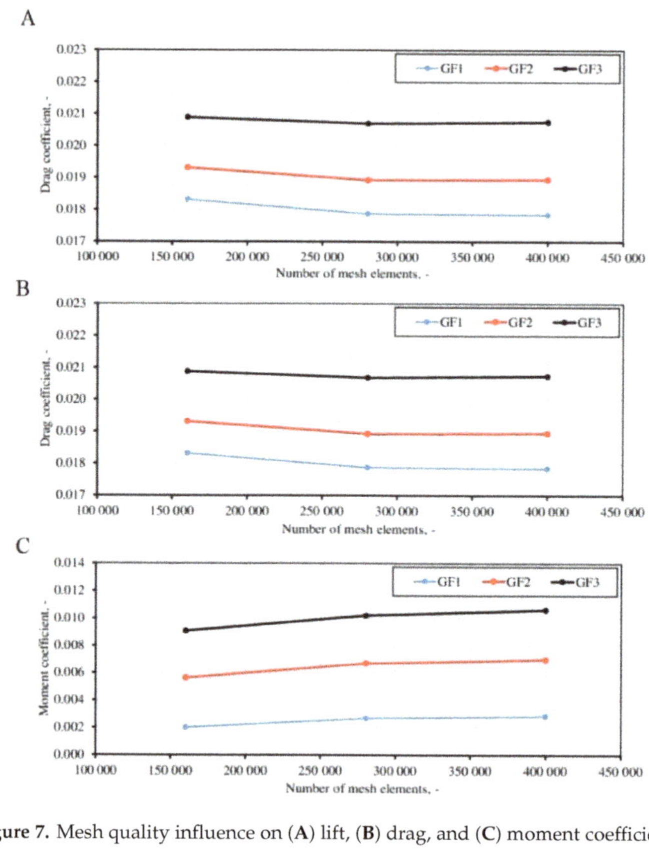

Figure 7. Mesh quality influence on (**A**) lift, (**B**) drag, and (**C**) moment coefficients.

It can be observed that asymptotic convergence is reached as the mesh is refined. The mesh with a number of elements equal to 400,000 was considered as not affected by the mesh size and further computations were carried out using this mesh. The values of the coefficients presented in Figure 7 have been averaged over 500 time steps after getting a converged solution. As extensively discussed in [22], the influence of the time step size also has to be considered carefully in order to obtain the desired accuracy of the computational model. In order to perform reliable dependence studies of temporal discretization, both the one-side GF and the fish tail GF were tested, since they were thought to induce a quite different vortex shedding at the trailing edge for low AoAs (see [6]), thus leading to different characteristic Strouhal numbers. Tests were carried out using a time-step of 10^{-3}, 10^{-4}, and 10^{-5} s, respectively. Upon examination of the results, it was apparent that a time-step of 10^{-4} s was sufficient to achieve independent results (differences in the absolute value lower than 0.1%).

3.2. Pitching Movements and Conventions

The cycloidal motion of each blade on board a Darrieus VAWT generates a continuous variation of the angle of attack on the blade itself as a function of the relative positioning of the chord and the oncoming wind. This, in turn, leads to a variable intensity of the relative speed and to discontinuous forces exerted by the airfoils.

For the conventions used in the study, please refer to Figure 8. The overall torque T produced by the blade is given by Equation (2), where L and D are the lift and drag forces, respectively, and M it the moment around the blade-spoke connection point, which does not always correspond to the aerodynamic centre [23].

$$T = (L \sin \alpha - D \cos \alpha)R + M \tag{2}$$

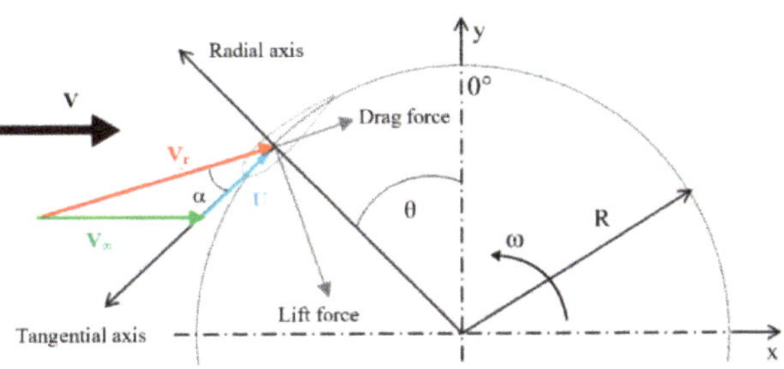

Figure 8. Forces acting on the turbine blade.

The dependency of the AoA on the relative positioning between the blade and the wind, as well as on the force really exerted by the blade, which induces a reduction of the oncoming wind, lead to the well-known variation trends of the AoA that are non-symmetrical between the upwind and the downwind halves of the revolution and characterized by very steep variation rates. As a result, recent research works (e.g., [17]) showed that simulating a blade in pure pitching motion is not sufficient to give reliable estimation of the functioning of cycloidal motion.

To this end, in the present study, the AoA trends depicted in Figure 4 were applied directly to the blades. An average value of the Reynolds number (calculated based on the actual ones taken from CFD) was imposed at the inlet boundary, while the actual forces (lift, drag, and pitching motion) in motion were reconstructed point-by-point by means of the relative speed value also presented in the same figure, in order to have a more precise estimation. By doing so, the variation of the airfoil performance with the Reynolds number is unfortunately neglected, but this cannot be avoided in the case of a pitching approach like the one presented here. However, since each simulation is carried out with each specific equivalent TSR, this variation is thought to be sufficiently small to not compromise the accuracy of the results. In order to compare airfoils and turbines working in different conditions, the introduction of dimensionless coefficients is needed. The main coefficients used in the following of the study are the torque coefficient (Equation (3)) and the power coefficient (Equation (4)):

$$C_T = \frac{T}{\frac{1}{2}\rho c^2 V_\infty^2} \tag{3}$$

$$C_p = \frac{P}{\frac{1}{2}\rho A V_\infty^3} \tag{4}$$

3.3. Radial Basis Functions (RBFs) Interpolation for Data Reduction

The parametric analysis carried out on the airfoil performance with different GFs provided only a finite set of points. In order to obtain a continuous response surface, the multivariate radial basis functions (RBFs) interpolation was applied. This method is very efficient for interpolation of large scattered data sets. It also has some drawbacks connected with unstable solutions and fast growth of the computational cost for large data series and also non-negligible error connected with Runge's phenomenon at the boundary of the domain. Drawbacks notwithstanding, it seems to be a suitable choice for the considered case of data reduction [24].

A multivariate function Φ is called radial (RBF) if its value at each point depends only on its distance from the base point, what is written in mathematical notation as:

$$\Phi(r) = \Phi(\|r - r_0\|) \tag{5}$$

where $\|\cdot\|$ is the Euclidian norm in the R^n space and r_0 is the vector describing the position of the base point. The radial function based interpolation takes the form of a linear combination of base functions attached to all N collocation nodes giving following equation:

$$\mathrm{u}(r) = \sum_{i=1}^{N} \alpha_\mathrm{i}\Phi(\|r - r_0\|) \tag{6}$$

where α_i is an unknown interpolation coefficient. The values of interpolation coefficients can be found by collocating the interpolation function of Equation (6) and then solving the resultant linear set of equations which can be briefly written as:

$$\alpha_\mathrm{i} = \Phi^{-1}\cdot\mathrm{u}(r) \tag{7}$$

where the interpolation (or collocation) matrix is computed as:

$$\Phi = \left\{\Phi_{ij}\right\} = \left\{\Phi\left(\|r_i - r_j\|\right)\right\} \text{ where } i, j = 1, \ldots, N \tag{8}$$

The radial function form has to be chosen adequately with respect to the considered problem. Thus, to find the most suitable function, a dedicated analysis was carried out for different common types of radial functions. The functions are shown in Table 3.

Table 3. Utilized radial basis functions.

Functions Form	
Multiquadric (MQ)	$\varphi(r) = \sqrt{r^2 + c^2}$
Inverse Quadratic (IQ)	$\varphi(r) = \frac{1}{r^2+c^2}$
Inverse Multiquadric (IMQ)	$\varphi(r) = \frac{1}{\sqrt{r^2+c^2}}$

The shape parameter c, which appears in the definition of different radial basis functions, is a parameter that controls the shape of the basis function and hence the size of the region of influence of a given basis function around the collocation point. The higher the value of shape parameter, the bigger is the region of influence of the basis function; unfortunately, this also causes deterioration of conditioning of the approximation problem.

It needs to be pointed out that the interpolation matrix Φ is symmetric and positively defined, hence it is always invertible. However, by incrementing the number of nodes and consequently the number of base functions, the matrix conditioning becomes worse. The condition number of a matrix measures error is introduced by the finite arithmetic of computations on computers [25]. The matrix condition number for inversion is given by following equation:

$$(\Phi) = ||\Phi||||\Phi^{-1}|| \tag{9}$$

where is the conditioning number. In case of an interpolation based on the RBFs, the conditioning number value of the interpolation matrix is strictly connected with the shape of applied interpolation function. In the case of the functions presented in Table 3, this is controlled by the shape parameter [26]. For every interpolation problem it is possible to find optimal value of the shape parameter value; see [26]. To this end, two additional simulations, regarding the geometrical parameters of the airfoil thickness and the Gurney flap length, which had not been covered during the case studies, were done for each interpolation data set. Further, based on the additional simulations, the root main square (RMS) method was used to assess the interpolation accuracy, which is given by the following equation:

$$RMS = \sqrt{\frac{\sum_{j=1}^{N}\left(\widetilde{f}(p_j) - f(p_j)\right)}{N}} \tag{10}$$

where p is the vector of points, which in the interpolation corresponds with the collocation points, $\widetilde{f}$ indicates the interpolation function, and f is the numerical values of the original function. Figure 9 shows the values for the matrix of conditioning number and resulting RMS error as a function of the shape parameter.

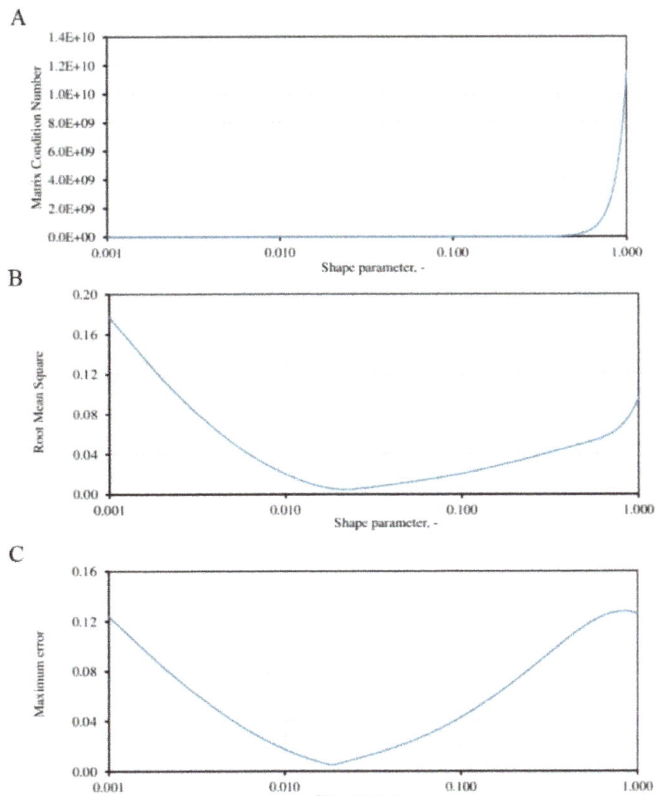

Figure 9. Shape parameter influence on: (**A**) matrix conditioning number, (**B**) root mean square (RMS) error, (**C**) maximum error.

4. Results and Discussions

4.1. Experimental Assessment of the Numerical Model

As discussed, an extensive validation of the numerical model was carried out preliminary upon comparison with experimental data from the Hermann Föttinger Institute (HFI) of the TU Berlin [17].

The test case was a NACA0021 airfoil under a Reynolds number of 180 k. The numerical model was tested both in terms of static polars and in dynamic pitching motion. For the sake of brevity, only the results with a one-sided GF (2.5%c) are reported here: Figure 10 displays the comparison of static polars and Figure 11 the performance in sinusoidal movement within an AoA range between 10 and 30 deg and a reduced frequency k of 0.05.

$$k = \frac{\omega}{u_0} \cdot \frac{c}{2} = \frac{\pi f c}{u_0} \tag{11}$$

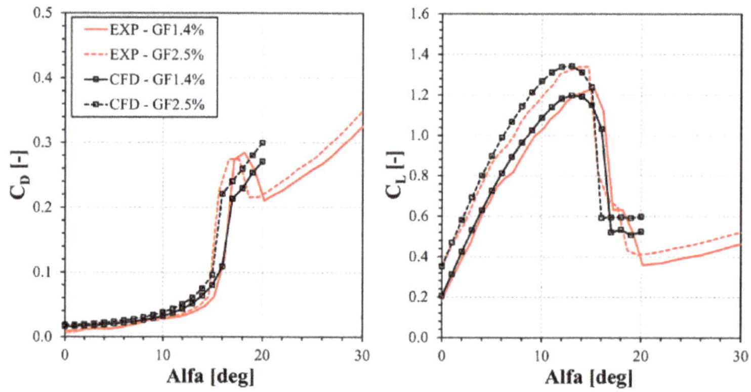

Figure 10. Comparison experimental results and CFD computations for static flow around airfoil equipped with GF at different values of angle of attack.

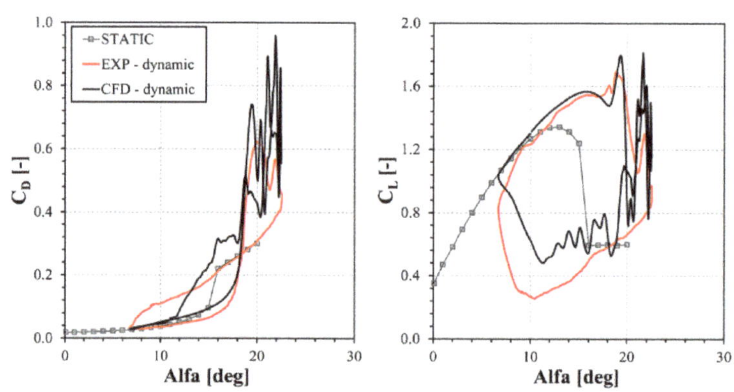

Figure 11. Comparison experimental results and CFD computations for dynamically changed angle of attack.

It has to be remarked that the numerical data in Figure 10 were run in unsteady RANS mode for several incidence angles (see Figure 12). Figure 12, in particular, shows the contours of normalized velocity, i.e., the local velocity divided by the undisturbed one. In particular, as common practice, the use of unsteady simulations was mandatory for high AoAs after a stall, where the analysis of the residuals showed issues in convergence history, characterized by the intense fluctuations of the calculated aerodynamic forces already shown by [6].

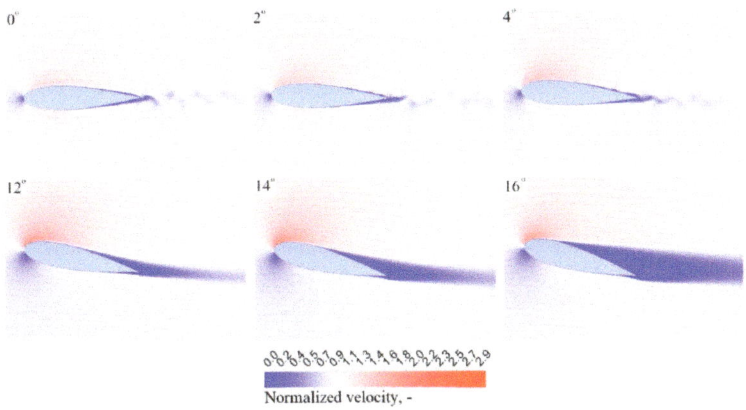

Figure 12. Normalized velocity contours for 0°, 2°, 4° and 12°, 14°, 16° angles of attack.

In thte case of GFs, the use of unsteady simulations was also necessary for very low AoAs (between 0° and 4°), where the GF itself produces some vortices that detach alternatively from the corners and then are convected downstream. In those cases, the timestep for the simulation was based on the previous experience of [6]. Upon examination of both comparisons, sound agreement can be noted overall between experiments and simulations, proving the effectiveness of the method.

4.2. GF Effects in Cycloidal Motion: Impact on Torque Profiles

If the final expected outcome of the application of GFs to Darrieus turbines is the power enhancement (that will be discussed in detail in the next section), it is worth analyzing from a physical point of view their impact on the functioning of the airfoils during a revolution. The balance of the energy extraction between the upwind and the downwind halves of the revolution is in fact very important not only for the overall efficiency, but also for the possible creation of stresses and vibrations of the turbine.

Figure 13 shows the influence of the Gurney flap length and configuration (i.e., the single facing outward or inward, and the fish tail GF, respectively) on the torque coefficient as a function of the turbine position. Displayed data do refer to the first test case only (solidity equal to 0.057 and NACA0018), even though the physical behavior was of general validity in the case of blades in cycloidal motion. The values indicated as "No GF" show the torque distributions with the smooth airfoil.

Upon examination of the figure, it is apparent that the outward pointing Gurney flap tends to increase the unbalance of the torque distribution. The increment of torque in the upwind part is significant and it is connected with the lift-to-drag ratio increment induced by the GF, which is particularly relevant for the higher AoAs reached upwind. On the other hand, the torque reduction downwind is related to the increased drag at those low AoAs that derive from the low wind speed going through the rotor. The inward pointing Gurney flap instead leads to an increment of the torque coefficient along the downwind part of turbine, leading to a more balanced torque profile, even though the extracted work (i.e., the area below the curve) is pretty much the same. The fish tail configuration finally confirmed the hypothesized change in performance, providing a relative increase of the torque coefficient for both the upwind and the downwind parts of the revolution.

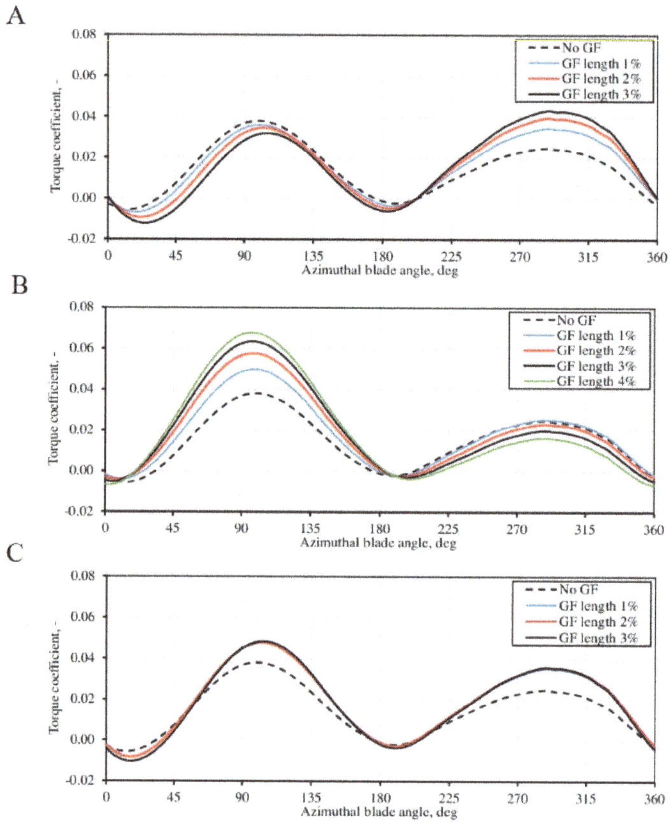

Figure 13. (**A**) Outward pointing one-sided, (**B**) inward pointing one-sided, and (**C**) fish tail Gurney flap length influence on the torque distribution in function of turbine position.

The relative impact to the produced power of the two halves of the machine is even more apparent from Figure 14. It is very interesting to notice that the fish tail not only provides an increase of the performance on both halves of the machine, but also a very balanced power between the two.

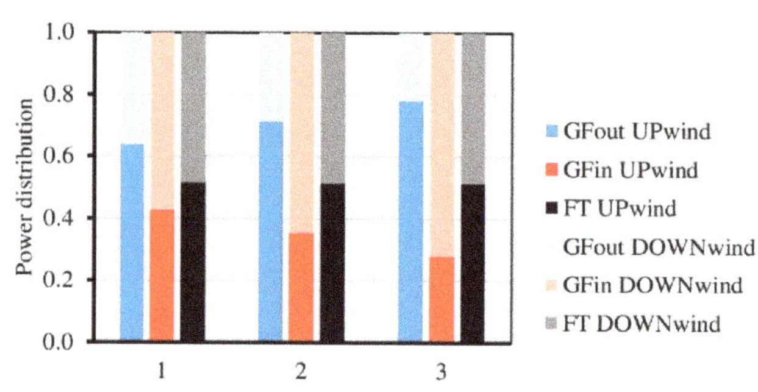

Figure 14. Gurney flap length and configuration influence on the power extraction distribution.

4.3. GF Effects in Cycloidal Motion: Sensitivity Analysis on GF Characteristics

As discussed, the scope of the present analysis was to study the prospects of different GF configuration in terms of power augmentation of Darrieus VAWTs using symmetric airfoils. To this end, a large number of simulations were carried out. Figure 15A–C and Figure 16A–C show the results of all the studied cases in a way which allows the reader to have an overall outlook on the main effects. In further detail, the two figures show the influence of the one-sided Gurney flap (Figure 15) and the fish tail configuration (Figure 16), respectively, in terms of power coefficient variation for different airfoils thickness. Graphs A–C refer to the three different equivalent turbine solidities (increasing from A to C). The white color in the color palette indicates the reference value of the power coefficient with the smooth airfoil, the red color indicates an incremented one, and the blue color a decreased one.

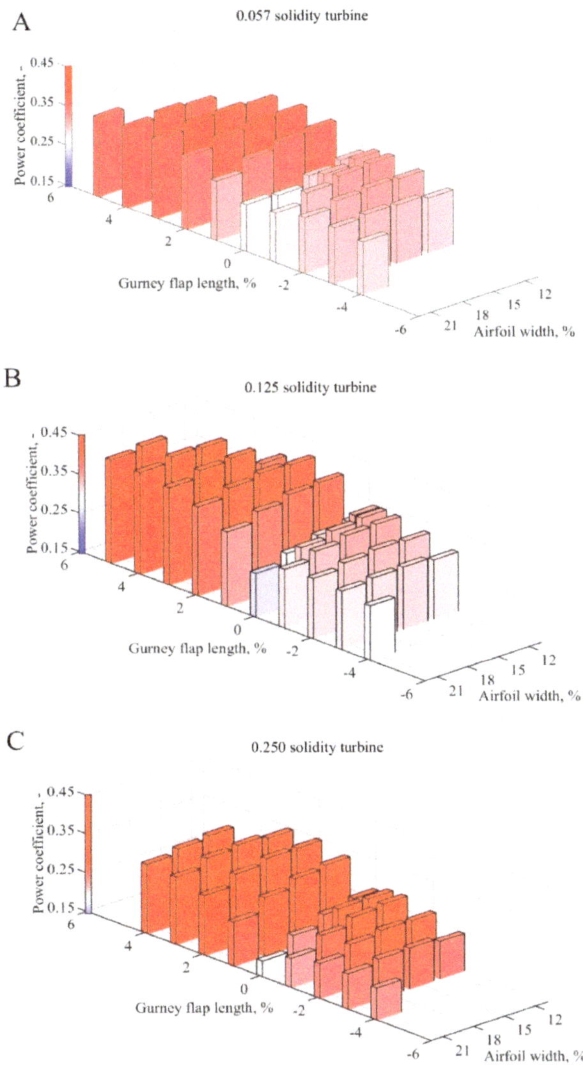

Figure 15. Numerical model results for one-sided Gurney flap for (**A**) 0.057, (**B**) 0.125, and (**C**) 0.25 equivalent solidity turbines.

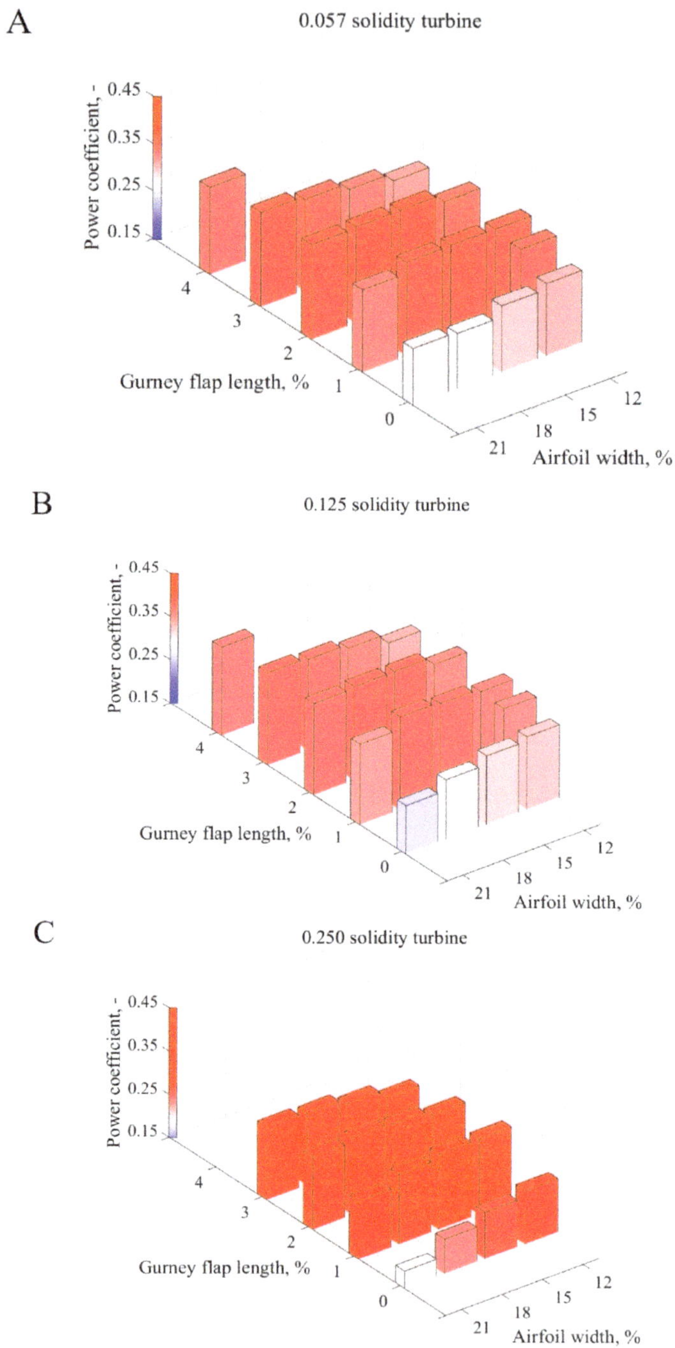

Figure 16. Numerical model results for the fish tail Gurney flap for (**A**) 0.057, (**B**) 0.125, and (**C**) 0.25 equivalent solidity turbines.

Finally, the positive and negative values of Gurney flap length in Figure 15 indicate its outward and inward pointing directions, respectively, with respect to the turbine axis.

Upon examination of the graphs, some interesting observations can be noted:

- The outward positioning of the GF (if one-sided) always provided the largest power increase in a blade in cycloidal motion; however, as soon as the solidity increased, the possibility of having a more balanced energy extraction (i.e., with the inward positioning) became attractive;
- For higher solidities, the application of the GF seems to provide a constant increase of performance. This can be explained as follows: a) in case of the inward positioning, this is due to the discussed re-distribution of the energy extraction between the upwind and downwind halves. In a very solid turbine, indeed, the wind velocity oncoming to the downwind half of the revolution is very low, thus leading to small AoAs and then to a reduced torque production. In this view, adding a GF that is able to increase the performance in this region (where it acts on the pressure side of the airfoils) leads to potential benefits; b) in case of the outward positioning, the torque extraction is maximized in the upwind half (where the flow is more energized), sacrificing the performance downwind;
- For low solidities, this latter approach is the only one providing significant benefits. In this case, the torque profile is sufficiently balanced even in the original configuration and then it is more convenient to maximize the impact of the GF upwind, where the flow speeds are higher.

Due to the complexity of analyzing so much data at a glance, some relevant trends have been extracted and reported in Figures 17 and 18.

Figure 17 first reports the extracted power for the four tested airfoils (i.e., as a function of the thickness-to-chord ratio of the airfoils) as a function of the GF length in the case of the outward (A), inward (B), mounting, and fish tail configurations (C). The high-solidity test case was selected, even if the same considerations can be repeated for the other three cases. Upon examination of the figure, one can notice that for the inward mounting (i.e., the one privileging the downwind side), the thinner the airfoil, the higher the performance that can be achieved. Also, the optimal GF length decreases monotonically with the airfoil thickness. On the other hand, the thinner NACA0012 airfoil is more sensitive to the GF length, with steeper variation curves. This is due to the larger impact of the GF additional drag on the thin airfoil. Overall, the thicker NACA0021 airfoil shows a quite different behavior than the other ones, with flatter response curves and much larger optimal GF lengths.

On the other hand, in case of the outward mounting, the best performance is achieved for medium-solidity airfoils, where the application of a GF to very thin or very thick ones does not provide benefits. The optimal GF length keeps shifting to lower values as soon as the airfoil thickness decreases. The same analysis is repeated in case of the fish tail configuration (see Figure 17C). One can notice that the fish tail configuration provides consistent benefits for almost all the airfoils, with only the exception of the very thin one, where the draft increase is probably not compensated for by the additional lift.

Figures 18 and 19 compare the optimal configurations found among the tested turbines with different values of solidity.

Upon examination of the figure, some of the relevant trends discussed before are still clearly noticeable. In addition, it is worth noticing that:

- In the case of the inward mounting, the maximum performance (presented as a torque increment with respect to no GF configuration) in the case of low solidity is monotonically increasing with the airfoil thickness-to-chord ratio, while for the high solidity, better performance is obtained with the medium NACA00018. The optimal GF height generally increases with the airfoil thickness.
- An opposite trend than the one above was noticed for the outward mounting of the Gurney flap. However, the optimal GF height kept increasing with the airfoil thickness;
- In the cases of the fish tail configurations quite thick airfoils are preferred, with the optimal GF height also increasing with the airfoil thickness.

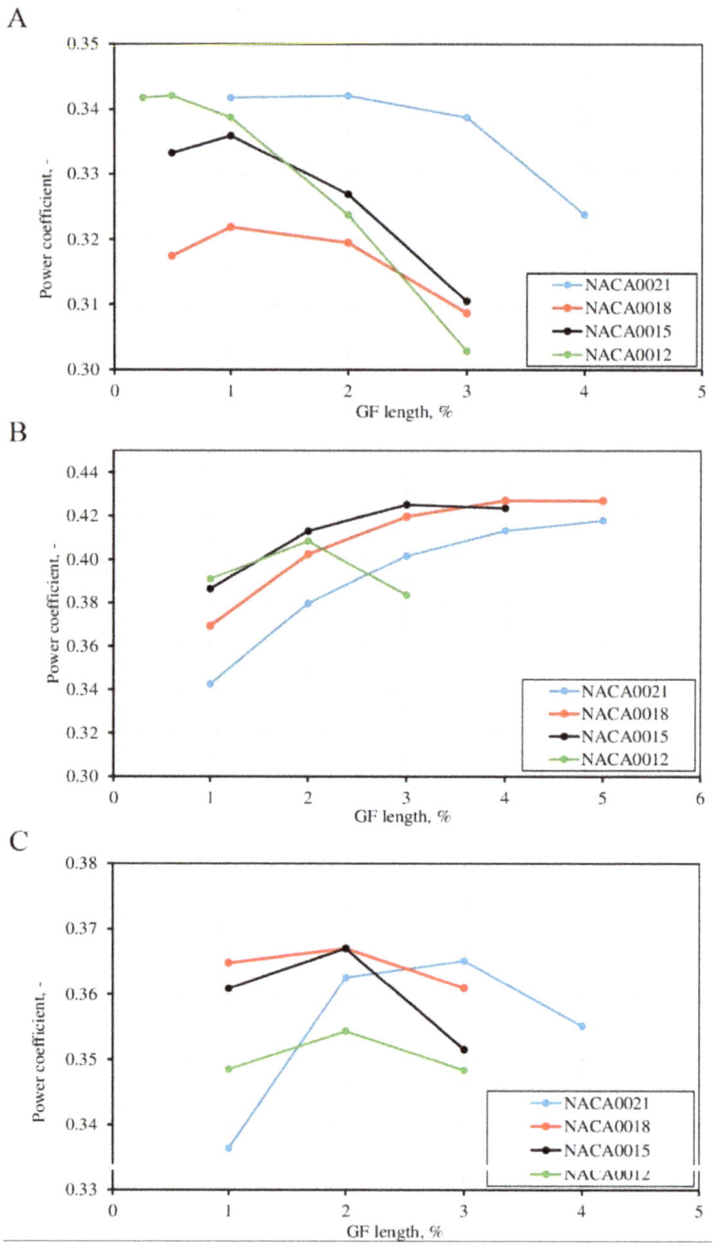

Figure 17. Turbine power coefficient as a function of the GF length and airfoil thickness for equivalent turbine solidity 0.25: (**A**) inward GF, (**B**) outward GF, (**C**) fish tail.

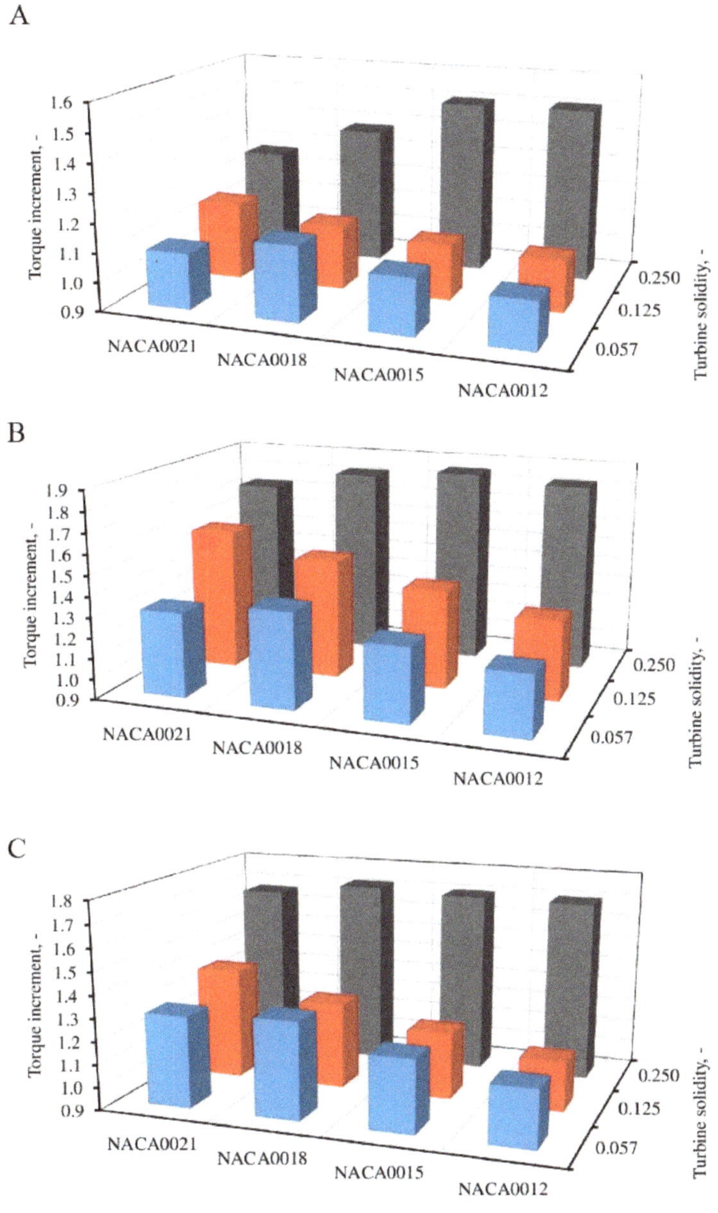

Figure 18. Increment of the torque value for optimal length of GFs for all investigated turbine configurations: (**A**) inward GF, (**B**) outward GF, (**C**) fish tail.

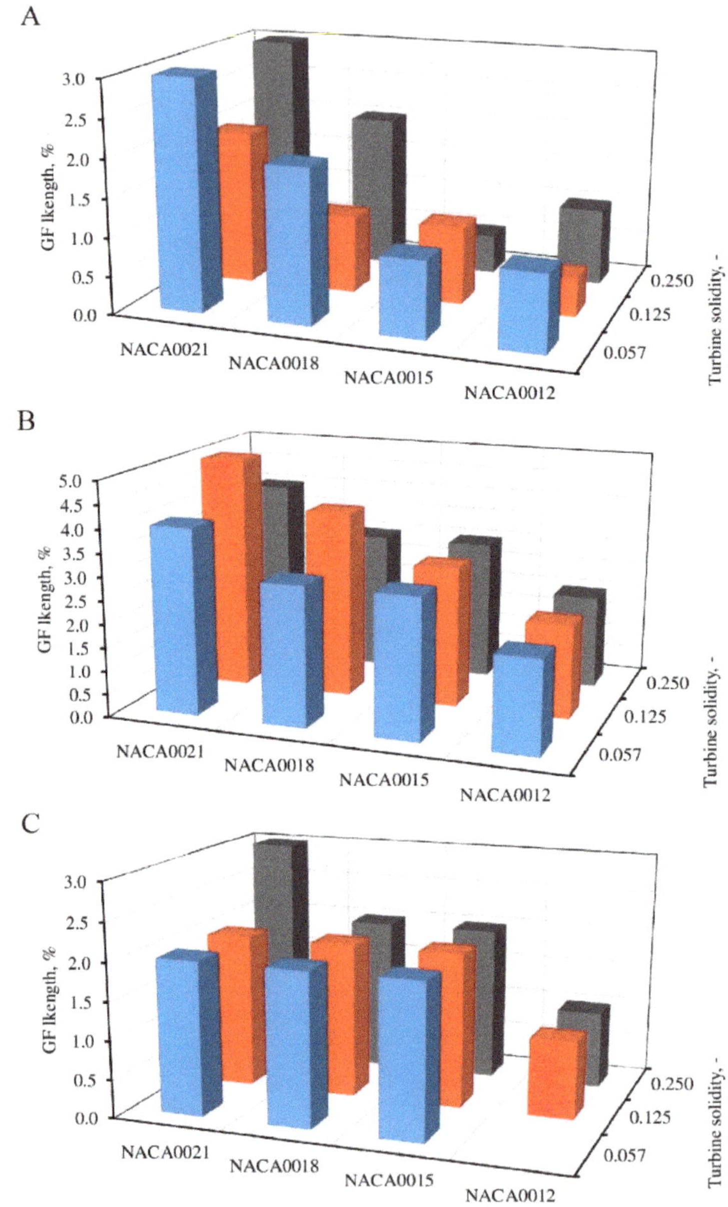

Figure 19. Optimal value of the GFs length for all investigated turbine configurations: (**A**) inward GF, (**B**) outward GF, (**C**) fish tail.

4.4. Response Surfaces

Even though scattered data coming from the simulations already provided some interesting indications about the relevant trends, more detailed results were needed to find optima with a sufficient accuracy. To this end, a radial basic interpolation was carried out on the data. This analysis provided a three dimensional solution of the surface response. As a result, the power coefficient as a function of the airfoil thickness and the Gurney flap length for different configurations are shown in Figures 20 and 21 for the single-side GF and the fish tail GF, respectively. These plots were obtained for interpolation with use of the Inverse Multiquadric (IMQ) basis functions.

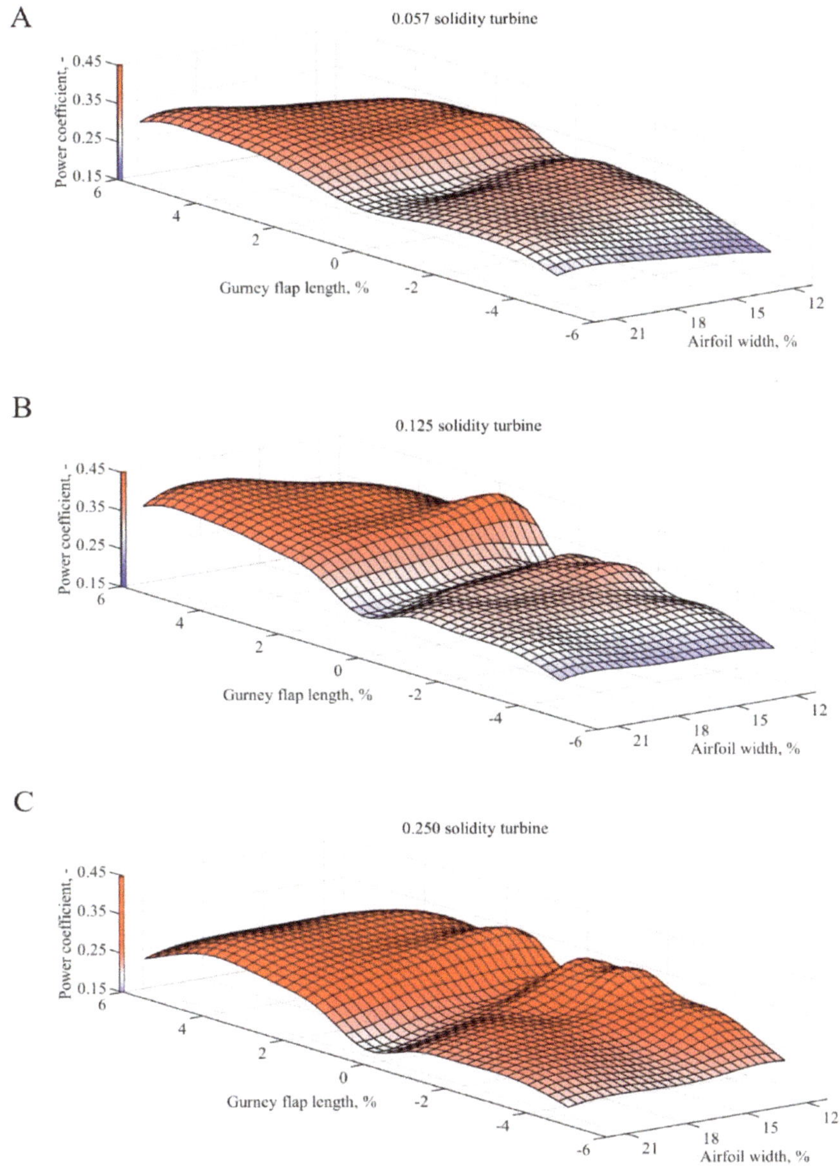

Figure 20. Power coefficient for one-sided Gurney flap using IMQ functions for (**A**) 0.057, (**B**) 0.125, and (**C**) 0.25 equivalent solidity turbines.

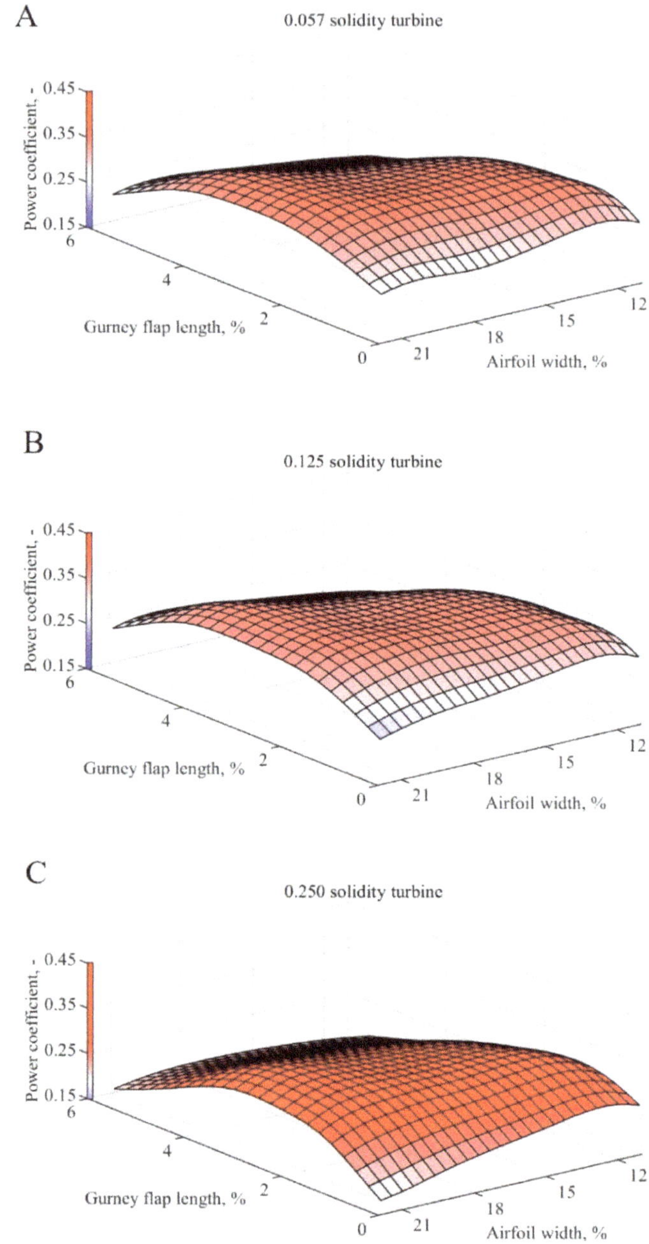

Figure 21. Power coefficient for the fish tail configuration using IMQ functions for (**A**) 0.057, (**B**) 0.125, and (**C**) 0.25 equivalent solidity turbines.

Optima were then found on the surfaces. Table 4 reports the values of the power coefficient for the configurations equipped with the most efficient blade and the reference blade, respectively. The power coefficient increment was defined as a ratio of the power coefficient of the blade with changed thickness and of the Gurney flap to the power coefficient of the reference blade.

Table 4. Power coefficient and geometrical parameters of the most efficient turbine and the reference one using IMQ functions.

					IMQ				
Turbine solidity		0.057			0.125			0.250	
				Geometrical Parameters					
	Ref.	One sided	Fish tail	Ref.	One sided	Fish tail	Ref.	One sided	Fish Tail
Airfoil Thickness	18.00	18.38	19.87	18.00	15.52	19.94	21.00	14.06	19.74
Gurney flap length	0.00	3.48	2.08	0.00	3.50	2.84	0.00	3.42	2.42
				Results					
Power coefficient	0.27	0.37	0.36	0.29	0.43	0.37	0.19	0.36	0.33
Increment, %	-	37.04%	33.33%		48.28%	27.59%	-	89.47%	73.68%

It can be observed that the potential increment of the power coefficient due to the introduction of the Gurney flap can be really significant (up to +89.5%), especially when higher-solidity turbines are considered.

Overall, the increment provided by the fish tail configuration is lower than that of a one-sided GF. The fish tail shape is more suitable for turbines with lower solidity, while the one-sided Gurney flap configuration significantly influences the performance of turbines with higher solidity values. It has to be pointed out, however, that the quantitative results reported in Table 3 only refer to the selected study cases (some of them quite theoretical) and operating conditions. Performance increases to be expected for real rotors are probably lower than the reported values. However, the tendencies are thought to be of general application and they clearly highlight the potential of GFs for use in Darrieus VAWTs. Finally, it is worth noticing that the introduction of the Gurney flap in case of medium-high solidity turbines would suggest that the airfoil thickness should be slightly changed, going toward notably thinner airfoils in the case of the one-sided GF and to slightly thinner ones for the fish tail configuration. In case of low-solidity machines, on the other hand, a medium-thickness airfoil is still the best choice.

5. Conclusions

In this study, the possible benefits provided by Gurney flaps when applied to blades subject to continuous and non-sinusoidal variations of the angle of attack have been analyzed by means of an extended sensitivity analysis. These analyses are propaedeutic for understanding the possible use of GFs as power augmentation devices in Darrieus wind turbines. Symmetric NACA 4-digits airfoils were considered. The airfoil thickness which represents one of the investigated parameters had values of 12% chord (NACA0012), 15% chord (NACA0015), 18% chord (NACA0018), and 21% chord (NACA0021). The GF height and mounting (angle of inclination with respect to the chord and airfoil side with respect to the revolution axis) represented the other main variables. The different test cases were analyzed by means of unsteady CFD simulations. Then, radial basis functions were used to interpolate the results and provide detailed response surfaces describing the impact of the aforementioned variables.

Upon examination of the results, it was shown that:

- A proper GF selection can indeed provide performance increases when used in a Darrieus wind turbine (possible benefits up to +89.5% when added in combination with the correct thickness of airfoil);
- The potential benefits are higher in case of more solid turbines;
- The introduction of a GF should be coupled with a re-optimization of the airfoil thickness to obtain the maximum performance. In case of medium-high solidity turbines, this would imply a reduction in the thickness-to-chord ratio;

- The one-sided Gurney flap allows obtaining the highest turbine efficiency, but it leads to a significant imbalance of the torque distribution between the upwind and the downwind part of the turbine;
- The Fish Tail Gurney flap configuration (both-sided GF with an inclination angle of 45°) provides a lower increment of the turbine power coefficient compared to the one-sided Gurney flap, but it results in a more balanced torque output.

Further developments of the present model could consider higher dimensionality of the model itself, viz., increasing the number of variables considered in the surrogate model. Additional variables that could be considered are, for example, the airfoil type, additional Gurney flap angles relative to the chord, or combination of differently shaped Gurney flaps in inward or outward pointing directions.

Author Contributions: Conceptualization, A.B. and F.B.; Methodology, A.B., F.B. and Z.B.; Software, F.B.; Validation, P.W.; Formal Analysis, P.W. and A.B.; Investigation, P.W.; Resources, A.B. and Z.B.; Data Curation, P.W. and A.B.; Writing-Original Draft Preparation, P.W.; Writing-Review & Editing, A.B.; Visualization, F.B.; Supervision, A.B. and Z.B.; Project Administration, Z.B.; Funding Acquisition, Z.B. All authors have read and agreed to the published version of the manuscript.

Funding: This research received the financial support by National Science Centre within the OPUS scheme under contract UMO-2017/27/B/ST8/02298.

Acknowledgments: Thanks are due to Giampaolo Manfrida from the Università degli Studi di Firenze for supporting the visiting period of MSc. Wiśniewski in Firenze and to Giovanni Ferrara of the same university for providing the facilities for hosting him. All the authors would like also to thank Holst at TU Berlin for providing the experimental data for validation.

Conflicts of Interest: The authors declare no conflict of interest.

References

1. Bianchini, A. Trends, prospects, and R&D directions in wind turbine technology. In *Reference Module in Earth Systems and Environmental Sciences*; Elsevier: Amsterdam, The Netherlands, 2019; ISBN 978-0-12-409548-9.
2. Paraschivoiu, I. *Wind Turbine Design: With Emphasis on Darrieus Concept*; Presses Internat Polytechnique: Montreal, QC, Canada, 2009; Reprinted; ISBN 978-2-553-00931-0.
3. Molina, A.C.; Massai, T.; Balduzzi, F.; Bianchini, A.; Ferrara, G.; De Troyer, T.; Bartoli, G. Combined experimental and numerical study on the near wake of a Darrieus VAWT under turbulent flows. *J. Phys. Conf. Ser.* **2018**, *1037*, 072052. [CrossRef]
4. Simão Ferreira, C.J.; Van Zuijlen, A.; Bijl, H.; Van Bussel, G.; Van Kuik, G. Simulating dynamic stall in a two-dimensional vertical-axis wind turbine: Verification and validation with particle image velocimetry data. *Wind Energy* **2010**, *13*, 1–17. [CrossRef]
5. Islam, M.R.; Mekhilef, S.; Saidur, R. Progress and recent trends of wind energy technology. *Renew. Sustain. Energy Rev.* **2013**, *21*, 456–468. [CrossRef]
6. Bianchini, A.; Balduzzi, F.; Di Rosa, D.; Ferrara, G. On the use of Gurney Flaps for the aerodynamic performance augmentation of Darrieus wind turbines. *Energy Convers. Manag.* **2019**, *184*, 402–415. [CrossRef]
7. Amini, Y.; Liravi, M.; Izadpanah, E. The effects of Gurney flap on the aerodynamic performance of NACA 0012 airfoil in the rarefied gas flow. *Comput. Fluids* **2018**, *170*, 93–105. [CrossRef]
8. Han, D.; Dong, C.; Barakos, G.N. Performance improvement of variable speed rotors by Gurney flaps. *Aerosp. Sci. Technol.* **2018**, *81*, 118–127. [CrossRef]
9. Zhu, H.; Hao, W.; Li, C.; Ding, Q. Numerical study of effect of solidity on vertical axis wind turbine with Gurney flap. *J. Wind Eng. Ind. Aerodyn.* **2019**, *186*, 17–31. [CrossRef]
10. Zhu, B.; Huang, Y.; Zhang, Y. Energy harvesting properties of a flapping wing with an adaptive Gurney flap. *Energy* **2018**, *152*, 119–128. [CrossRef]
11. Aramendia, I.; Fernandez-Gamiz, U.; Ramos-Hernanz, J.A.; Sancho, J.; Lopez-Guede, J.M.; Zulueta, E. Flow control devices for wind turbines. In *Energy Harvesting and Energy Efficiency*; Bizon, N., Mahdavi Tabatabaei, N., Blaabjerg, F., Kurt, E., Eds.; Springer International Publishing: Cham, Switzerland, 2017; Volume 37, pp. 629–655. ISBN 978-3-319-49874-4.

12. Saenz-Aguirre, A.; Fernandez-Gamiz, U.; Zulueta, E.; Ulazia, A.; Martinez-Rico, J. Optimal wind turbine operation by artificial neural network-based active gurney flap flow control. *Sustainability* **2019**, *11*, 2809. [CrossRef]
13. Islam, M.; Ting, D.S.-K.; Fartaj, A. Desirable airfoil features for smaller-capacity straight-bladed VAWT. *Wind Eng.* **2007**, *31*, 165–196. [CrossRef]
14. Bianchini, A.; Balduzzi, F.; Ferrara, G.; Ferrari, L. Critical analysis of dynamic stall models in low-order simulation models for vertical-axis wind turbines. *Energy Procedia* **2016**, *101*, 488–495. [CrossRef]
15. Fernandez-Gamiz, U.; Gomez-Mármol, M.; Chacón-Rebollo, T. Computational modeling of gurney flaps and microtabs by POD method. *Energies* **2018**, *11*, 2091. [CrossRef]
16. Holst, D.; Balduzzi, F.; Bianchini, A.; Church, B.; Wegner, F.; Pechlivanoglou, G.; Ferrari, L.; Ferrara, G.; Nayeri, C.N.; Paschereit, C.O. Static and dynamic analysis of a NACA 0021 airfoil section at low reynolds numbers based on experiments and computational fluid dynamics. *J. Eng. Gas Turbines Power* **2019**, *141*. [CrossRef]
17. Balduzzi, F.; Holst, D.; Melani, P.F.; Wegner, F.; Nayeri, C.N.; Ferrara, G.; Paschereit, C.O.; Bianchini, A. Combined numerical and experimental study on the use of Gurney flaps for the performance enhancement of NACA0021 airfoil in static and dynamic conditions. In Proceedings of the ASME Turbo Expo, London, UK, 22–26 June 2020.
18. Bianchini, A.; Balduzzi, F.; Ferrara, G.; Ferrari, L. A computational procedure to define the incidence angle on airfoils rotating around an axis orthogonal to flow direction. *Energy Convers. Manag.* **2016**, *126*, 790–798. [CrossRef]
19. Bianchini, A.; Balduzzi, F.; Ferrara, G.; Persico, G.; Dossena, V.; Ferrari, L. A critical analysis on low-order simulation models for darrieus vawts: How much do they pertain to the real flow? *J. Eng. Gas Turbines Power* **2019**, *141*. [CrossRef]
20. Balduzzi, F.; Bianchini, A.; Maleci, R.; Ferrara, G.; Ferrari, L. Critical issues in the CFD simulation of Darrieus wind turbines. *Renew. Energy* **2016**, *85*, 419–435. [CrossRef]
21. Menter, F.R. Two-equation eddy-viscosity turbulence models for engineering applications. *AIAA J.* **1994**, *32*, 1598–1605. [CrossRef]
22. Balduzzi, F.; Bianchini, A.; Ferrara, G.; Ferrari, L. Dimensionless numbers for the assessment of mesh and timestep requirements in CFD simulations of Darrieus wind turbines. *Energy* **2016**, *97*, 246–261. [CrossRef]
23. Bianchini, A.; Balduzzi, F.; Ferrara, G.; Ferrari, L. Aerodynamics of Darrieus wind turbines airfoils: The impact of pitching moment. *J. Eng. Gas Turbines Power* **2017**, *139*. [CrossRef]
24. Lazzaro, D.; Montefusco, L.B. Radial basis functions for the multivariate interpolation of large scattered data sets. *J. Comput. Appl. Math.* **2002**, *140*, 521–536. [CrossRef]
25. Golub, G.H.; Loan, C.F.V. *Matrix Computations*; JHU Press: Baltimore, MD, USA, 2013; ISBN 978-1-4214-0794-4.
26. Buliński, Z.; Nowak, A.J.; Šarler, B. *Numerical Experiments with the Local RBF Collocation*; International Conference on Computational Experimental Engineering and Science Special Symposium on Meshless Methods ICCES MM: Dubrovnik, Croatia, 2006.

MDPI
St. Alban-Anlage 66
4052 Basel
Switzerland
Tel. +41 61 683 77 34
Fax +41 61 302 89 18
www.mdpi.com

Energies Editorial Office
E-mail: energies@mdpi.com
www.mdpi.com/journal/energies

www.ingramcontent.com/pod-product-compliance
Ingram Content Group UK Ltd.
Pitfield, Milton Keynes, MK11 3LW, UK
UKHW060123010826
14089UKWH00006B/113

9 783036 511658